G. Lubec (ed.)

Advances in Down Syndrome Research

SpringerWienNewYork

Prof. Dr. G. Lubec
Department of Pediatrics
University of Vienna
Währinger Gürtel 18–20
A-1090 Vienna, Austria

Printed in Austria

Springer-Verlag Wien New York is a part of Springer Science+Business Media
springeronline.com

Typesetting: SNP Best-Set Typesetter Ltd., Hong Kong
Printing: Holzhausen Druck & Medien GmbH, A-1140 Wien

Printed on acid-free and chlorine-free bleached paper

SPIN: 10953891

CIP data applied for

With numerous (partly coloured) Figures

ISBN 3-211-40781-2 (hard cover) Springer-Verlag Wien New York
ISBN 3-211-40776-6 Journal of Neural Transmission [Suppl 67]
(soft cover) Springer-Verlag Wien New York

Preface

In this publication information on biology, clinics and morphology of the Down Syndrome (DS) phenotype and mouse models thereof is provided in one volume in order to show the necessity of a common approach to solve the enigma, DS.

And indeed, the DS phenotype may be seen as the consequence of a series of biochemical phenomena with aberrant RNA or protein expression of pivotal structures as e.g. transcription factors BACH1 or ERG, with different expression patterns in fetal and adult DS. It is also these structures that may be leading to the development of Alzheimer's disease neuropathology, inevitably supervening in adult DS, making studying DS a useful approach to investigate Alzheimer's disease.

We learn that in early fetal life neuropathology does not show a specific histopathological pattern and we understand that molecular changes are preceding morphological alterations.

The report of a new animal model forms the basis along with the already existing ones, to carry out biochemical and pharmacological experiments enabling testing current hypotheses, comparison with the human system and much more.

Aberrant folate metabolism in mothers is a risk factor for developing DS and there is now biochemical evidence for impaired folate handling structures in fetal DS brain. Last not least overexpression of SOD-1, proposed to lead to the generation of reactive oxygen species and possibly responsible for neuronal loss is back to stage and the antioxidant system still is a matter of discussion, although literature indicates involvement of this pathway from early life.

Many pathways, channels, metabolic, signaling, chaperone, cytoskeleton, etc.etc. are involved in the pathomechanisms leading to the DS phenotype and each may well explain impaired function or neuronal death, but the etiological question is still open and many findings may only reflect byphenomena or consequences of deranged different systems. There are also no definitive therapeutical targets so far and even existing tentative pharmacological targets have not been tested in humans due to restrictions by law or Ethical Committees or by cutting budgets for science or maybe even by (experimental) therapeutic nihilism.

The editor of this (now third) booklet on DS is highly indebted to the Red Bull Company, Salzburg, Austria for generous support for many years and to the many partners worldwide that cooperated at different levels from providing antibodies to carrying out analyses or fruitful discussions. This is also to thank contributors to this volume as well as parents and patients who participated in studies published herein.

Vienna, November 2003 **G. Lubec**

This book is dedicated to Dietrich Mateschitz,
Salzburg, Austria

Contents

A new mouse model for Down syndrome

Y. Kazuki[1,2], **T. C. Schulz**[2,3], **T. Shinohara**[2], **M. Kadota**[2], **R. Nishigaki**[2], **T. Inoue**[4], **M. Kimura**[2], **Y. Kai**[1], **S. Abe**[2], **Y. Shirayoshi**[2], and **M. Oshimura**[1,2]

[1] Department of Biomedical Science, Institute of Regenerative Medicine and Biofunction, Graduate School of Medicine, and
[2] Department of Molecular and Cell Genetics, School of Life Sciences, Faculty of Medicine, Tottori University, Tottori, Japan
[3] BresaGen, University of Georgia, Athens, GA, U.S.A.
[4] Department of Human Genome Science, Graduate School of Medicine, Tottori University, Tottori, Japan

Summary. Trisomy 21 (Ts21) is the most common live-born human aneuploidy and results in a constellation of features known as Down syndrome (DS). Ts21 is a frequent cause of congenital heart defects and the leading genetic cause of mental retardation. Although overexpression of a gene(s) or gene cluster on human chromosome 21 (Chr 21) or the genome imbalance by Ts21 has been suggested to play a key role in bringing about the diverse DS phenotypes, little is known about the molecular mechanisms underlying the various phenotypes associated with DS. Four approaches have been used to model DS to investigate the gene dosage effects of an extra copy of Chr 21 on various phenotypes; 1) Transgenic mice overexpressing a single gene from Chr 21, 2) YAC/BAC/PAC transgenic mice containing a single gene or genes on Chr 21, 3) Mice with intact/partial trisomy 16, a region with homology to human Chr 21 and 4) Human Chr 21 transchromosomal (Tc) mice.

Here we review our new model system for the study of DS using the Tc technology, including the biological effects of an additional Chr 21 in vivo and in vitro.

Overview

Trisomy 21 is the most frequent live-born aneuploidy and occurs at approximately one in 700 in humans. This trisomy produces DS, which includes a variety of developmental anomalies including facial dysmorphology, congenital defects of the heart (DS-CHD) and gut, infertility, immunodeficiencies, the early onset of Alzheimer's disease (DS-Alz) and an increased incidence of leukemia (Epstein et al., 1986). A major goal of DS research is to correlate the dosage imbalance of specific genes from Chr 21 with different clinical aspects of the syndrome. This goal has been facilitated by the recent report of the sequence of 99.7% of the long arm of Chr 21 (33.46 Mb) (Hattori et al., 2000).

A gene catalogue of expressed sequences was developed from the initial annotation, including 127 known genes and 98 predicted genes. Furthermore, the gene expression atlas on Chr 21 in mice facilitated the understanding of gene function and of the pathogenetic mechanisms in DS (Reymond et al., 2002; Gitton et al., 2002). Comparison of the phenotype and genotype of individuals who are partially trisomic for Chr 21 has been used to localize genes whose dosage imbalance contributes to these features (McCormick et al., 1989; Korenberg et al., 1994; Rahmani et al., 1989). However, as there are a number of genes on Chr 21 included in the DS critical region (DSCR), it is difficult to correlate the genes to phenotypes in DS exactly. Furthermore, the number of these individuals is limited and many aspects of the DS phenotype are in place at birth, and the presentation of DS, even with full trisomy 21, is highly variable, limiting the resolution of phenotype maps. Systematic studies of prenatal development cannot be carried out in humans (Reeves et al., 2001).

Thus, mouse models are powerful tools to study phenotype-genotype correlations in DS (Kola et al., 1997). The merits and demerits of the four approaches for DS models as shown in Fig. 1 are outlined as follows. Firstly, transgenic mice over-expressing a single gene on Chr 21 have been produced by a number of groups (Sumarsono et al., 1996; Ema et al., 1999; Epstein et al., 1987; Elson et al., 1994; Bustin et al., 1995; Altafaj et al., 2001; Gerlai et al., 1993). These transgenic mice showed parts of the DS phenotype, for example, *Ets2* over-expressing mice developed skeletal/bone defects (Sumarsono et al., 1996) and *Sim2* over-expressing mice developed mild impairment of learning and memory (Ema et al., 1999). However, as these vectors for over-expression are integrated into the host genome, the integration event mutates the host genome and as these vectors are present at greater than one copy per cell and the promoters used in the vectors aren't original ones of the genome, the gene expression cannot be physiologically controlled. Thus the transgenic over-expression approach may not result in a complete mouse model for DS. Secondly, some YAC/BAC/PAC transgenic mice have been made (Smith et al., 1997; Chrast et al., 2000; Lamb et al., 1997), for example, YACs spanning 2 Mb of Chr 21q22.2 were introduced into the mouse genome and one of the four YACs, containing the *Dyrkla* (*Minibrain*) gene, was implicated in causing subtle behavioral defects when present in three copies (Smith et al. 1997). Although YAC can have multiple genes or a large gene up to about 2 Mb in length, the potential for insertional mutagenesis and transgene silencing following integration of the DNA vector into the host genome has to be addressed as per plasmid vector over-expression. Thirdly, four mouse models with complete or segmental trisomy for mouse chromosome 16 have been established (Gropp et al., 1975; Reeves et al., 1995; Sago et al., 1998, 2000). Chr 21 has conserved syntenies to mouse chromosome (MMU) 16, 17, 10. Most of the Chr 21 region shares homology with MMU16 and the distal end of Chr 21 has syntenies to MMU10 and MMU17. Trisomy 16 (Ts16) mice that have three copies of MMU16 show a number of phenotypic characteristics similar to those of DS subjects, including cardiac defects and anasarca (Epstein et al., 1985). However, as they die in utero, the research on this

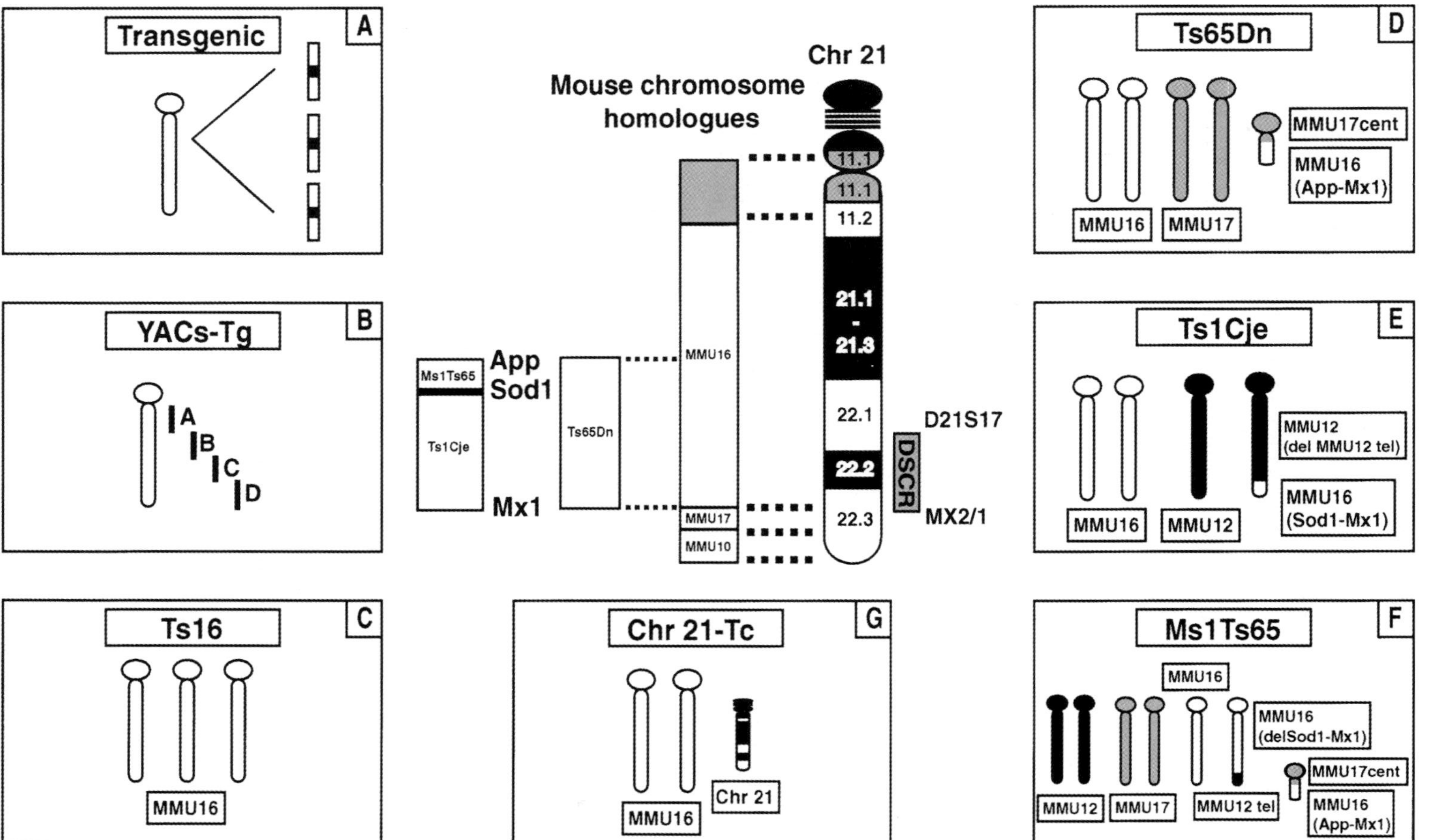

Fig. 1. Summary of mouse models for DS [modification based on Dierssen et al. (2001)]. **a** transgenic mice overexpressing a single gene from Chr 21 **b** YAC transgenic mice containing a single gene or some genes on Chr 21 **c** Trisomy 16 mice with three copy of MMU16 **d** Ts65Dn mice, partial trisomy 16 (App-Mx1) **e** Ts1Cje, partial trisomy 16 (Sod1-Mx1) **f** Ms1Ts65 mice, partial trisomy 16 (App-Sod1) **g** Chr 21 transchromosomal (Tc) mice. Human chromosome 21 (Chr 21) has homology with MMU 10, 16, and 17. MMU, mouse chromosome. *DSCR* Down syndrome critical region

model has been restricted to the fetal stage of development. Furthermore, as MMU16 has homology to other human chromosomes, including human chromosome 3, 16, 22, it is unknown whether these phenotypes observed in Ts16 mice are caused by mouse homologues of genes on Chr 21 (Davisson et al., 1990). Thus these issues limit this model approach. Three mouse strains with segmental trisomy 16, Ts65Dn, Ts1Cje, and Ms1Ts65, carry an extra copy of the segment of mouse Chr 16 extending from *Gabpa* to *Mx1* (~16 Mb), *Sod1* to *Mx1* (~10 Mb), and *Gabpa* to *Sodl* (~6 Mb), respectively (Reeves et al., 1995; Sago et al., 1998, 2000). These mice have deficits in learning and memory. Ts65Dn mice show precise parallels to anomalies of the craniofacial skeleton (Richtsmeier et al., 2000) and the cerebellum (Baxter et al., 2000) that are seen in DS. However, one limitation of partial trisomy 16 mouse models is that they are trisomic for a subset of the genes on human Chr 21. As the heart defect characteristic of DS observed in Ts16 has not been observed in live born segmental trisomy 16 mice, nor do these mice show the histopathology of early-onset Alzheimer disease, these mice might not be a model for DS-CHD or DS-Alz.

Finally, our approach has been to produce mice containing an additional, entire or partial human Chr 21, using microcell-mediated chromosome transfer (MMCT) (Shinohara et al., 2001). This approach has also been reported by Hernandez et al., (1999). Because human chromosomes or their fragments are used as a vector, they can carry greater than megabase-sized genomic DNA compared to BACs, PACs and YACs that have a cloning capacity of ~300 kb, ~100 kb and ~2 Mb respectively. Moreover, because they replicate and segregate as a normal chromosome, they can evade the transgene silencing observed in single gene over-expressing or YAC/BAC/PAC transgenic mice, although the efficiency of introduction of DNA by MMCT approach is lower than other approaches and the deletion or rearrangement of transferred chromosomes sometimes may occur.

Here, we present insights learned from the generation of chimeric mice via MMCT and in vitro studies with ES cells containing partial Chr 21 as an independent chromosome, and demonstrating specific parallels to developmental anomalies in DS.

Analyses of mice containing a Chr 21

Construction of ES cells containing Chr 21

Construction of a mouse A9 cell library containing different human chromosomes tagged with the neomycin-resistance gene is described previously (Inoue et al., 2001). After screening the library (Tomizuka et al., 1997) to identify A9 hybrid clones carrying an independent, intact human Chr 21, microcells prepared from these donor A9 cells were fused with mouse TT2F ES cells (39, XO) (Uchida et al., 1995). The hybrids containing Chr 21 were selected with G418 and the transferred Chr 21 in each line was characterized by PCR and FISH analyses (Fig. 2). Chimera production was carried out as

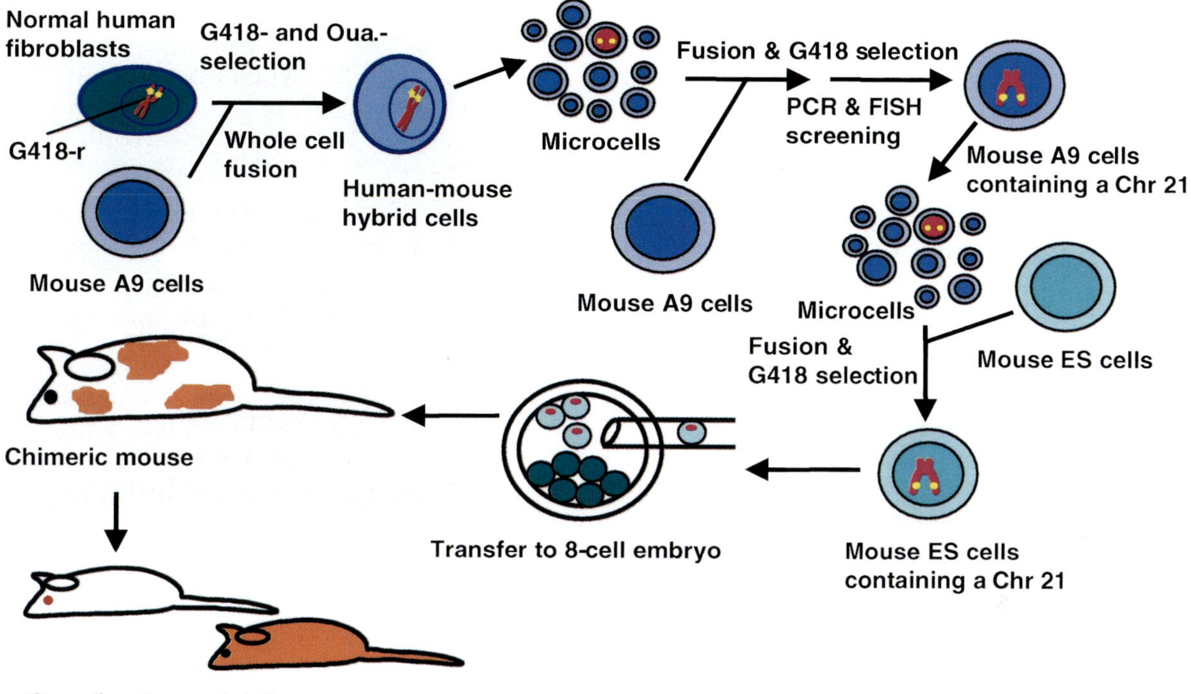

Fig. 2. A schematic diagram showing the construction of ES cells containing a human Chr 21 to produce chimeric mice and Tc mice expressing human genes on Chr 21. *G418*-r G418 resistance. *Oua* ouabain

described previously (Tomizuka et al., 1997), i.e., ES cells were injected into 8-cell stage embryos derived from jcl: MCH (ICR) mice (Crea Japan, Inc.), and transferred into pseudopregnant jcl: MCH (ICR) females.

Retention of Chr 21 in chimeric mice

PCR analysis was used to determine the transferred region of Chr 21 in each chimeric mouse, using 13 Chr 21 markers (Fig. 3A). All of the 11 markers retained in the ES (#21)-10 cell line were present in the ES (#21)-10 chimeras (C10) (Shinohara et al., 2001). In contrast, the high percentage chimeras (C11) produced with the ES (#21)-11 line showed deletions of *SIM2* and *PWP*, even though the parental cell line retained all 13 markers by PCR (Shinohara et al., 2001). This ES cell line may represent a mixed population of cells, some of which contain an intact Chr 21 and others, a partially deleted Chr 21. Chimeras from ES (#21)-10 were examined in further analyses including FISH, gene expression, behavior and two-dimensional electrophoresis (2-DE). On the other hand, chimeric fetuses from ES (#21)-11 were analysed histologically. None of the sixteen chimeric mice with >50% Chr 21 retention contained an intact Chr 21, possibly suggesting that the presence of an entire copy of the chromosome might be severely deleterious during mouse development, as is the case for trisomy 21 in human beings.

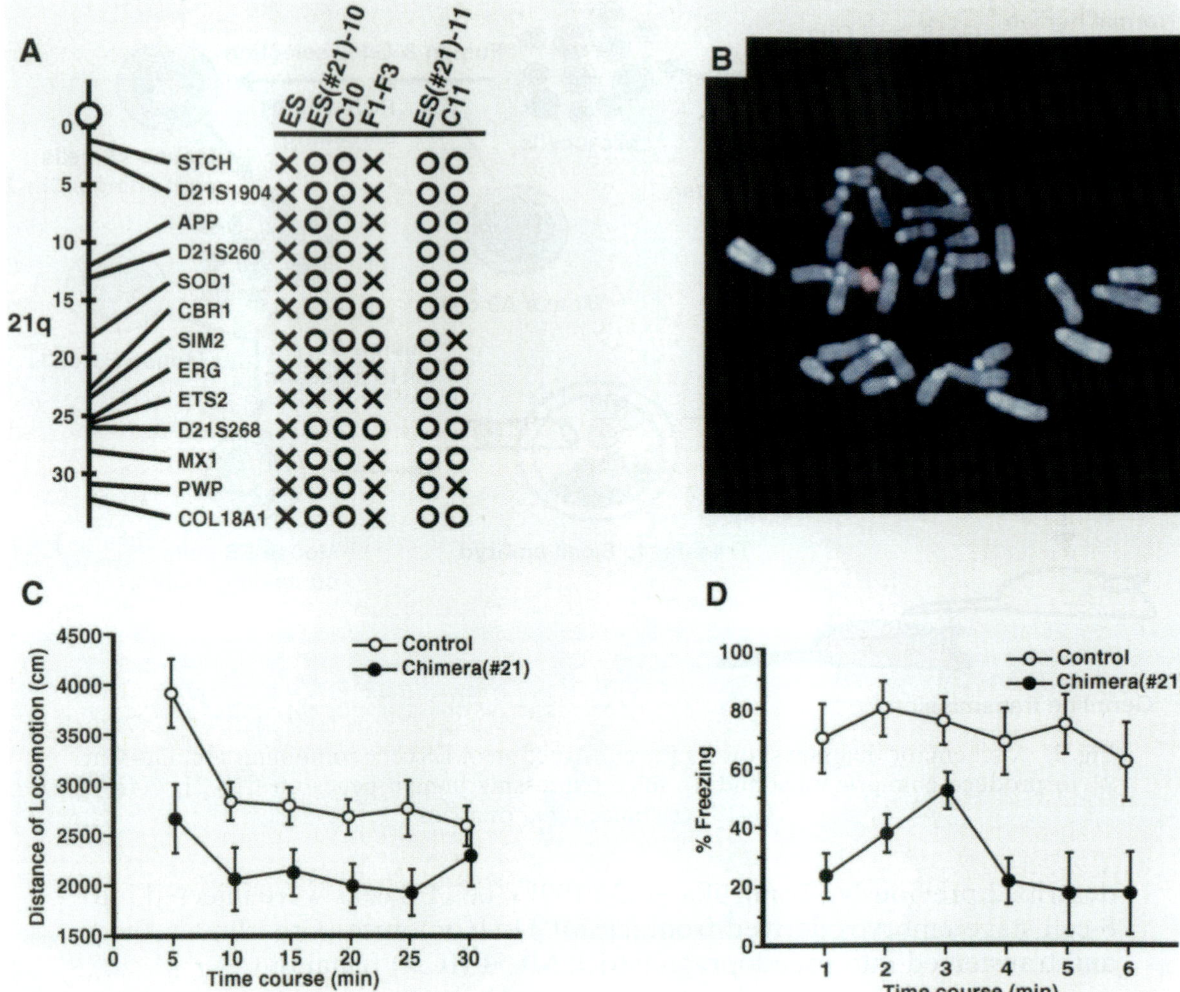

Fig. 3. Molecular and cytogenetic analyses of the retention of the transferred Chr 21 in chimeric mice and behavioral analyses for general activity and learning in chimeric mice [modification based on Shinohara et al. (2001)] **a** Summary of PCR analyses of ES cells containing Chr 21 using primers for the Chr 21-specific STS markers and genes (Shinohara et al., 2001). Chr 21 genomic DNA in tail fibroblasts of chimeric and Tc mice were detected by PCR, with ES cells and each ES cell line containing Chr 21 as controls. The left panel represents the approximate physical order of STS or EST markers and genes tested, and the distance along q-arm from centromere is indicated. In the right panel, open circles and crosses represent presence and absence of tested loci, respecitvely. **b** FISH with human COT-1 DNA to metaphase chromosomes prepared from tail fibroblasts in a Chr 21 chimera (Shinohara et al., 2001). On metaphase spreads, a single Chr 21 (red signal) was detected in addition to mouse chromosomes (blue). **c** Hypoactivity in exploratory behavior in ES (#21)-10 chimeras (n = 4) and control mice (n = 5) (Shinohara et al., 2001). Distance traveled in locomotion was measured for 30 min in open-field in the ES (#21)-10 chimera and control mice. Trials were duplicated for each mouse at an interval of two days to attenuate effects of the estrous cycle. Mean values obtained from these trials are plotted for each 5 min. **d** Impairment in contextual fear conditioning in ES (#21)-10 chimeras (n = 4) and control mice (n = 5) (Shinohara et al., 2001). Conditioned freezing responses 24 hr after training. The mean percentage of time spent for freezing is plotted for each min over a 6-mine retention test in ES (#21)-10 chimera and control mice. The ES (#21)-10 chimeras showed a less conditioned freezing response through the test compared with the normal mice

FISH analysis was used to determine the retention rate of Chr 21 in cells from brain, heart, liver, spleen, bone-marrow, and in cultured tail-fibroblasts of adult chimeric mice. A human COT-1 probe detected Chr 21 as an independent extra copy in tail-fibroblasts, indicating that no recombination or rearrangement had taken place between human and mouse chromosomes (Fig. 3B) (Shinohara et al., 2001). The retention rate of the Chr 21 varied considerably, among different mice and even among different tissues of the same mouse (Shinohara et al., 2001). Notably, the retention rate of Chr 21 was relatively high in brain and low in heart compared to other tissues tested, suggesting a possibility that chimeras retaining a high proportion of Chr 21 in heart might die after birth because of severe heart defects as shown as follows. Chimeric mice produced offspring with the dominant agouti coat color, indicating that they originated from ES-derived germ cells. However, PCR analysis showed that only a Chr 21 fragment (hCF21), including the *CBR1*, and *SIM2* markers, was transmitted from the F1 to F3 generation at least (Fig. 3A) (Kazuki et al., 2001).

Transcripts of genes on Chr 21 in chimeric mice

To determine whether human genes on the transferred Chr 21 were transcribed appropriately, total RNA isolated from tissues of chimeric mice were analyzed by RT-PCR using specific primers for human expressed sequence tags (ESTs). The brain specific polypeptide, *PEP19* (*PCP4*) (Ziai et al., 1988), was detected only in brain, while the transcripts for the ubiquitously expressed human *SIM2* and *GIRK2* (Mjaatvedt et al., 1995; Chrast et al., 1997) were detected in all the tissues tested. Thus human genes on the transferred Chr 21 examined were expressed under appropriate tissue-specific transcriptional regulation in chimeric mice despite the human origin of the chromosome and its passage in a differentiated fibroblast cell line (Shinohara et al., 2001). This phenomenon may result from interaction between cis-elements on human chromosome sequences and mouse trans-acting factors, suggesting that these chimeras can be very useful for studying the capability of the regulation of gene expression between human and mouse.

Learning and behaviors in chimeras

Chimeric mice that had >90% ES contribution to coat color and C57BL/6xCBA-F1 mice were subjected to a battery of learning and behavioral tasks. Spontaneous motor activity in a novel environment was measured using an open-field box (Hogan et al., 1994). The chimeras showed less activity in exploratory behavior than control mice (Fig. 3C). This data demonstrated that the chimeras were hypoactive or hypokinetic in exploratory behavior compared to control mice (Shinohara et al., 2001). Furthermore, there was a significant correlation between the retention rate of Chr 21 and a behavioral decline, with a higher percentage of chimerism in the brain resulting in more anomalies (Shinohara et al., 2001).

The contextual fear conditioning task requires the integrated neural circuit of the hippocampus and amygdala. For the training trial, mice were placed in a conditioning chamber where they received three sequential foot shocks. When returned to the conditioning chamber 24 hr after the training session, the chimeras showed a less conditioned freezing response than control mice did (Fig. 3D) (Shinohara et al., 2001). Furthermore, the percentage of freezing in the chimeras and control mice correlated significantly with the contribution of Chr 21-containing cells in the brain (Shinohara et al., 2001). These results revealed that the chimeras had impairment in the associative learning of fear to contextual stimuli. We examined two behaviors, including the light-dark choice test and sensitivity test to shock, which might affect performance in the fear conditioning test, and found no difference between chimeras and control mice (Shinohara et al., 2001). This data demonstrated that pain sensitivity in the chimeras was normal, which suggests that the chimeras had impairment in associative learning of fear to contexual stimulus. The hippocampus and cortex play a role in exploratory behavior and novelty-seeking (Sandin et al., 1997), so these results imply that this trisomy affects development and/or function of these critical brain regions.

The forced swim test has been widely used for screening of antidepressant drugs (Porsolt et al., 1978). Mice that spend a longer time immobile (floating) in an inescapable and stressful situation are considered to reflect a "depressive-like behavior" or passive coping strategy. The stress response related with emotionality was examined in the chimeras. Two trials in the forced swim test were performed for two consecutive days. The chimeras showed a larger increase in immobility on the second trial, 24 hr after the first trial, than that of the control mice, suggesting that the chimeras are in a more depressive state than control mice (Shinohara et al., 2001).

Developmental anomalies in chimeric mouse fetuses

Live born chimeras were produced from the ES (#21)-11 cell line, which retained an intact Chr 21. Since 50% or more of human conceptuses with trisomy 21 do not survive until birth (Epstein et al., 1986) the persistence of an entire copy of Chr 21 might also be expected to disturb gestation in the mouse, as well. The development of 34 prenatal fetuses was examined at 18 days post-coitum (Shinohara et al., 2001). The retention rate of Chr 21 was determined in tail fibroblasts of twenty-one chimeras with pigmented eyes. Seven of the 21 chimeric fetuses (33.3%) showed some degree of developmental retardation, which tended to be more pronounced in fetuses with a retenion rate of Chr 21 of >78%. Hypoplastic thymus was observed in 6 fetuses, five of which also had conotruncal malformations of the heart (see below). Thymic hypoplasia occurred more frequently in fetuses with Chr 21 retention of >92% and did not correlate with body size. Conotruncal malformations were observed in 7 chimeric fetuses (33%) and four of these displayed hypoplastic pulmonary artery. Only the most severely affected fetus displayed remarkable pathology of the aorta, and the pulmonary semilunar valves. The fetus had a large ascending aorta originating from the right ventricle. There was a rudimentary main pulmonary artery extending to the ductus arteriosus on the left side of the aorta. Further observation revealed a thin main pulmonary artery that originated from the right ventricle without a pulmonary valve, and a narrow infundibulum. This was diagnosed as double outlet right ventricle. AV canal malformations were observed in 10 chimeric fetuses (48%). AV septal defects, which are frequently observed in DS patients, were not observed in the chimeric mice. Myocardial layers of heart ventricles showed characteristic phenotypes among chimeric fetuses. A fetus with Chr 21 retention under 10% had mostly normal development of ventricles, while the myocardial layers in fetuses with Chr 21 retention over 68% showed increasing structural change with increased Chr 21 retention. Why are so many phenotypic variations observed in the chimeras and DS patients? In general, more abnormalities of greater severity were seen in fetuses with a greater portion of cells retaining Chr 21. However, even among fetuses with more than 80% Chr 21 contribution, growth retardation, hypoplastic thymus and cardiovascular malformations were not associated uniformly. Although the cause of phenotypic variation in DS patients may be related to individual genetic differences, this would not be the case in our model system where the normal and Tc mice share the same genetic background. There may be several sources of phenotypic variation in these model mice. A first source of variation is the different Chr 21 retention rate in different tissues. The retention rate of the introduced Chr 21 could not be tested in heart and thymus, but was observed in tail fibroblasts. Differences between individual fetuses could have contributed to inconsistency in the severity of the phenotypes observed. A second source of variation may be the difference of expression of some genes on Chr 21 or other mouse chromosomes. This may be examined in more detail in a new model shown in Fig. 4, where the retention of the transferred Chr 21 should be stable, but variable phenotypes are observed.

Proteomic analysis in chimeric mice and in human DS patients

To determine the gene dosage effects of an extra copy of Chr 21 *in vivo*, overall protein expression patterns were compared in whole heart and hippocampus of Chr 21 chimeric and wild type mice by 2-DE analysis (Nishigaki et al., 2002). Out of 843 protein spots resolved by 2-DE in the hippocampus, only one spot was observed to be absent in the wild type mice, but present in the Chr 21 chimeric mice. This spot was isolated from the 2-DE gel of hippocampus extract of chimeric mice and identified as human superoxide dismutase [Cu-Zn] (*SOD1*) using an amino acid sequencer. For whole heart, 495 protein spots were observed, with one spot exhibiting downregulation in Chr 21 chimeric mice compared to control mice. This spot was isolated from the 2-DE gel of heart extract of control mice and was identified as Myosin light chain-2 isoform mlc-2a (*mlc-2a*) using amino acid sequencing. To confirm differences in protein expression levels of mouse mlc-2a, proteins obtained from whole heart of chimeric mice were resolved by 2-DE, and then examined

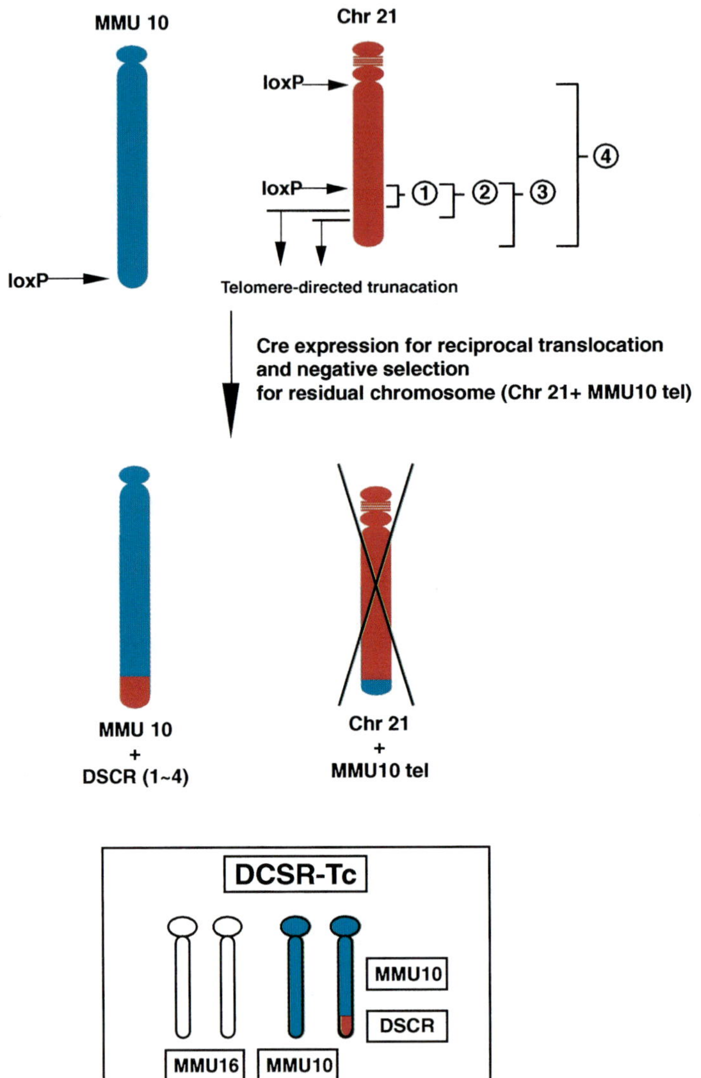

Fig. 4. The strategy to generate DSCR-Tc containing variable regions of Chr 21 (1–4). The strategy for reciprocal translocation was a modification of that described by Smith et al. LoxP-3′HPRT and loxP-5′HPRT is inserted into the terminal region of mouse chromosome 10 in ES cells and into a defined region on Chr 21 in DT40 cells, respectively. Furthermore the Chr 21 is modified by telomere-directed chromosome truncation at a desired site. The modified Chr 21 is transferred to the modified ES cells via MMCT. Following MMCT, chromosome translocation is mediated by transient expression of Cre recombinase and HAT selection to detect reconstitution of the HPRT cassette. After the translocation, negative selection by 6-thioguanine can be applied to remove the bulk of Chr 21 containing the mouse chromosome 10 terminal region, resulting in ES cells containing the translocated variable DSCR regions (1–4). These ES cells can be used for the generation of chimeric mice and attempted germ-line transmission of the translocated DCSR fragments

A

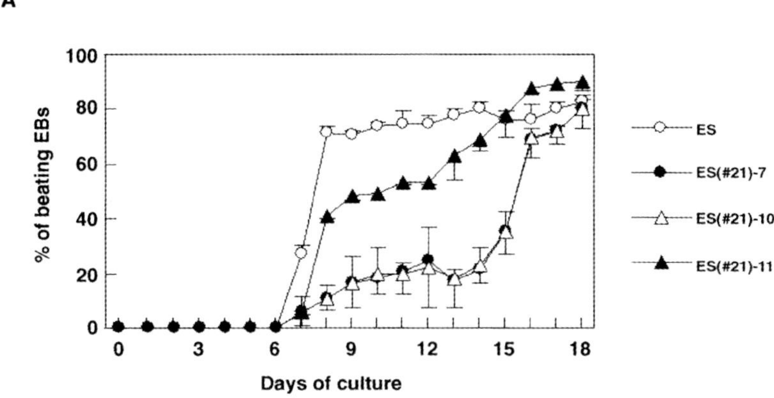

B

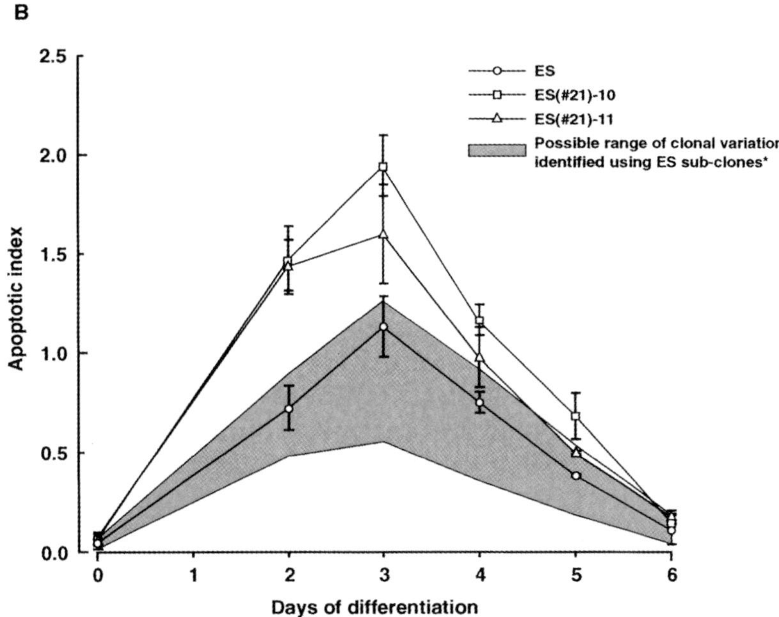

Fig. 5. Impairment of cardiogenesis and elevated apoptosis in neurogenesis during in vitro differentiation of ES (#21). **a** Percentages of EBs exhibiting cardiomyocyte differentiation (Inoue et al., 2000). The EBs derived from wild type ES cells and ES cells containing a Chr 21 [ES (#21)-7, ES (#21)-10, and ES (#21)-11] were used to monitor differentiation into cardiomyocytes. The data points are shown as means (bars, standard deviation) ($n = 3$). **b** Time course study of occurrence of apoptosis in the ES and ES (#21) cells (Kadota et al., 2002). Cells were subjected to staining with Annexin-V (FITC). The apoptotic cell index was determined as the ratio between Annexin-V positive and negative cells. Data are represented as mean ± S.D.; n = three independent experiments, $p < 0.05$ relative to control by the student's t test at days 2–5. *10 sub-clones from parental ES cells were selected and subjected to FACS scanning analysis with Annexin-V. Range was identified by plotting the highest and the lowest value among the sub-clones

by western blotting using an anti-myosin monoclonal antibody. The mouse *mlc-2a* protein was downregulated in Chr 21 chimeric mice, confirming the 2-DE silver staining data (Nishigaki et al., 2002).

To confirm quantitatively the level of protein expression of human atrial myosin light chain-2 (*MLC-2A*), we examined atrium and ventricle samples from DS neonate and normal controls using the anti-myosin light chain monoclonal antibody by 2-DE and western blot analysis. Consequently, the expression of human *MLC-2A* from the atrium of DS neonates was suppressed three-fold compared to the normal control (Nishigaki et al., 2002).

In both human and mouse, *mlc-2a* is a cardiac-specific gene expressed in the adult atrium and during the earliest stages of cardiac development. In situ hybridization studies during mouse embryogenesis showed that *mlc-2a* was expressed specifically in cardiac tissue throughout days 8–16 postcoitum, with atrial-restricted expression from day 12 and qualitatively greater expression in the atrium than the ventricles from day 9 (Kubalak et al., 1994). The preferential expression of this gene in the atrium occurs prior to chamber septation, and negative-regulation of this gene in the ventricles is likely to be an important step for the maintenance of normal ventricular development (Grubr et al., 1998). Thus, *mlc-2a* expression corresponds to the activation of the earliest known markers of the cardiac muscle gene program. Consequently, this data suggest that downregulation of *mlc-2a* observed in the heart of Chr 21 chimeric mice and DS patients has distinct effects on developmental anomalies in early stages of cardiogenesis and altered expression of the *mlc-2a* gene plays a key role in CHD observed in DS. The cause of reduction of *mlc-2a* might be the change of gene expression that controls the *mlc-2a* gene expression, whose candidates on human Chr 21 are *ADARB1*, *KCNJ15*, *PFKL*, *PWP2H* and *SH3BGR*, because these genes are expressed during cardiogenesis in mice (Reymond et al., 2002). Taken together, this suggests that Chr 21 chimeric mice is a good model for DS and also a useful model for identification of genes responsible for DS-CHD.

In vitro differentiation analyses of ES cells containing a Chr 21

Impaired differentiation of ES cells containing a Chr 21 into beating cardiac myocytes

Mouse ES cells differentiate spontaneously into a variety of cell types including beating cardiac myocytes *in vitro*. We investigated whether the presence of Chr 21 in mouse ES cells had any effects on formation of beating cardiac myocytes using the hanging drop culture method according to the method of Oyamada et al. (1996; Inoue et al., 2000). For the quantitative estimation of differentiation, cultures were monitored every day under a microscope to detect the presence of rhythmically beating cardiac myocytes in the dense center of embryoid bodies (EBs) and comparisons were made among three independent ES cell lines containing a Chr 21, ES (#21)-7, ES (#21)-10 and ES

(#21)-11, and wild type ES by *in vitro* differentiation analysis. ES (#21)-10 chimeras carried a Chr 21 in approximately 90% of their cells in many organs except the heart and some of ES (#21)-11 chimeras exhibited cardiovascular malformations as described above. All cell lines gave rise to EBs of similar size and gross morphology. After 7 days in culture, rhythmically beating cardiac myocytes appeared at the dense center of EBs derived from all cell lines, but the ratio of cardiogenic differentiation in EBs derived from the ES (#21) clones was lower than that of EBs derived from wild type ES cells (Fig. 5A). Approximately 80% of EBs derived from TT2F cells were beating by day 8. In EBs derived from ES (#21) clones, the ratio of beating EBs remained significantly lower than this from days 7 to 12. From days 12 to 17, the ratio of beating EBs derived from ES (#21) cells reached approximately 80%. To test whether the retardation of differentiation in ES cells containing a Chr 21 is specific to cardiac myocytes, we analyzed EBs derived from each ES cell line for the differentiation into cartilage, using the same system as before. Differentiation into cartilage occurred with a similar timing and essentially to the same extent in all the cell lines. Taken together, we concluded that the Chr 21 specifically delays the maximum differentiation of beating cardiac myocytes in this *in vitro* system.

We speculate that a delay in cardiogenesis during development caused by trisomy of Chr 21 could be one of the causes that leads to cardiovascular malformations and heart disease in babies with DS, and death of embryos with DS during pregnancy. In this hypothesis, we assume that the genes responsible for this phenotype on Chr 21 are adjusted to restore normal expression levels by epigenetic modifications or deletion, in one of the three copies of Chr 21 in the hearts of DS patients that do not exhibit congenital heart disease. It should be noted however, that further study of *in vivo* differentiation in chimeric mice is required to demonstrate how a delay in the differentiation into cardiomyocytes leads to these phenotypes.

The next question to be addressed is the mechanism of how the Chr 21 in ES cells retards the maximum differentiation into cardiac myocytes. Although we found that the difference in the differentiation between wild type ES cells and ES cells containing a Chr 21 is not likely caused by soluble factors, the mechanism is largely unknown. Because EBs derived from equal numbers of wild type ES cells and ES (#21) cells could normally differentiate into cardiac myocytes, we speculate that the factors that are different between ES and ES (#21) cells may be signalling molecules that control neighboring cells in the development of cardiac myocytes, such as paracrine signaling factors or adhesion molecules. These genes could be on Chr 21 or could be downstream genes located on mouse chromosomes that are up- or down-regulated because of the presence of Chr 21. In this context, the gap junction protein *connexin 43* is a candidate because this molecule is known to be essential for normal heart formation and function (Reaume et al., 1995; Hagendorff et al., 1999). Several transcription factor genes expressed in precardiac mesoderm, such as *Cript-1*, *Nkx2.5*, *Mef2C*, *eHAND*, and *dHAND* were predicted to be responsible for the commitment of mesodermal cells to the cardiac lineage (Xu et al., 1998; Baker et al., 1996).

Neuronal differentiation of ES cells containing a Chr 21 and differentiation associated apoptosis

Two independently established Chr 21 lines, ES (#21)-10 and ES (#21)-11 were used for neuronal differentiation by the SDIA method as described by Kawasaki et al. (2000). The neuronal differentiation profiles of ES (#21) cells and the parental ES cells was examined by RT-PCR for several neuronal differentiation markers, which exhibited the same expression patterns (Kadota et al., 2002).

ES (#21) cells showed about 20–60% reduction in both the number of colonies and the sizes of the colonies growing on PA6 cells. To clarify the cause of an observed reduction of the number and size of the differentiating colonies, the occurrence of apoptosis was analyzed in this system. TUNEL detection revealed that a high proportion of cells within the colony of both parental ES and ES (#21) at day 3 of differentiation were undergoing apoptosis, while the proportion of apoptotic cells at day 6 of differentiation was reduced. Since it was difficult to quantify the rate of apoptosis during differentiation of parental ES and ES (#21) cells by TUNEL, quantitative analysis was performed by FACS scanning. After extensive washes with PBS, the number of cells at the early phase of apoptosis was examined quantitatively at day 0, 2, 3, 4, 5 and 6 of differentiation using Annexin-V-FITC, which binds to the phosphatidylserine on the surface of apoptotic cells. Massive cell apoptosis was detected during neuronal differentiation of both ES and ES (#21) cells at days 2–5 of differentiation, but few apoptotic cells were observed at days 0 and 6 of differentiation (Fig. 5B). This was consistent with the result of TUNEL analysis. The time course of the induction of apoptosis in this system was the same between parental ES and ES (#21) cells, however the proportion of apoptotic cells was 1.5–2 fold higher in ES (#21) cells than in parental ES cells (Fig. 5B). There was no difference at day 0 of differentiation, and at day 6 of differentiation (Fig. 5B), suggesting that the elevated apoptosis observed in ES (#21) cells was specific to days 2–5 of neuronal differentiation patterns (Kadota et al., 2002).

Although mature neurons are present in the DS brain, a reduction in the brain volume and total number and concentration of neurons are the common symptoms patterns (Wisniewski et al., 1986). Moreover, pathological studies of the Ts65Dn mouse model, which exhibits reduced number of neurons in the granule cell layer of the dentate gyrus, also reveals a reduction in the population or defects in the proliferation of neural stem cells patterns (Insausti et al., 1998). Bahn et al. described a striking reduction in neurogenesis of DS brain samples in vitro patterns (Bahn et al., 2002). Furthermore, apoptosis in differentiated neurons is thought to be the cause of Alzheimer disease with amyloid plaques and neurofibrillary tangles that is commonly observed in all individuals with DS over the age of 40 (Wisniewski et al., 1985). Thus, increased apoptosis in DS neurons, including apoptosis in neural stem cells as shown in this study, probably contributes to alterations in the DS brain such as a reduced total brain volume, reduced neuronal number and density (Wisniewski et al., 1986) and Alzheimer disease, which may result in mental retardation or dementia. We are currently studying the cause of this elevated apoptosis using microarray and proteomic analysis.

Therefore, from these in vitro differentiation results, ES cells containing a Chr 21 have the potential to allow examination of early DS development because studying this early developmental process is impossible in humans and is also difficult even with use of mouse models. The abnormalities in vitro differentiation were observed specifically in neurogenesis and cardiogenesis, suggesting that the DS phenotypes in heart and brain might be caused by the developmental anomalies including impaired cardiogenesis and increased apoptosis during neurogenesis. Moreover, because we are likely to be able to introduce Chr 21 to human ES cells using MMCT, the in vitro differentiation studies of neuronal or cardiac differentiation using Ts 21 ES cells might help the investigation of DS during early development.

Perspectives

In the study described above, the most significant problem is that the introduced Chr 21 is not mitotically stable in mice, making further analyses difficult. To resolve this and enable stable inheritance, we are developing mice that have Chr 21, or defined fragments of Chr 21, translocated to a mouse chromosome using Cre-loxP chromosome engineering techniques (Ramirez-Solis et al., 1995; Smith et al., 1995; Kuroiwa et al., 2000). Our strategy for generating chromosomal translocations of different Chr 21 regions containing the DSCR to mouse chromosome 10 is summarised in Fig. 4. The translocation is carried out using a modification to the system described by Smith et al. (1995). These mice will hopefully allow stable germ-line transmission of the included human genetic information, circumventing the inherent difficulties in analysis of chimeras. The generation of these mice may enable to identify a gene or combination of genes responsible for DS phenotype and to examine the hypothesis that DS (Ts21) may be caused by genome imbalance rather than overexpression of genes on Chr 21.

Conclusions

The trigger for DS phenotypes is genome imbalance or over-expression of a gene or combination of genes on Chr 21. As DS is caused by Ts 21, this might affect other gene expression at the transcriptional, post-transcripitional, translational, or post-translational levels, contributing to variable DS phenotypes.

To elucidate the relationship between the variable phenotypes and a gene or combination of genes on Chr 21, mouse models for systematic analyses have been used. In this study, chimeric mice were produced using ES cell clones containing a Chr 21 using MMCT, for the systematic analysis of DS phenotypes at developmental and adult stages. These mice demonstrated specific developmental anomalies including learning or behavioral impairments and cardiac anomalies, similar to the clinical manifestations of DS. Furthermore, differentiation using Chr 21 ES cells showed impaired cardiogenesis and increased apoptosis in pre-mature neurons. To date, as there are few DS studies using ES cells, our model demonstrates a unique and powerful tool. Although the models described above may or may not be the best for the study of DS, ES cells and mice containing Chr 21 are anticipated to be useful for elucidation of DS developmental mechanisms and therapies for DS symptoms.

As discribed here, the Tc mouse approach via MMCT to ES cells was demonstrated to be a very useful tool for analysis of DS. A potential application of the Tc approach is the construction of animals carrying human genetic elements to model specific diseases including DS (Ts 21) and other human aneuploidy syndromes. Another Tc application, for biomedical purposes, is to produce various therapeutic products including human Ig, HLA, and p450 whose genes are contained in Mb sized-clusters, by a combination of Cre/loxP-mediated chromosome translocation and telomere-directed chromosome truncation in homologous recombination-proficient chicken DT40 cells.

Acknowledgements

These studies were performed in collaboration with K. Tomizuka, A. Ohguma, I. Ishida, Pharmaceutical Research Laboratory KIRIN Brewery CO. Ltd, S. Miyabara, Department of Pathology, Saga Medical School, H. Nakane, A. Iino, First Department of Anatomy, Faculty of Medicine, Tottori University, S. Ikegami, K. Inokuchi, Mitsubishi Kasei Institute of Life Sciences, RH. Reeves, Department of Physiology, Johns Hopkins University School of Medicine, T. Toda, Department of Gene Regulation and Protein Function, Tokyo Metropolitan Institute of Gerontology, A. Omori, S. Ichinose, Laboratory of Protein structure analysis section, Mitsubishi Kagaku Institute of Life Sciences, M. Itoh, Department of Mental Retardation and Birth Defect Research, National Institute of Neuroscience, National Center of Neurology and Psychiatry. We thank J. Inoue, S. Takehara and C. Okita for technical assistance and M. Katoh, H. Kugoh and A. Kurimasa, Tottori University for valuable comments. Y. K. is a Research Fellow of the Japan Society for the Promotion of Science. The studies were supported by grants from the Ministry of Education, Culture, Sports, Science, and Technology of Japan and from the Ministry of Health, Labour and Welfare of Japan.

References

Altafaj X, Dierssen M, Baamonde C, Marti E, Visa J, Guimera J, Oset M, Gonzalez JR, Florez J, Fillat C, Estivill X (2001) Neurodevelopmental delay, motor abnormalities and cognitive deficits in transgenic mice overexpressing Dyrk1A (minibrain), a murine model of Down's syndrome. Hum Mol Genet 10: 1915–1923

Baker RK, Lyons GE (1996) Embryonic stem cells and in vitro muscle development. Curr Top Dev Biol 33: 263–279

Bahn S, Mimmack M, Ryan M, Caldwell MA, Jauniaux E, Starkey M, Svendsen CN, Emson P (2002) Neuronal target genes of the neuron-restrictive silencer factor in neurospheres derived from fetuses with Down's syndrome: a gene expression study. Lancet 359: 310–315

Baxter LL, Moran TH, Richtsmeier JT, Troncoso J, Reeves RH (2000) Discovery and genetic localization of Down syndrome cerebellar phenotypes using the Ts65Dn mouse. Hum Mol Genet 22: 195–202

Bustin M, Alfonso PJ, Pash JM, Ward JM, Gearhart JD, Reeves RH (1995) Characterization of transgenic mice with an increased content of chromosomal protein HMG-14 in their chromatin. DNA Cell Biol 14: 997–1005

Chrast R, Scott HS, Chen H, Kudoh J, Rossier C, Minoshima S, Wang Y, Shimizu N, Antonarakis SE (1997) Cloning of two human homologs of the Drosophila single-minded gene SIM1 on chromosome 6q and SIM2 on 21q within the Down syndrome chromosomal region. Genome Res 7: 615–624

Chrast R, Scott HS, Madani R, Huber L, Wolfer DP, Prinz M, Aguzzi A, Lipp HP, Antonarakis SE (2000) Mice trisomic for a bacterial artificial chromosome with the single-minded 2 gene (Sim2) show phenotypes similar to some of those present in the partial trisomy 16 mouse models of Down syndrome. Hum Mol Genet 9: 1853–1864

Davisson MT, Lalley PA, Peters J, Doolittle DP, Hillyard AL, Searle AG (1990) Report of the comparative subcommittee for human and mouse homologies. Cytogenet Cell Genet 55: 434–456

Dierssen M, Fillat C, Crnic L, Arbone's M, Flo'rez J, Xavier Estivill (2001) Murine models for Down syndrome. Physiol Behav 73: 859–871

Elson A, Levanon D, Weiss Y, Groner Y (1994) Overexpression of liver-type phosphofructokinase (PFKL) in transgenic-PFKL mice: implication for gene dosage in trisomy 21. Biochem J 299: 409–415

Ema M, IkegamiS, Hosoya T, Mimura J, Ohtani H, Nakao K, Inokuchi K, Katsuki M, Fujii-Kuriyama Y (1999) Mild impairment of learning and memory in mice overexpressing the mSim2 gene located on chromosome 16: an animal model of Down's syndrome. Hum Mol Genet 8: 1409–1415

Epstein CJ (1986) Consequences of chromosome imbalance: principals, mechanisms and models. Cambridge University Press, New York

Epstein CJ, Cox DR, Epstein LB (1985) Mouse trisomy 16: an animal model of human trisomy 21 (Down syndrome). Ann NY Acad Sci 450: 157–168

Epstein CJ, Avraham KB, Lovett M, Smith S, Elroy-Stein O, Rotman G, Bry C, Groner Y (1987) Transgenic mice with increased Cu/Zn-superoxide dismutase activity: animal model of dosage effects in Down syndrome. Proc Natl Acad Sci USA 84: 8044–8048

Gerlai R, Friend W, Becker L, O'Hanlon D, Marks A, Roder J (1993) Female transgenic mice carrying multiple copies of the human gene for S100 beta are hyperactive. Behav Brain Res 31: 51–59

Gitton Y, Dahmane N, Baik S, Ruiz i Altaba A, Neidhardt L, Scholze M, Herrmann BG, Kahlem P, Benkahla A, Schrinner S, Yildirimman R, Herwig R, Lehrach H, Yaspo ML (2002) A gene expression map of human chromosome 21 orthologues in the mouse. Nature 420: 586–590

Gropp A, Kolbus U, Giers D (1975) Systematic approach to the study of trisomy in the mouse. II. Cytogenet Cell Genet 14: 42–62

Gruber PJ, Kubalak SW, Chien KR (1998) Downregulation of atrial markers during cardiac chamber morphogenesis is irreversible in murine embryos. Development 125: 4427–4438

Hagendorff A, Schumacher B, Kirchhoff S, Luderitz B, Willecke K (1999) Conduction disturbances and increased atrial vulnerability in Connexin40-deficient mice analyzed by transesophageal stimulation. Circulation 99: 1508–1515

Hattori M, Fujiyama A, Taylor TD, Watanabe H, Yada T, Park HS, Toyoda A, Ishii K, Totoki Y, Choi DK et al (2000) DNA sequence of human chromosome 21. Nature 405: 311–319

Hernandez D, Mee PJ, Martin JE, Tybulewicz VL, Fisher EM (1999) Transchromosomal mouse embryonic stem cell lines and chimeric mice that contain freely segregating segments of human chromosome 21. Hum Mol Genet 8: 923–933

Hogan B, Beddington R, Costantini F, Lacy E (1994) Manipulating the mouse embryo: a laboratory manual. Cold Spring Harbor Laboratory Press, Cold Spring Harbor, New York, NY, pp 379–380

Inoue T, Shinohara T, Takehara S, Inoue J, Kamino H, Kugoh H, Oshimura M (2000) Specific impairment of cardiogenesis in mouse ES cells containing a human chromosome 21. Biochem Biophys Res Commun 24: 219–224

Insausti AM, Megias M, Crespo D, Cruz-Orive LM, Dierssen M, Vallina IF, Insausti R, Florez J, Vallina TF, Vallina TF (1998) Hippocampal volume and neuronal number in Ts65Dn mice: a murine model of Down syndrome. Neurosci Lett 253: 175–178

Kadota M, Shirayoshi Y, Oshimura M (2002) Elevated apoptosis in pre-mature neurons differentiated from mouse ES cells containing a single human chromosome 21. Biochem Biophys Res Commun 299: 599–605

Kawasaki H, Mizuseki K, Nishikawa S, Kaneko S, Kuwana Y, Nakanishi S, Nishikawa SI, Sasai Y (2000) Induction of midbrain dopaminergic neurons from ES cells by stromal cell-derived inducing activity. Neuron 28: 31–40

Kazuki Y, Shinohara T, Tomizuka K, Katoh M, Ohguma A, Ishida I, Oshimura M (2001) Germline transmission of a transferred human chromosome 21 fragment in transchromosomal mice. J Hum Genet 46: 600–603

Kuroiwa Y, Tomizuka K, Shinohara T, Kazuki Y, Yoshida H, Ohguma A, Yamamoto T, Tanaka S, Oshimura M, Ishida I (2000) Manipulation of human minichromosomes to carry greater than megabase-sized chromosome inserts. Nat Biotechnol 18: 1086–1090

Kola I, Hertzog PJ (1997) Animal models in the study of the biological function of genes on human chromosome 21 and their role in the pathophysiology of Down syndrome. Hum Mol Genet 6: 1713–1727

Korenberg JR, Chen XN, Schipper R, Sun Z, Gonsky R, Gerwehr S, Carpenter N, Daumer C, Dignan P, Disteche C et al (1994) Down syndrome phenotypes: the consequences of chromosomal imbalance. Proc Natl Acad Sci USA 24: 4997–5001

Kubalak SW, Miller-Hance WC, O'Brien TX, Dyson E, Chien KR (1994) Chamber specification of atrial myosin light chain-2 expression precedes septation during murine cardiogenesis. J Biol Chem 269: 16961–16970

Lamb BT, Call LM, Slunt HH, Bardel KA, Lawler AM, Eckman CB, Younkin SG, Holtz G, Wagner SL, Price DL, Sisodia SS, Gearhart JD (1997) Altered metabolism of familial Alzheimer's disease-linked amyloid precursor protein variants in yeast artificial chromosome transgenic mice. Hum Mol Genet 6: 1535–1541

McCormick MK, Schinzel A, Petersen MB, Stetten G, Driscoll DJ, Cantu ES, Tranebjaerg L, Mikkelsen M, Watkins PC, Antonarakis SE (1989) Molecular genetic approach to the characterization of the "Down syndrome region" of chromosome 21. Genomics 5: 325–331

Mjaatvedt AE, Cabin DE, Cole SE, Long LJ, Breitwieser GE, Reeves RH (1995) Assessment of a mutation in the H5 domain of Girk2 as a candidate for the weaver mutation. Genome Res 5: 453–463

Nishigaki R, Shinohara T, Toda T, Omori A, Ichinose S, Itoh M, Shirayoshi Y, Kurimasa A, Oshimura M (2002) An extra human chromosome 21 reduces mlc-2a expression in chimeric mice and Down syndrome. Biochem Biophys Res Commun 295: 112–118

Oyamada Y, Komatsu K, Kimura H, Mori M, Oyamada M (1996) Differential regulation of gap junction protein (connexin) genes during cardiomyocytic differentiation of mouse embryonic stem cells in vitro. Exp Cell Res 229: 318–326

Porsolt RD, Anton G, Blavet N, Jalfre M (1978) Behavioural despair in rats: a new model sensitive to antidepressant treatments. Eur J Pharmacol 47: 379–391

Rahmani Z, Blouin JL, Creau-Goldberg N, Watkins PC, Mattei JF, Poissonnier M, Prieur M, Chettouh Z, Nicole A, Aurias A et al (1989) Critical role of the D21S55 region on chromosome 21 in the pathogenesis of Down syndrome. Proc Natl Acad Sci USA 86: 5958–5962

Ramirez-Solis R, Liu P, Bradley A (1995) Chromosome engineering in mice. Nature 378: 720–724

Reaume AG, de Sousa PA, Kulkarni S, Langille BL, Zhu D, Davies TC, Juneja SC, Kidder GM, Rossant J (1995) Cardiac malformation in neonatal mice lacking connexin43. Science 267: 1831–1834

Reeves RH, Irving NG, Moran TH, Wohn A, Kitt C, Sisodia SS, Schmidt C, Bronson RT, Davisson MT (1995) A mouse model for Down Syndrome exhibits learning and behaviour deficits. Nat Genet 11: 177–184

Reeves RH, Baxter LL, Richtsmeier JT (2001) Too much of a good thing: mechanisms of gene action in Down syndrome. Trends Genet 17: 79–84

Reymond A, Marigo V, Yaylaoglu MB, Leoni A, Ucla C, Scamuffa N, Caccioppoli C, Dermitzakis ET, Lyle R, Banfi S, Eichele G, Antonarakis SE, Ballabio A (2002) A human chromosome 21 gene expression atlas in the mouse. Nature 420: 582–586

Richtsmeier JT, Baxter LL, Reeves RH (2000) Parallels of craniofacial maldevelopment in Down Syndrome and Ts65Dn mice. Dev Dyn 217: 137–145

Sago H, Carlson EJ, Smith DJ, Kilbridge J, Rubin EM, Mobley WC, Epstein CJ, Huang TT (1998) Ts1Cje, a partial trisomy 16 mouse model for Down syndrome, exhibits learning and behavioral abnormalities. Proc Natl Acad Sci USA 95: 6256–6261

Sago H, Carlson EJ, Smith DJ, Rubin EM, Crnic LS, Huang TT, Epstein CJ (2000) Genetic dissection of region associated with behavioral abnormalities in mouse models for Down syndrome. Pediatr Res 48: 606–613

Sandin J, Georgieva J, Schott PA, Ogren SO, Terenius L (1997) Nociceptin/orphanin FQ microinjected into hippocampus impairs spatial learning in rats. Eur J Neurosci 9: 194–197

Shinohara T, Tomizuka K, Miyabara S, Takehara S, Kazuki Y, Inoue J, Katoh M, Nakane H, Iino A, Ohguma A, Ikegami S, Inokuchi K, Ishida I, Reeves RH, Oshimura M (2001) Mice containing a human chromosome 21 model behavioral impairment and cardiac anomalies of Down's syndrome. Hum Mol Genet 15: 1163–1175

Smith AJ, De Sousa MA, Kwabi-Addo B, Heppell-Parton A, Impey H, Rabbitts P (1995) A site-directed chromosomal translocation induced in embryonic stem cells by Cre-loxP recombination. Nat Genet 9: 376–385

Smith DJ, Stevens ME, Sudanagunta SP, Bronson RT, Makhinson M, Watabe AM, O'Dell TJ, Fung J, Weier HU, Cheng JF, Rubin EM (1997) Functional screening of 2 Mb of human chromosome 21q22.2 in transgenic mice implicates minibrain in learning defects associated with Down syndrome. Nature Genet 16: 28–36

Sumarsono SH, Wilson TJ, Tymms MJ, Venter DJ, Corrick CM, Kola R, Lahoud MH, Papas TS, Seth A, Kola I (1996) Down's syndrome-like skeletal abnormalities in Ets2 transgenic mice. Nature 379: 534–537

Tomizuka K, Yoshida H, Uejima H, Kugoh H, Sato K, Ohguma A, Hayasaka M, Hanaoka K, Oshimura M, Ishida I (1997) Functional expression and germline transmission of a human chromosome fragment in chimaeric mice. Nat Genet 16: 133–143

Uchida M, Tokunaga T, Aizawa S, Niwa K, Imai H (1995) Isolation and nature of mouse embryonic stem cell line efficiently producing female germ-line chimeras. Anim Sci Technol (Jpn) 66: 361–367

Wisniewski KE, Wisniewski HM, Wen GY (1985) Occurrence of neuropathological changes and dementia of Alzheimer's disease in Down's syndrome. Ann Neurol 17: 278–282

Wisniewski KE, Laure-Kamionowska M, Connel F, Wen GY (1986) Neuronal density and synaptogenesis in the postnatal stage of brain maturation in Down syndrome. In: Epstein CJ (ed) The neurobiology of Down syndrome. Raven Press, New York, pp 29–45

Xu C, Liguori G, Adamson ED, Persico MG (1998) Specific arrest of cardiogenesis in cultured embryonic stem cells lacking Cripto-1. Dev Biol 196: 237–247

Ziai MR, Sangameswaran L, Hempstead JL, Danho W, Morgan JI (1988) An immunochemical analysis of the distribution of a brain-specific polypeptide, PEP-19. J Neurochem 51: 1771–1776

Authors' address: Dr. M. Oshimura, Department of Biomedical Science, Institute of Regenerative Medicine and Biofunction, Graduate School of Medicine, Tottori University, Nishimachi 86, Yonago, Tottori 683–8503, Japan, e-mail: oshimura@grape.med.tottori-u.ac.jp

Predicting pathway perturbations in Down syndrome

K. Gardiner

Eleanor Roosevelt Institute, University of Denver, and
Department of Biochemistry and Genetics, University of Colorado Health Sciences Center, Denver, CO, U.S.A.

Summary. Comparative annotation of human chromosome 21 genomic sequence with homologous regions of mouse chromosomes 16, 17 and 10 has identified 170 orthologous gene pairs. Functional annotation of these genes, based on literature reports and computationally-derived predictions, shows that a broad range of cellular processes are represented. A goal of Down syndrome research is to determine which of these processes are perturbed by overexpression of chromosome 21 genes, and which may, therefore, contribute to the cognitive deficits that characterize Down syndrome. Eleven chromosome 21 genes are annotated to interact with or be affected by components of the MAP Kinase pathway and eight are involved in Ca^{2+}/calcineurin signaling. Both pathways are critical for normal neurological function, and consequently their perturbations are proposed as candidates for phenotypic relevance. We present evidence suggesting that the MAP Kinase pathway is perturbed in the Ts65Dn mouse model of Down syndrome at 4–6 months of age. Analysis is complicated by the observation that overexpression of chromosome 21 genes in trisomy may be affected by method of detection, organism, tissue or brain region, and/or developmental age.

Introduction

With an incidence of one in 700 live births, Down syndrome (DS) remains the leading genetic cause of mental retardation (Hassold and Jacobs, 1984). The majority of DS is associated with trisomy of the complete long arm of human chromosome 21 and is due to overexpression of the otherwise normal genes encoded within it. Because analysis of cases of DS due to partial trisomy 21 has not eliminated any significant region of 21q from containing genes relevant to the cognitive deficits (Korenberg et al., 1994), the candidate region and gene numbers remain large. The generation of the complete finished sequence of 21q (Hattori et al., 2000) has resulted, after several iterations of computationally-based gene identification and experimental verification (Reymond et al., 2001, 2002; Gardiner et al., 2002), in a catalogue that currently contains approximately 350 genes and gene models (Gardiner et al., 2003). Determining which of these genes contribute to features of mental retardation and through what mechanisms is now a critical focus of DS research. Such research is facilitated by a number of resources. First, the recent availability of the draft sequence of the mouse genome has allowed the comparative analysis of chromosome 21 with homologous regions of mouse chromosomes 16, 17 and 10, and the identification of a set of 170 highly conserved, likely orthologous genes (Gardiner et al., 2003). Second, the mouse chromosome 16 segmental trisomies, Ts65Dn and Ts1Cje (reviewed in Davisson and Costa, 1999), provide model systems trisomic for 94 and 71 genes orthologous to those on human chromosome 21. These mice display DS-relevant phenotypic features (Davisson and Costa, 1999; Crnic and Pennington, 2000); notably, the Ts65Dn displays deficits in hippocampal function and degeneration of basal forebrain cholinergic neuron function similar to those seen in DS (Holtzman et al., 1996; Granholm et al., 2000; Hyde et al., 2001; Pennington et al., 2003). Third, there is a steady increase in the number of chromosome 21 genes for which functional information is being generated and in the detail in which it is being generated. Such data have derived from overexpression of chromosome 21 cDNAs and cDNA fragments in cell culture, determination of protein interactions partners in yeast two hybrid and large scale mass spectrometry/proteomics experiments, and creation of mouse models using single gene knockouts, overexpression and mutation.

Together these developing resources permit a re-evaluation of the strengths and limitations of current and potential mouse models. Importantly they also provide the framework for the beginnings of a systems biology approach to the study of DS.

The particular challenge of DS is one of subtleties. A gene dosage-defined alteration will be a modest 50% increase in the expression of any one chromosome 21 gene. While it is intuitively obvious that no chromosome 21 gene acts independently of other genes in the genome, it is also becoming increasingly clear that many chromosome 21 genes mutually interact within individual complexes or pathways. To understand their role as it is played in DS it is therefore necessary to analyze the complete complex or pathway as it is represented in DS, i.e. all appropriate components overexpressed and others not. The perturbations of any pathway may be subtle, because they are initiated by a 50% increase in expression of one or more components, and this may present particular challenges for quantitation, sensitivity and reproducibility of measurements of perturbations.

As a first step in a systems biological approach to the study of DS, here we first present a functional classification of the identified human chromosome 21/mouse orthologous gene pairs. We next compile data from the literature that illustrate the interactions of chromosome 21 genes in two important pathways, the MAP Kinase and the Ca-calcineurin. Lastly, we present initial experiments on overexpression of some of the pathway-relevant chromosome 21 genes in the Ts65Dn and illustrate the perturbation of the MAP kinase pathway.

Materials and methods

Orthologous gene identification and functional classification

Chromosome 21 finished sequence and homologous mouse chromosome 16, 17 and 10 draft sequence were annotated as described (Fortna and Gardiner, 2001; Gardiner et al., 2003). Gene functional annotations were defined from literature searches and by computational analysis, particularly SMART and TM-PRED (http://us.expasy.org/tools) to identify domains and motifs. All annotations were reviewed and curated. Annotations and gene lists are available at http://eri.uchsc.edu/chromosome21/index.html.

Western analysis

Ts65Dn trisomy and normal control mice, bred as described (Davisson et al., 1993), were generously provided by Dr. Linda Crnic (University of Colorado Health Sciences Center) and Dr. Alberto Costa (Eleanor Roosevelt Institute). Mice were sacrificed by cervical dislocation according to approved protocols. Whole brain minus cerebellum or intact hippocampi were dissected out and immediately frozen in liquid nitrogen. Protein lysates were prepared by standard methods in the presence of protease and phosphatase inhibitors. For analysis of chromosome 21-encoded proteins and MAP kinase components, respectively, 40 μg and 15μg of total protein were loaded per lane on gels of 7% for APP detection; 15% for SOD1 detection and otherwise 10% polyacrylamide. Proteins were electrotransferred to nitrocellulose; antibodies were purchased from Santa Cruz Biotechnologies, with the exceptions of tERK and APP (BD Biosciences), ITSN1 (Ese1, Transduction Laboratories), SYNJ1 (produced by Bio-Synthesis Inc, Texas) and GART (gift from David Patterson, Eleanor Roosevelt Institute). Signals were detected either with ECL-Plus (Amersham) or CDP-Star (Tropix-ABI). RIP 140 or tubulin antibodies were included with other chromosome 21 and the MAP kinase antibodies to serve as an internal control for protein load. Signals were visualized and quantitated on a phosphoimager (Molecular Dynamics) or on the DIANA III chemiluminescence imager (Raytest) using appropriate software. Ratios of trisomy to normal signal were calculated. For chromosome 21 proteins, except as indicated, three pairs of 4–6 month normal and trisomic littermate pairs were analyzed. For MAP kinase pathway components, hippocampi from five 4–6 month old normal and five trisomic mice were analyzed; non-littermates were used.

Results

Chromosome 21/mouse orthologous genes: distribution, functional annotation and associated pathways

A recent comparative analysis of the genomic sequence of human chromosome 21 with homologous regions of mouse chromosomes 16, 17 and 10 identified 170 orthologous gene pairs (Gardiner et al., 2003). Given their high sequence conservation and their syntenic mapping, these genes are assumed to serve the same functions in both organisms. Table 1 lists the 140 orthologous genes that can be associated with some functional information, either experimentally defined (from the literature) or computationally predicted from domain and motif analyses. While often limited and sometimes

Table 1. Chromosome 21/mouse orthologous genes: functional categories

Functional categories	Functional assignments
Transcription factors, regulators, modulators (17)	GABPA, BACH1, OLIG1, OLIG2, RUNX1, SIM2, ERG, ETS2 (transcription factors); ZNF294, ZNF295, ZNF298, APECED (zinc fingers); KIAA0136 (leucine zipper); SON (DNA binding domain); PKNOX1 (homeobox); HSF2BP (heat shock transcription factor binding protein); RIP140 (modulator of transcriptional activation by hormone receptors)
Chromatin structure (4):	HMG14 (high mobility group), CHAF1B (chromatin assembly factor), PCNT (pericentrin, an integral component of the percentriolar matrix of the centrosome); MCM3AP (DNA replication-associated factor)
Proteases/inhibitors (6):	BACE (beta-site APP cleaving enzyme); TMPRSS2, TMPRSS3 (transmembrane serine proteases); ADAMTS1, ADAMTS5 (metalloproteinases), CSTB (protease inhibitor)
Ubiquitin pathway (4):	USP25, USP16 (ubiquitin proteases); UBE2G2 (ubiquitin conjugating enzyme); SMT3A (ubiquitin-like)
Interferon/immune response (8):	IFNAR1, IFNAR2, IL10RB, IFNGR2 (receptors/auxilliary factors); MX1, MX2 (interferon induced); CCT8 (T-complex subunit); C21Orf7 (TAB2 binding domain of Tak1; interleukin signaling?)
Kinases (7):	PRSS7 (enterokinase); HUNK, DYRK1A, SNF1LK (serine/threonine); PDXK (pyridoxal kinase), PFKL (phosphofructokinase); ANKRD3 (ankirin-like with kinase domains)
Phosphatases (2):	SYNJ1 (polyphosphinositide phosphatase); PDE9A (cyclic-phosphodiesterase)
Spliceosome (7):	SFRS15 (SR protein), U2AF35 (splicing factor), ADAR2 (pre-mRNA adenosine deaminase), PCBP3 (poly C binding protein); RBM11 (RNA binding motif); C21orf70; C21orf66 (spliceosome associated)
Adhesion molecules (9):	NCAM2 (neural cell), DSCAM; ITGB2 (lymphocyte); JAM2 (similar to endothelial tight junction molecule); CXADR (Coxackie/adenovirus receptor; tight junctions); Claudin 8, 14, 17 (tight junction, pore complex); TSPEAR (thrombospondin and epilepsy repeats)
Channels (7):	GRIK1 (glutamate receptor, calcium channel); KCNE1, KCNE2, KNCJ6, KCNJ15 (potassium); CHCL4 (chloride); TRPC7 (calcium)
Receptors (1):	CHODL (chondrolectin; manose receptor)
Transporters (2):	SLC5A3 (Na-myoinositol); ABCG1 (ATP-binding cassette)

Table 1. *Continued*

Functional categories	Functional assignments
Mitochondrial (6):	ATP50 (ATP synthase oligomycin sensitivity conferral protein); ATP5J (ATPase coupling factor 6); NDUFV3 (NADH-ubiquinone oxoreductase subunit precursor); CRYZL1 (quinone oxidoreductase); MRPL39, MRPS6 (mitochondrial ribosomal proteins)
Structural (4):	CRYA (lens protein); COL18, COL6A1, COL6A2 (collagens)
Methyl transferases (3):	DNMT3L (cytosine methyl transferase), HRMT111 (protein arginine methyl transferase); N6AMT (N6-DNA methyltransferase)
SH3 domain (3):	SAMSN1, SH3BGR, UBASH3A
One carbon metabolism (4):	GART (purine biosynthesis), CBS (cystathionine-b synthetase), FTCD (formiminotransferase cyclodeaminase), SLC19A1 (reduced folate carrier)
AA composition (9):	CYRR1 (cys-tyr rich); C21orf2, C21orf102, DSCR2 (Leu rich); PWP2 (periodic tryptophan protein); WRB (tryptophan rich); WDR4, WDR9, C21orf6 (WD repeats)
Transmembrane domains (#) (6)	C21orf61 (1); C21orf108 (6); C21orf4 (4); DSCR5 (2); C21orf1 (1); TMEM1 (4)
Signal sequence (2)	C21orf62; KIAA0958
Oxygen metabolism (3):	SOD1 (superoxide dismutase); CBR1, CBR3 (carbonyl reductases);
Miscellaneous domains (9)	KIAA0653, IGSF5 (Ig domains); C21orf63 (Galactose binding lectin); C21orf55 (DnaJ); S100β, C21orf25 (Ca binding); KIAA0184 (AMP binding); C21orf33 (ThiJ); TTC3 (tetratricopeptide repeat containing)
Miscellaneous functions (17):	HLCS (holocarboxylase synthase); LSS (lanosterol synthetase); B3GALT5 (galactosyl transferase)); STCH (microsomal stress protein); BTG3 (cell cycle control); APP (Alzheimers amyloid precursor); TFF1, 2, 3 (trefoil proteins); Pred5 (lipase); SCL37A1(glycerol phosphate permease); AGPAT3 (lysophosphatidic acid acyl transferase); DSCR1 (calcineurin inhibitor); PCP4 (camstatin; inhibitor of calmodulin); ITSN (Rho-GEF); TIAM1 (RAC GTPase); DSCR3 (vacuolar protein-sorting)

only inferred, such functional annotations are nevertheless useful. Genes have been grouped in general categories, clearly representing a broad range of cellular processes, protein complexes and pathways. If expression levels of chromosome 21 genes are increased by 50% in DS, as much but not all data in the literature suggest, and depending upon mechanism, it is possible that many of these processes will be perturbed.

Table 2 reorganizes subsets of the orthologous gene pairs according to location within chromosome 21 relative to homologous regions of mouse chromosomes 16, 17 and 10, and, importantly, relative to the trisomic regions of the mouse chromosome 16 segmental trisomy models of DS, Ts65Dn and Ts1Cje. An important observation from Table 2 is that for no process are all components trisomic within either of the segmental trisomy models. Therefore, if RNA processing, protein degradation or mitochondrial function, for example, are altered in DS due to increased expression of chromosome 21 encoded components, it cannot be assumed that current mouse models are completely recapitulating the phenotypic consequences.

Table 2 also reorganizes two sets of genes that are listed with disparate function in Table 1 but that, nevertheless, function within two important pathways, MAP Kinase and calcium/calcineurin signaling. A functional MAPK pathway has been shown to be required for many types of cognitive and behavioral tasks (reviewed in Adams and Sweatt, 2002). Pharmacological treatments or genetic modifications that decrease or increase flux through the MAPK pathway have been shown to result in deficits in spatial learning (Costa et al., 2002; Blum et al., 1999), similar to those seen in the Ts65Dn and in DS. In addition, nerve growth factor (NGF) signals through MAPK, and abnormalities in NGF concentrations, distributions and transport have been demonstrated in the Ts65Dn (Cooper et al., 2001). Together with the location on chromosome 21 of the MAPK-relevant genes listed in Table 2, these data support the hypothesis that perturbation of this pathway may contribute to the cognitive phenotype of DS.

Calcineurin is a Ca^{2+}/calmodulin-dependent serine-threonine protein phosphatase. In mammals, it has been shown to function in T cell activation, muscle hypertrophy and, especially relevant to Down syndrome, synaptic plasticity/memory development (reviewed in Sugiura et al., 2001). Mice overexpressing calcineurin showed impaired LTP and deficits in long term spatial and visual memory (Mansuy et al., 1998). Mice in which calcineurin had been knockout showed no alteration in either LTP or LTD but depotentiation was completely abolished (Zhuo et al., 1999). Calcineurin is inhibited by DSCR1 (Down syndrome critical region 1 gene, also known as Adapt78 and MCIP1) (Ermak et al., 2002) and therefore activity would be predicted to be decreased in DS (but see also Vega et al., 2003).

MAP Kinase and chromosome 21 genes

MAPK pathways are activated by stimuli that include binding of growth factors to receptor tyrosine kinases followed by endocytosis. Initial MAPK

Table 2. Distribution of functionally related genes among mouse genomic regions

Mouse genomic region	RNA Processing	Ubiquitin Proteasome	Methylation	MAP kinase/ Endocytosis	Ca/ CaN	Cell adhesion/Tight junctions	Mitochondria
Chr 16; − Ts650Dn	RBM11	USP25		RIP140	RIP140	CXADR	
Chr 16; +Ts65Dn, −Ts1Cje		USP16	N6AMT1	APP TIAM1 GABPA	APP, SOD1	JAM2 CLDN17 CLDN8	MRPL39 ATP5J
Chr 16; +Ts65Dn, +Ts1Cje	SFRS15 C21orf66			SYNJ1, ITSN DYRK1A DSCR1, RUNX1, ETS2	DSCR1 PCP4 C21orf25	CLDN14	ATP50 MRPS6
Chr 17	U2AF1		CBS				NDUVF3
Chr 10	C21orf70 ADAR2 PCBP3	UBE2G2 SMT3H1	DNMT3L SCL19A1 FTCD HRMT1L1	ITGB2	TRPC7 S100β		

components are the trio of kinases that effect a cascade of activations through phosphorylation of their respective targets. As shown in Fig. 1, these trios include Raf-Mek1/2-Erk1/2, Mekk1-Mkk4/7-JNK and Mekk3-Mek5-Erk5 (reviewed in Kaplan and Millar, 2000; Watson et al., 2001). Targets of MAPK include transcription factors, with downstream consequences for activation or repression of transcription of numerous target genes. Figure 1 shows the relationships of eleven chromosome 21 genes with components of the pathway.

Five chromosome 21 genes interact with proteins functioning in endocytosis and initial MAPK signaling. Intersectin (ITSN1) is a multi-functional protein containing two EH, five SH3, a RhoGEF, pleckstrin and Ca interaction domains (Hussain et al., 1999). It is involved in endocytosis through direct, possibly competitive, interactions with Sos, dynamin and other scaffolding proteins (McPherson et al., 2000). Independent of endocytosis, when overexpressed, ITSN1 also inhibits the activation of Ras, an initial step in the MAPK cascade (Tong et al., 2000a,b). TIAM1 is a guanine nucleotide exchange factor with specificity for Rac activation (Michiels et al., 1995); it also functions through PI3K-dependent and independent pathways (Fleming et al., 2000). Synaptojanin (SYNJ1) is a phosphoinositol-phosphatase that directly interacts with ITSN and dynamin, and its substrates affect TIAM1 activity (Slepnev and DeCamilli, 2000). DYRK1a (Minibrain) is a dual specificity tyrosine phosphorylated and regulated kinase, whose substrates include dynamin (Chen-Hwang et al., 2002). Interestingly, DYRK1A activates or inhibits dynamin functions in a concentration dependent fashion. A fifth chromosome 21 gene relevant to MAPK is DSCR1/MCIP1, an inhibitor of calcineurin, whose targets include SYNJ1, dynamin and additional endocytic proteins (see below).

Chromosome 21 encoded proteins are also affected downstream of MAPK activity. APP acts synergistically with NGF and, in a concentration dependent fashion, can increase or decrease Erk1/2 activation (Wallace et al., 1997; Mills et al., 1997; Rossner et al., 1999). The MAPK targets, transcription factors CREB and the estrogen receptor, are also targets of the chromosome 21 proteins DYRK1A and RIP140, respectively (Yang et al., 2001; Cavailles et al., 1995). Lastly, three chromosome 21 encoded transcription factors, GABPA, RUNX1 and ETS2 are also targets of MAPK (reviewed in Lewis et al., 1998). Predicting effects of overexpression of these genes due to trisomy is therefore complicated by possible additional alterations in their levels of activation caused by perturbations of MAPK activity. The chromosome 21 gene ITGB2 is a target of GABPA; analysis of its transcription level in trisomy may be indicative of the status of GABPA activation (Rosmarin et al., 1998).

While not involved in MAPK, it is noteworthy that DYRK1A also phosphorylates the Tau protein, at the site that is hyperphosphorylated in AD-associated plaques (Woods et al., 2000).

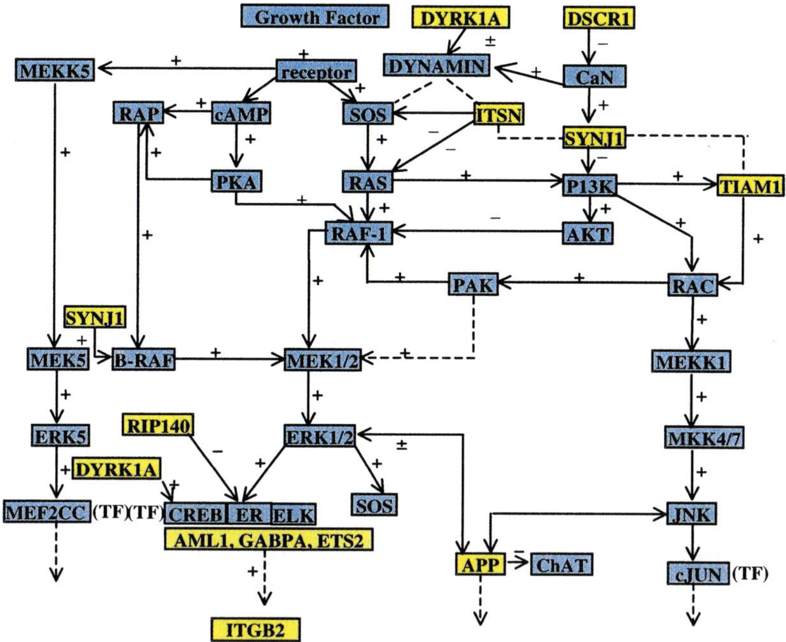

Fig. 1. Interactions of chromosome 21 genes with the MAP Kinase pathway. Standard MAP Kinase pathways, with Erk5, Erk1/2 and Jnk, are shown. *CaN* calcineurin. Growth factors initiate signaling by binding to receptors. Arrows indicate the effect of pathway components; +, activation; −, inhibition. Dashed lines indicate a protein-protein interaction. Chromosome 21 genes that impact the pathway are shown in yellow e.g. ITSN blocks activation of Ras; TIAM1 activates Rac. Chromosome 21 transcription factors that are targets of the pathway are shown in yellow. *TF* transcription factor

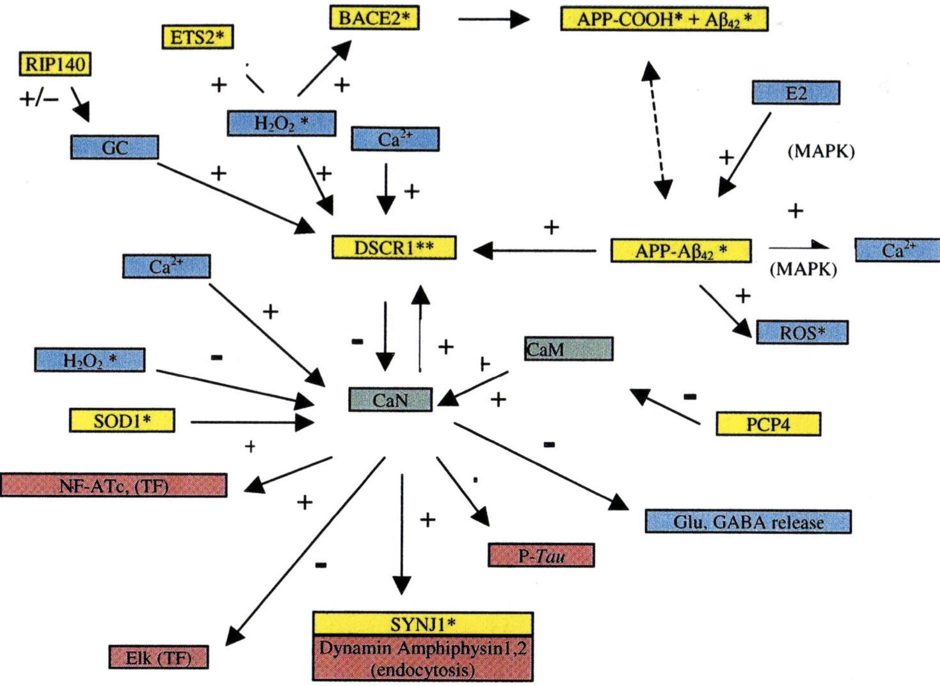

Fig. 2. The DSCR1-Calcineurin pathway and Down syndrome. *, increased in DS; **, also increased in AD; *CaN* calcineurin; *CaM* calmodulin; E2 estrogen; *GC* glucocorticoid; *TF* transcription factor; + activation; −, inhibition; *yellow* chromosome 21 genes; *pink* proteins regulated by CaN; *blue* small molecules

Calcium signaling, calcineurin and chromosome 21 genes

Calcineurin is activated by Ca^{2+} influx through L-type voltage gated calcium channels, and when activated in the hippocampus, calcineurin inhibits glutamate and γ-amino butyric acid (GABA) release and desensitization of post synaptic NMDA-receptor coupled Ca^{2+} channels (see Klee et al., 1998 and references therein). As mentioned above and indicated in Fig. 1, targets of calcineurin include the endocytic proteins dynamin, amphiphysins 1 and 2, and the chromosome 21 gene, synaptojanin; all four are activated by calcineurin-mediated dephosphorylation (Marks and McMahon, 1998; Cousin et al., 2001). Calcineurin also dephosphorylates the Tau protein, a component of the neurofibrillary tangles that are a hall mark of Alzheimers Disease (AD); decreased calcineurin activity and increased DYRK1A (see above) could combine to cause the hyperphosphorylation of Tau seen in AD, or at least in the DS-associated AD. Calcineurin positively regulates the transcription factor NF-ATc where dephosphorylation results in target gene transcriptional activation (Graef et al., 1999). Conversely, calcineurin-mediated dephosphorylation of the transcription factor Elk results in its inactivation (Sugimoto et al., 1997).

Figure 2 illustrates the interconnections of chromosome 21 genes with calcineurin. DSCR1 inhibits calcineurin activity by binding to the catalytic domain (for review of DSCR1/calcineurin see Rothermel et al., 2003). Levels of DSCR1 are increased in DS 2–3 fold, i.e. more than predicted by gene dosage; interestingly, levels are also increased in AD (Ermak et al., 2001). DSCR1 is induced by calcineurin, implying a negative feedback mechanism (Yang et al., 2000), and by Ca^{2+}, by H_2O_2, and by the Alzheimers amyloid precursor protein (APP) $A\beta_{42}$ peptide (reviewed in Rothermel et al., 2003). SOD1, Superoxide Dismutase, produces H_2O_2; its increased production in DS is believed to contribute to the observed increased levels of oxidative damage which may also serve to increase levels of DSCR1. The transcription factor ETS2 and the APP cleavage protein, BACE2, are also induced by H_2O_2 (Tamagno et al., 2002; Sanji et al., 2001). Increased BACE2 activity has been shown to increase APP proteolytic cleavage, increasing proportions of the APP C-terminal fragments, and possibly producing increased levels of APP-$A\beta_{42}$, which may in turn increase DSCR1 expression. The transcription modulator, RIP140, inhibits, in a concentration dependent fashion, the activity of the Glucocorticoid Receptor transcription factor; glucocorticoids have been shown to activate DSCR1 (Yoshida et al., 2002; Subramanian et al., 1999; Windahl et al., 1999). Lastly, calcineurin is calmodulin dependent; the chromosome 21 gene PCP4 is an inhibitor of calmodulin, and therefore the overexpression of PCP4 may also contribute to decreases in calcineurin activity (Slemmon et al., 1996).

As with the MAPK-chromosome 21 gene interactions, precisely how increased expression of chromosome 21 genes would affect this network cannot be predicted. It is however reasonable to hypothesize that the network will be perturbed in DS. Downstream effects again involve several transcription factors, allowing for a cascade effect of the perturbations.

Table 2 includes three additional genes involved in or predicted to be involved in Ca^{2+} regulation. TRPC7 is a novel M-type calcium channel and C21orf25 encodes a predicted Ca^{2+} binding domain. Lastly, S100β, a Ca^{2+} binding protein, modulates long term synaptic plasticity (Nishiyama et al., 2002) and is affected by pharmacological inhibitors of calcineurin (Alexanian and Bamburg, 1999).

Gene dosage effects in trisomy

Figure 3 shows results of Western analysis of several chromosome 21 encoded proteins in brains of Ts65Dn normal and trisomic littermate pairs. The proteins App, Tiam1, Synj1, Itsn1 and Gart show the predicted gene dosage increase of ~50% in the trisomic tissues, although not all isoforms are equally

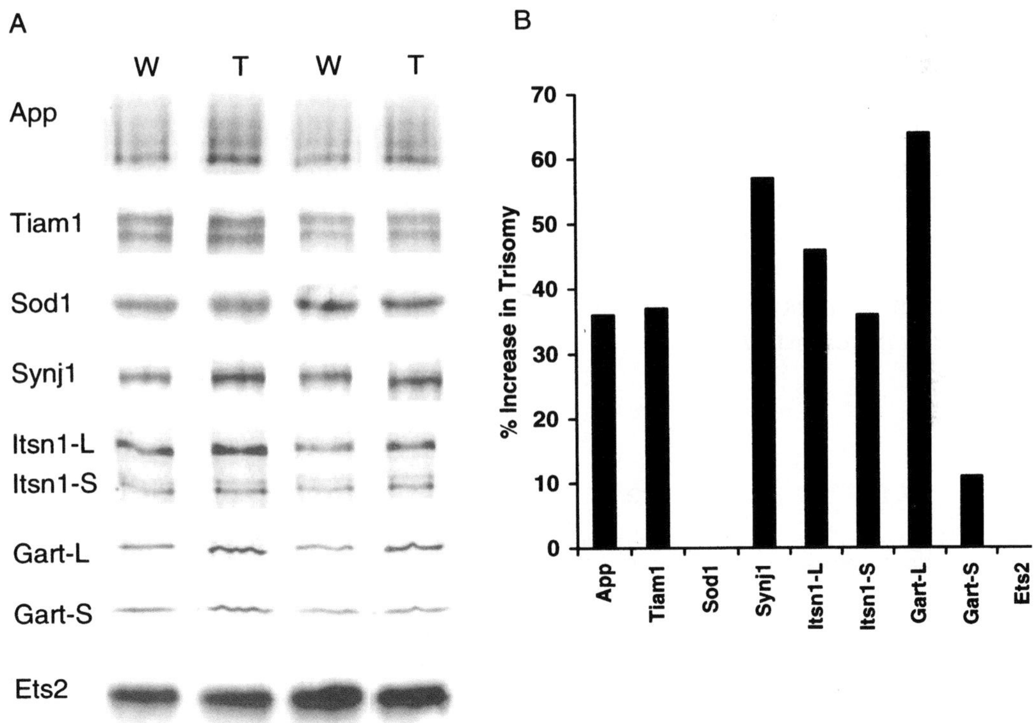

Fig. 3. Expression of chromosome 21 orthologous proteins in Ts65Dn mice. **A** Protein lysates from brains (minus cerebellum) of Ts65Dn (T) and normal littermate controls (W) were analyzed with antibodies to the indicated chromosome 21-encoded proteins. Isoforms of App and Tiam1 appear to be equally increased. For the isoforms of Itsn1 and Gart, indicated as long (L) and short (S), the short forms show less of an increase. Sod1 and Ets2 do not appear to be increased in any pair. Three littermate pairs (aged 4–6 months, except for Gart where mice were 12 months) were analyzed in 3–5 replicates. Signals were normalized for protein load using antibodies to tubulin or Rip140 (not shown). **B** Protein levels in Ts65Dn vs normal controls were quantitated from the Westerns in **A**. Consistent with the images, Sod1 and Ets2 showed no increases in trisomy. The short isoforms of Itsn1 and Gart show lesser increases than the long forms

affected. In contrast, Sod1 and Ets2 protein levels are unchanged in the same trisomy samples.

Abnormalities in MAPK in trisomy

Figure 4a shows Western analysis of the activation of MAPK components Erk1/2 in cerebrums and hippocampi dissected from five normal and five trisomic Ts65Dn brains. Visually, levels of phosphorylated Erk1/2, pErk1 and pErk2, are clearly decreased in the Ts65Dn relative to normal controls. Quantitation, after normalization to tubulin which is unchanged in trisomy (data not shown), shows that in Ts65Dn, levels of pErk1 and pErk2 are 60% and 54%, respectively, of their values in normal mice (Fig. 4b,c). This decrease is significant ($p < 0.0001$). These data show that the MAPK pathway is indeed perturbed in the trisomy mice.

Discussion

The availability of the complete sequence of human chromosome 21 and the draft sequence of the homologous mouse chromosomal regions has allowed detailed comparative annotation of gene contents. Of 170 orthologous gene pairs, 140 can be associated with some functional annotation, either from reports of direct experimental analysis, or by computational methods that identify similarities to protein domains and motifs of known function. Several points are noteworthy.

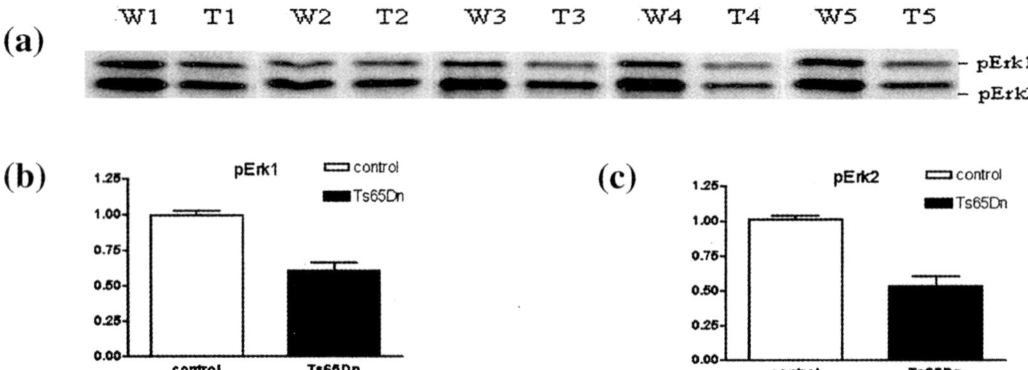

Fig. 4. Abnormalities of MAPK in Ts65Dn. Protein lysates were prepared from cerebrums or hippocampi dissected from five Ts65Dn and five normal controls, of approximately 4–6 months of age. **a** Western blots were probed with antibodies recognizing only the phosphorylated forms of Erk1 and Erk2. W1–W5, wild type (normal) mice; T1–T5, trisomic mice. **b,c** Levels of pErk1 and pErk2 in the Ts65Dn were compared to the corresponding average levels in normal mice. Decreases to 60% and 54% of normal are significant ($p < 0.0001$) by the t-test. Data are from 3–5 replicates

i) While the analysis shows that chromosome 21 contains genes predicted to function in a very broad range of cellular processes, only 94 and 71 of these are trisomic in the mouse chromosome 16 segmental trisomies, Ts65Dn and Ts1Cje. In addition, for subsets of genes functioning within a common pathway or process, again not all are trisomic in the mouse models. These points suggest that phenotypic effects of trisomy on these pathways cannot be completely reproduced in current mouse models.

ii) Simple gene dosage effects, i.e. 50% increases in activity of all trisomic genes, cannot be assumed. Of the seven chromosome 21 encoded proteins evaluated here, five showed gene dosage-related increases in 4–6 (or 12) month mouse brain and two, Ets2 and, surprisingly, Sod1, showed no increases. Some of these data are consistent with many previous reports of these genes at the RNA, protein and activity level in DS samples; others conflict (Cheon et al., 2003a,b; see also Engidawok and Lubec, 2003, for some summaries). Indeed, a review of the literature shows that data on overexpression are frequently conflicting. Possible causes for differences among experiments include the use of RNA vs protein, mouse vs human samples, adult vs fetal brain, and whole brain vs different brain regions vs cell lines. In addition, for a single gene, isoforms resulting from alternative splice variants may not show equal dosage effects (consider the short vs long forms of Itsn1 and Gart, Fig. 3). Careful and comprehensive quantitation of levels of chromosome 21 encoded proteins and their mouse orthologues in trisomy vs normal are required. Failure to observe gene dosage effects could be due to disregulation of processes involved in transcription, translation and/or processing, all of which use components encoded on chromosome 21.

iii) A number of chromosome 21 genes function within the same complex or pathway. Eleven genes are associated with MAPK and eight with Ca-calcineurin signaling. Functional data strongly predict perturbation of these pathways by overexpression of the chromosome 21 components. Preliminary data for the MAPK pathway suggest it is indeed perturbed modestly in unstressed Ts65Dn mice.

iv) Even MAPK and Ca/calcineurin pathways are not independent with respect to chromosome 21 genes. DSCR1, APP and SYNJ1 proteins function within both. In addition, DSCR1 is activated by phosphorylation by MAPK (Rothermel et al., 2003); if MAPK activity is perturbed, DSCR1 relative activity could be affected in addition to protein levels being increased.

v) Predicting effects of extra copies of transcription factors is not trivial. For example, GABPA functions in a complex with the non-chromosome 21 GABPB, which is required to enhance GABPA binding to DNA. In addition, GABPA must be phosphorylated by MAPK to be capable of transcription activation of target genes (Fromm and Burden, 2001). Thus, disruption of complex stoichiometry due to overexpression of GABPA and altered levels of phosphorylation due to perturbed MAPK both could contribute to disregulation of transcription of GABPA target genes.

Lastly, the number of chromosome 21 genes and the number and complexity of their interactions, responses and effects, limited as current knowledge is, suggests that it is unlikely that there are any "single gene-one phenotypic feature" correlations to be discovered; rather the emphasis needs to include "one pathway-one phenotypic feature" correlations. Furthermore, it is unlikely that experimental systems alone, whether cell culture or mouse models, will be sufficient to unravel gene-pathway-phenotype correlations in DS. The presence of modest perturbations, in pathways, in both positive and negative feedback loops, and in downstream transcriptional networks, will require the incorporation into DS research of mathematical modeling. Happily, such approaches to systems biology are being developed (Neves and Iyenga, 2002; Bolouri and Davidson, 2002).

Acknowledgements

This work was supported by grants from the Fondation Jerome Lejeune and the National Down Syndrome Society, and by HD17449 from the National Institutes of Health. We thank L. Crnic (University of Colorado Health Sciences Center) and A. Costa (Eleanor Roosevelt Institute) for Ts65Dn mice; D. Patterson (Eleanor Roosevelt Institute) for GART antibodies; and A. Fortna and K. O'Brien for technical assistance.

References

Adams JP, Sweatt JD (2002) Molecular psychology: roles for the ERK MAP kinase cascade in memory. Annu Rev Pharmacol Toxicol 42: 135–163

Alexanian AR, Bamburg JR (1999) Neuronal survival activity of S100ββ is enhanced by calcineurin inhibitors and requires activation of NF-κB. FASEB J 13: 1611–1620

Blum S, Moore AN, Adams F, Dash PK (1999) A mitogen-activated protein kinase cascade in the CA1/CA2 subfield of the dorsal hippocampus is essential for long-term spatial memory. J Neurosci 19: 3535–3544

Bolouri H, Davidson EH (2002) Modeling transcriptional regulatory networks. BioEssays 24: 1118–1129

Cavaillès V, Dauvois S, L'Horset F, Lopez G, Hoare S, Kushner PJ, Parker MG (1995) Nuclear factor RIP140 modulates transcriptional activation by the estrogen receptor. EMBO J 14: 3741–3751

Chen-Hwang M-C, Chen H-R, Elzinga M, Hwang Y-W (2002) Dynamin is a minibrain kinase/dual specificity Yak1-related kinase 1A substrate. J Biol Chem 277: 17597–17604

Cheon MS, Kim SH, Yaspo ML, Blasi F, Aoki Y, Melen K, Lubec G (2003) Protein levels of genes encoded on chromosome 21 in fetal Down syndrome brain: challenging the gene dosage effect hypothesis, part I. Amino Acids 24: 111–117

Cooper JD, Salehi A, Delcroix J-D, Hower CL, Belichenko PV, Cua-Couzens J, Kilbridge JR, Carlson EJ, Epstein CJ, Mobley WC (2001) Failed retrograde transport of NGF in a mouse model of Down's syndrome: reversal of cholinergic neurodegenerative phenotypes following NGF infusion. Proc Natl Acad Sci USA 98: 10439–10444

Costa RM, Federov NB, Kogan JH, Murphy GG, Stern J, Ohno M, Kucherlapati R, Jacks T, Silva AJ (2002) Mechanism for the learning deficits in a mouse model of neurofibromatosis type 1. Nature 415: 526–530

Cousin MA, Tan TC, Robinson PJ (2001) Protein phosphorylation is required for endocytosis in nerve terminals: potential role for the dephosphins dynamin I and synaptojanin, but not AP180 or amphiphysin. J Neurochem 76: 105–116

Crnic LS, Pennington BF (2000) Down syndrome: neuropsychology and animal models. Progr Infancy Res 1: 69–111

Davisson MT, Costa ACS (1999) Mouse models of Down syndrome. In: Mouse models in the study of genetic neurological disorders. Plenum Press, New York, pp 297–327 (Adv Neurochem)

Davisson MT, Schmidt C, Reeves RH, Irving NG, Akeson EC, Harris BS, Bronson RT (1993) Segmental trisomy as a mouse model for Down syndrome. Prog Clin Biol Res 384: 117–133

Engidawork E, Lubec G (2003) Molecular changes in fetal Down syndrome brain. J Neurochem 84: 895–904

Ermak G, Morgan TE, Davies KJA (2001) Chronic overexpression of the calcineurin inhibitory gene *DSCR1* (*Adapt78*) is associated with Alzheimer's disease. J Biol Chem 276: 38787–38794

Ermak G, Harris CD, Davies KJA (2002) The *DSCR1* (*Adapt78*) isoform 1 protein calcipressin 1 inhibits calcineurin and protects against acute calcium-mediated stress damage, including transient oxidative stress. FASEB J 16: 814–824

Fleming IN, Gray A, Downes CP (2000) Regulation of the Rac1-specific exchange factor Tiam1 involves both phosphoinositide 3-kinase-dependent and -independent components. Biochem J 351: 173–182

Fortna A, Gardiner K (2001) Genomic sequence analysis tools: a user's guide. Trends Genet 17: 158–164

Fromm L, Burden SJ (2001) Neuregulin 1-stimulated phosphorylation of GABP in skeletal muscle cells. Biochem 40: 5306–5312

Gardiner K, Slavov D, Bechtel L, Davisson MT (2002) Annotation of human chromosome 21 for relevance to Down syndrome: gene structure and expression analysis. Genomics 79: 833–843

Gardiner K, Fortna A, Bechtel L, Davisson M (2003) Mouse models of Down syndrome: how useful can they be: Comparison of the gene content of human chromosome 21 with homologous mouse genomic regions. Gene (in press)

Graef IA, Mermelstein PG, Stankunas K, Neilson JR, Deisseroth K, Tsien RW, Crabtree GR (1999) L-type calcium channels and GSK-3 regulate the activity of NF-ATc4 in hippocampal neurons. Nature 401: 703–708

Granholm AC, Sanders LA, Crnic LS (2000) Loss of cholinergic phenotype in basal forebrain coincides with cognitive decline in a mouse model of Down's syndrome. Exp Neurol 161: 647–663

Hassold TJ, Jacobs PA (1984) Trisomy in man. Annu Rev Genet 18: 69–97

Hattori M, Fujiyama A, Taylor TD, Watanabe H, Yada T, Park HS, Toyoda A, Ishii K, Totoki Y, Choi DK, Groner Y, Soeda E, Ohki M, Takagi T, Sakaki Y, Taudien S, Blechschmidt K, Polley A, Menzel U, Delabar J, Kumpf K, Lehmann R, Patterson D, Reichwald K, Rump A, Schillhabel M, Schudy A, Zimmermann W, Rosenthal A, Kudoh J, Schibuya K, Kawasaki K, Asakawa S, Shintani A, Sasaki T, Nagamine K, Mitsuyama S, Antonarakis SE, Minoshima S, Shimizu N, Nordsiek G, Hornischer K, Brant P, Scharfe M, Schon O, Desario A, Reichelt J, Kauer G, Blocker H, Ramser J, Beck A, Klages S, Hennig S, Riesselmann L, Dagand El, Haaf T, Wehrmeyer S, Borzym K, Gardiner K, Nizetic D, Francis F, Lehrach H, Reinhardt R, Yaspo ML (2000) The DNA sequence of human chromosome 21. Nature 405: 311–319

Holtzman DM, Santucci D, Kilbridge J, Chuacouzens J, Fontana DJ, Daniels SE, Johnson RM, Chen K, Sun YL, Carlson E, Alleva E, Epstein CJ, Mobley WC (1996) Developmental abnormalities and age-related neurodegeneration in a mouse model of Down syndrome. Proc Natl Acad Sci USA 93: 13333–13338

Hussain NK, Yamabhai M, Ramjaun AR, Guy AM, Baranes D, O'Bryan JP, Der CJ, Kay BK, McPherson PS (1999) Splice variants of Intersectin are components of the endocytic machinery in neurons and nonneuronal cells. J Biol Chem 274: 15671–15677

Hyde LA, Frisone DF, Crnic LS (2001) Ts65Dn mice, a model for Down syndrome, have deficits in context discrimination learning suggesting impaired hippocampal function. Behav Brain Res 118: 53–60

Kaplan DR, Miller FD (2000) Neurotrophin signal transduction in the nervous system. Curr Opin Neurobiol 10: 381–391

Klee CB, Ren H, Wang X (1998) Regulation of the calmodulin-stimulated protein phosphatase, calcineurin. J Biol Chem 273: 13367–13370

Korenberg JR, Chen X-N, Schipper R, Sun Z, Gonsky R, Gerwehr S, Carpenter N, Daumer C, Dignan P, Disteche C, et al (1994) Down syndrome phenotypes: the consequences of chromosomal imbalance. Proc Natl Acad Sci USA 91: 4997–5001

Lewis TS, Shapiro PS, Ahn NG (1998) Signal transduction through MAP kinase cascades. Adv Canc Re 74: 49–139

Mansuy IM, Mayford M, Jacob B, Kandel ER, Bach ME (1998) Restricted and regulated overexpression reveals calcineurin as a key component in the transition from short-term to long-term memory. Cell 92: 39–49

Marks B, McMahon HT (1998) Calcium triggers calcineurin-dependent synaptic vesicle recycling in mammalian nerve terminals. Curr Biol 8: 740–749

McPherson PS, Kay BK, Hussain NK (2001) Signaling on the endocytic pathway. Traffic 2: 375–384

Michels F, Stam JC, Hordijk PL, van der Kammen RA, Ruuls-Van Stalle L, Feltkamp CA, Collard JG (1997) Regulated membrane localization of Tiam1, mediated by the NH_2-terminal Pleckstrin homology domain, is required for Rac-dependent membrane ruffling and c-Jun NH_2-terminal kinase activation. J Cell Biol 137: 387–398

Mills J, Charest DL, Lam F, Beyreuther K, Ida N, Pelech SL, Reiner PB (1997) Regulation of amyloid precursor protein catabolism involves the mitogen-activated protein kinase signal transduction pathway. J Neurosci 17: 9415–9422

Neves SR, Iyengar R (2002) Modeling of signaling networks. BioEssays 24: 1110–1117

Nishiyhama H, Knopfel T, Endo S, Itohara S (2002) Glial protein S100B modulates long-term neuronal synaptic plasticity. Proc Natl Acad Sci USA 99: 4037–4042

Pennington BF, Moon J, Edgin J, Stedron J, Nadel L (2003) The neuropsychology of Down syndrome: evidence for hippocampal dysfunction. Child Dev 74: 75–93

Reymond A, Friedli M, Henrichsen CN, Chapo F, Deutsch S, Ucla C, Rosier C, Lyle R, Guipponi M, Antonarakis SE (2001) From PREDs and open reading frames to cDNA isolation: revisiting the human chromosome 21 transcription map. Genomics 78: 46–54

Reymond A, Camargo AA, Deutsch S, Stevenson BJ, Parmigiani RB, Ucla C, Bettoni F, Rossier C, Lyle R, Guipponi M, de Souza S, Iseli C, Jongeneel CV, Bucher P, Simpson AJ, Antonarakis SE (2002) Nineteen additional unpredicted transcripts from human chromosome 21. Genomics 79: 824–832

Rosmarin AG, Luo M, Caprio DG, Shang J, Simkevich CP (1998) Sp1 cooperates with the ets transcription factor, GABP, to activate the CD18 (beta2 leukocyte integrin) promoter. J Biol Chem 273: 13097–13103

Roßner S, Ueberham U, Schliebs R, Perez-Polo JR, Bigl V (1999) Regulated secretion of amyloid precursor protein by TrkA receptor stimulation in rat pheochromocytoma-12 cells is mitogen activated protein kinase sensitive. Neurosci Lett 271: 97–100

Rothermel BA, Vega RB, Williams RS (2003) The role of modulatory calcineurin-interacting proteins in calcineurin signaling. Trends Cardiovasc Med 13: 15–21

Sanij E, Hatzistavrou T, Hertzog P, Kola I, Wolvetang E-J (2001) Ets-2 is induced by oxidative stress and sensitizes cells to H_2O_2-induced apoptosis: implications for Down's syndrome. Biochem Biophys Res Commun 287: 1003–1008

Slemmon JR, Morgan JI, Fullerton SM, Danho W, Hilbush BS, Wengenack TM (1996) Camstatins are peptide antagonists of calmodulin based upon a conserved structural motif in PEP-19, neurogranin and neuromodulin. J Biol Chem 271: 15911–15917

Slepnev VI, De Camilli P (2000) Accessory factors in clathrin-dependent synaptic vesicle endocytosis. Nat Rev Neurosci 1: 161–72

Subramaniam N, Treuter E, Okret S (1999) Receptor interacting protein RIP140 inhibits both positive and negative gene regulation by glucocorticoids. J Biol Chem 274: 18121–18127

Sugimoto T, Stewart S, Guan K-L (1997) The calcium/calmodulin-dependent protein phosphatase calcineurin is the major Elk-1 phosphatase. J Biol Chem 272: 29415–29418

Sugiura R, Sio SO, Shuntoh H, Kuno T (2001) Molecular genetic analysis of the calcineurin signaling pathways. Cell Mol Life Sci 58: 278–288

Tamagno E, Bardini P, Obbili A, Vitali A, Borghi R, Zaccheo D, Pronzato MA, Danni O, Smith MA, Perry G, Tabaton M (2002) Oxidative stress increases expression and activity of BACE in NT_2 neurons. Neurobiol Dis 10: 279–288

Tong X-K, Hussain NK, de Heuvel E, Kurakin A, Abi-Jaoude E, Quinn CC, Olson MF, Marais R, Baranes D, Kay BK, McPherson PS (2000a) The endocytic protein Intersectin is a major binding partner for the Ras exchange factor mSos1 in rat brain. EMBO J 19: 1263–1271

Tong X-K, Hussain NK, Adams AG, O'Bryan JP, McPherson PS (2000b) Intersectin can regulate the Ras/MAP kinase pathway independent of its role in endocytosis. J Biol Chem 275: 29892–29899

Vega RB, Rothermel BA, Weinheimer CJ, Kovacs A, Naseem RH, Bassel-Duby R, Williams RS, Olson EN (2003) Dual roles of modulatory calcineurin-interacting protein 1 in cardiac hypertrophy. Proc Natl Acad Sci USA 100: 669–674

Wallace WC, Akar CA, Lyons WE (1997) Amyloid precursor protein potentiates the neurotrophic activity of NGF. Mol Brain Res 52: 201–212

Watson FL, Heerssen HM, Bhattacharyya A, Klesse L, Lin MZ, Segal RA (2001) Neurotrophins use the Erk5 pathway to mediate a retrograde survival response. Nat Neurosci 4: 981–988

Windahl SH, Treuter E, Ford J, Zilliacus J, Gustafsson J-Å, McEwan IJ (1999) The nuclear-receptor interacting protein (RIP) 140 binds to the human glucocorticoid receptor and modulates hormone-dependent transactivation. J Steroid Biochem Mol Biol 71: 93–102

Woods YL, Cohen P, Becker W, Jakes R, Goedert M, Wang X, Proud CG (2001) The kinase DYRK phosphorylates protein-synthesis initiation factor eIF2Bε at Ser^{539} and the microtubule-associated protein tau at Thr^{212}: potential role for DYRK as a glycogen synthase kinase 3-priming kinase. Biochem J 355: 609–615

Yang J, Rothermel B, Vega RB, Frey N, McKinsey TA, Olson EN, Bassel-Duby R, Williams RS (2000) Independent signals control expression of the calcineurin inhibitory proteins MCIP1 and MCIP2 in striated muscles. Circ Res 87: E61–68

Yang EJ, Ahn YS, Chung KC (2001) Protein kinase Dyrk1 activates camp response element-binding protein during neuronal differentiation in hippocampal progenitor cells. J Biol Chem 276: 39819–39824

Yoshida NL, Miyashita T, UM, Yamada M, Reed JC, Sugita Y, Oshida T (2002) Analysis of gene expression patterns during glucocorticoid-induced apoptosis using oligonucleotide arrays. Biochem Biophys Res Commun 293: 1254–1261

Zhuo M, Zhang W, Son H, Mansuy I, Sobel RA, Seidman J et al (1999) A selective role of calcineurin Aα in synaptic depotentiation in hippocampus. Proc Natl Acad Sci USA 96: 4650–4655

Author's address: Dr. Kathleen Gardiner, Eleanor Roosevelt Institute at the University of Denver, and Department of Biochemistry and Genetics, University of Colorado Health Sciences Center, 1899 Gaylord Street, Denver Co 80206, U.S.A., e-mail: kgardine@du.edu

Aberrant protein expression of transcription factors BACH1 and ERG, both encoded on chromosome 21, in brains of patients with Down syndrome and Alzheimer's disease

K. S. Shim[1], **R. Ferrando-Miguel**[2], and **G. Lubec**[1]

[1] Department of Pediatrics, and
[2] Department of Neonatology, University of Vienna, Vienna, Austria

Summary. Down syndrome (DS; trisomy 21) is a genetic disorder associated with early mental retardation and patients inevitably develop Alzheimer's disease (AD)-like neuropathological changes. The molecular defects underlying the DS — phenotype may be due to overexpression of genes encoded on chromosome 21. This so-called gene dosage hypothesis is still controversial and demands systematic work on protein expression. A series of transcription factors (TF) are encoded on chromosome 21 and are considered to play a pathogenetic role in DS. We therefore decided to study brain expression of TF encoded on chromosome 21 in patients with DS and AD compared to controls: Frontal cortex of 6 male DS patients, 6 male patients with AD and 6 male controls were used for the experiments. Immunoblotting was used to determine protein levels of TF BACH1, ERG, SIM2 and RUNX1. SIM2 and RUNX1 were comparable between groups, while BACH1 was significantly reduced in DS, and ERG was increased in DS and AD as compared to controls. These findings may indicate that DS pathogenesis cannot be simply explained by the gene dosage effect hypothesis and that results of ERG expression in DS were paralleling those in AD probably reflecting a common pathogenetic mechanism possibly explaining why all DS patients develop AD like neuropathology from the fourth decade. We conclude that TF derangement is not only due to the process of neurodegeneration and propose that TFs BACH1 and ERG play a role for the development of AD — like neuropathology in DS and pathogenesis of AD per se and the manifold increase of ERG in both disorders may form a pivotal pathogenetic link.

Introduction

Down syndrome (DS) or trisomy 21 is the most frequent genetic cause of mental retardation, arising about once in every 700 life births. Adult DS is characterized by a reduced number of cortical neurons, malformed dendritic trees and spines, abnormal synapses and the development of Alzheimer's disease type neuropathology by age 30–40 years (Epstein, 1995). Overexpression of genes residing in chromosome 21 has been proposed to be a central point in the brain deficit of DS. However, normal, increased or even decreased expression of several gene products from chromosome 21 in DS such as e.g. T-oligomycin sensitivity conferring protein, peptide 19, SOD1, collagen type VI (Cheon et al., PART I–IV, 2003; Engidawork and Lubec, 2003; Gulesserian et al., 2001; Lubec and Engidawork, 2002) has been shown, suggesting that this hypothesis may not be sufficient to explain neuropathogenesis in DS. Recently, a converging line of evidence has shown that transcription factors (TF) play an important role in CNS development and maintenance of neuronal functions including learning or memory, thus TF could serve as an attractive candidate for the explanation of brain impairment in DS: More specifically, it has been shown that basic helix-loop-helix TF scleraxis and jun D, negative regulators of cell growth, were significantly decreased in brain regions of patients with adult and fetal DS (Labudova et al., 1998; Yeghiazaryan et al., 1999). In addition, a series of transcription related factors, CRK, elongation factor1-alpha1 and 2, PTB-associated splicing factor, were significantly downregulated in brain of DS fetuses as revealed by proteomic approaches (Freidl et al., 2001). Furthermore, overexpression of DNA excision-repair-cross-complementing (ERCC) gene products ERCC2 and ERCC3 which are identical to the p89 subunit of a general TF and associated with a class II TF in AD and DS (Fang-Kircher et al., 1999) was demonstrated.

Neuronal functions depend on coordinated patterns of gene activation or inactivation through TF and many TF classes are able to generate the vast diversity of transcription elements required for brain function (He and Rosenfeld, 1991). Therefore, imbalance of one TF or a component of a transcription factor complex may alter effectiveness of the activation or repression of transcription of target genes and is likely to have profound consequences for neuronal function and wiring of the brain.

The present study was undertaken to add data on candidate TF to the current body of literature by investigating the expression of different TF encoded on chromosome 21 in frontal cortex of adult DS and AD, complementing a previous report (Ferrando-Miguel and Lubec, 2003) on brain expression of BACH1, ERG, RUNX1 and SIM2 in fetal DS brain. The finding of significant reduction of BACH1 in adult DS in contrast to increased BACH1 in fetal DS (Ferrando-Miguel and Lubec, 2003) shows the important factor "age" which is of particular interest when controversial studies are compared or hypotheses have to be tested.

Materials and methods

Brain samples

Frontal cortex from patients with AD (n = 6; six males; 59.3 ± 6.4 years old), DS (n = 6; six males: 57.8 ± 8.2 years old) and controls (n = 6; six males; 60.2 ± 9.3 years old) were used in this study. Briefly, postmortem human brain samples were obtained from the MRC London Brain Bank for Neurodegenerative Diseases, Institute of Psychiatry,

King's College, UK. The AD patients fulfilled the National Institute of Neurological Disorders and Stroke and Alzheimer's disease and Related Disorders Association criteria for probable AD (Tierney et al., 1998). The neuropathological diagnosis of "definite AD" was confirmed using the CERAD criteria (Mirra et al., 1991). All the DS patients were karyotyped and possessed trisomy 21. A formal cognitive assessment of dementia in DS was not performed. In all DS brains there were abundant and extensive beta-amyloid deposits, neurofibrillary tangles and neuritic plaques. Normal brains obtained from individuals with no history of neurological or psychiatric illness were used as controls. The major cause of death was bronchopneumonia in AD and DS and heart disease in controls. Postmortem interval of brain dissection in AD, DS and controls was 36.5 $\pm$ 27.5, 57.8 $\pm$ 8.2, 34.0 $\pm$ 12.9 h. After dissection, coronal slices were snap frozen and stored at $-70°C$ until use and the freezing chain was never interrupted.

Antibodies

The polyclonal rabbit anti-Runx1 was provided from Prof. Yoram Groner (Department of Molecular Genetics, The Weizmann Institute of Science, Israel). Details of the preparation and characterization of antibody have been described previously (Aziz-Aloya et al., 1998). Five antibodies, against BACH1 (goat polyclonal antibody, Santa Cruz Biotechnology, USA), ERG (rabbit polyclonal antibody, Santa Cruz Biotechnology, USA), SIM2 (goat polyclonal antibody, Santa Cruz Biotechnology, USA), β-actin (mouse monoclonal antibody, Serotec Ltd., UK), and neuron specific enolase (NSE, rabbit polyclonal antibody, Chemicon, UK) were purchased.

Western blotting

Brain samples were ground with a mortar under liquid nitrogen, homogenized with cold homogenization buffer containing 10 mM Tris-HCI (pH 7.5), 150 mM NaCl, 0.05% (v/v) Tween 20, 1 mM PMSF (phenylmethylsulfonyl fluoride, Sigma, Austria) and 1 tablet of protease inhibitor cocktail (Roche, Austria). Homogenized samples were centrifuged at 8,000 $\times$ g for 10 minutes at 4°C. The BCA protein assay kit (Pierce, USA) was used to determine the concentration of protein in the supernatant. Samples were mixed with the same volume of sample buffer (60 mM Tris-HCl, 2% SDS, 0.1% bromophenol blue, 25% glycerol, 14.4 mM 2-mercaptoethanol, pH 6.8) and denatured at 95°C for 15 minutes. Brain samples (20 ug) were resolved by sodium dodecyl sulfate polyacrylamide gel electrophoresis (SDS-PAGE) using a 12.5% ExcelGel SDS homogeneous gel (Amersham Pharmacia Biotech, Sweden) in Multiphor II Electrophoresis System (Amersham Pharmacia Biotech, Sweden). Proteins separated on the gel were transferred onto PVDF membranes (Millipore, Bedford, USA) and membranes were blocked in blocking buffer (10 mM Tris-HCl, pH 7.5, 150 mM NaCl, 2% non-fat dried milk, 0.1% Tween 20). Membranes were incubated for 2 hours at room temperature with diluted primary antibodies (1:700 for BACH1 and RUNX1, 1:1,000 for SIM2 and ERG, 1:4,000 for beta-actin and NSE). After 4 times washing for 10 minutes with blocking buffer, membranes were probed with 1:2,000 diluted rabbit anti-goat IgG coupled to horseradish peroxidase (Southern Biotechnology Associates, Inc., Alabama, USA) for 1 hour. Subsequent washing 3 times for 10 minutes and developing with the Western blot Chemiluminescene reagent (NEN™ Life Science Products, Inc., USA) followed.

Analysis and statistics

The density of detected bands was measured using RFLP scan software version 2.1 (MWG, Germany). Between group differences were calculated by non-parametric Mann-

Whitney U test. The statistical analysis was performed by GraphPad Instat 2 software version 2.05 and the level of significance was set at $P < 0.05$. All results are presented as mean ± standard deviation (SD).

Results

We determined protein levels of four TF encoded on chromosome 21 (BACH1, ERG, SIM2, RUNX1) in adult brains (frontal cortex) of patients with DS and AD as compared to controls by western blot. As shown in Fig. 1, we detected a double band at 81 kDa using BACH1 antibody and one band at 50 kDa with the RUNX1 antibody. The two detected bands may represent posttranslational modifications including phosphorylation and glycosylation (Blom et al., 1999) or individual isoforms (Asou, 2003; Tsuji and Noda, 2000). ERG antibody reacted with one major band at 80 kDa which was used for quantification and five isoforms, ERG1, ERG2, ERG3/p55ERG, p49ERG and p38ERG have been identified produced by a combination of differential mRNA splicing, polyadenylation sites and alternative use of translational codons (Hewet et al., 2001). The antibody against SIM2 recognized one band at 40 kDa. We also determined NSE and actin levels as reference proteins for neuronal and cellular density, and showed comparable expression in DS, AD and controls (Table 1).

Density of immunoreactive bands of SIM2 and RUNX1 was comparable between DS, AD and controls. Protein levels of BACH1 were significantly and remarkably reduced in DS, and ERG levels were significantly and manifold increased in DS and AD as compared to controls (Table 1).

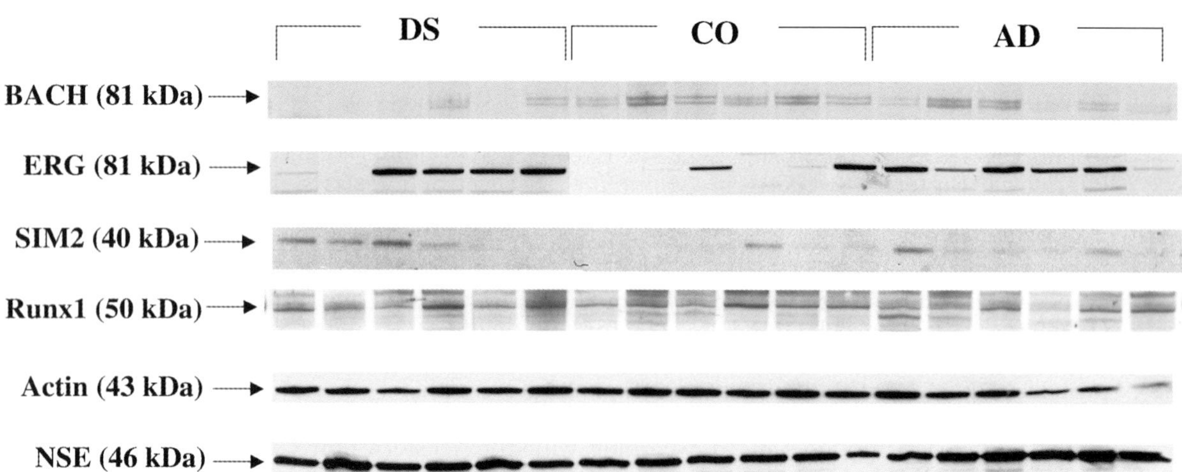

Fig. 1. Western blot of four proteins (BACH1, ERG, SIM2, RUNX1) in frontal cortex of patients with DS, AD and controls. Twenty micrograms of protein were separated on 12.5% homogeneous gel and transferred onto PVDF membranes. The membranes were immunoreacted with primary and secondary antibody as described in "materials and methods", and specific immunoreactive band (BACH1, 81 kDa; ERG, 80 kDa; SIM2, 40 kDa; RUNX1, 50 kDa; Actin, 43 kDa, NSE, 46 kDa) were detected

Table 1. Protein expression levels of four TF encoded on chromosome 21 in frontal cortex of patients with DS, AD and controls. Values are arbitrary units of densitometry. * $P < 0.05$

	BACH1	ERG	SIM	RUNX1	NSE	Actin
DS	1.66 ± 1.65*	9.96 ± 8.13*	3.68 ± 4.79	7.41 ± 3.31	18.42 ± 4.75	9.71 ± 2.10
CO	7.65 ± 6.63	2.76 ± 6.38	1.04 ± 1.37	5.72 ± 2.69	16.75 ± 4.96	11.58 ± 1.67
AD	4.46 ± 4.47	12.25 ± 8.55*	1.28 ± 0.88	4.22 ± 1.65	14.87 ± 4.21	7.29 ± 4.02

Table 2. The density of four TF was normalized with that of either actin or NSE levle in frontal cortex of patients with DS, AD and controls. * $P < 0.05$

		BACH1	ERG	SIM2	RUNX1
Protein density/Actin	DS	0.10 ± 0.11*	1.14 ± 1.16*	0.50 ± 0.75	0.75 ± 0.22
	CO	0.47 ± 0.19	0.27 ± 0.02	0.09 ± 0.13	0.51 ± 0.25
	AD	0.30 ± 0.27	1.99 ± 2.00	0.18 ± 0.18	0.93 ± 1.06
Protein density/NSE	DS	0.16 ± 0.13*	0.59 ± 0.49	0.23 ± 0.30	0.43 ± 0.24
	CO	0.65 ± 0.27	0.31 ± 0.73	0.06 ± 0.06	0.36 ± 0.16
	AD	0.58 ± 0.38	0.83 ± 0.56	0.09 ± 0.09	0.32 ± 0.19

When levels of TF proteins were normalised with those of actin, statistical reduction of BACH1 and the increase of ERG in DS persisted (Table 2). We also used NSE levels in order to normalise TF levels by neuronal density. The significant reduction of BACH1 was even more pronounced when related to NSE. NSE — normalised ERG levels still showed increased expression in DS and AD although not reaching statistical significance (Table 2). There was no statistically significant correlation between protein expression of BACH1 and ERG and either age or post-mortem interval of brain samples (data not shown).

Discussion

The major findings of the study are decreased expression of BACH1 protein in DS and increased expression of ERG protein in DS and AD frontal cortex.

BACH1 is a member of a novel family of the broad complex, tramtrack, bric-a-brac/poxvirus (BTB/POZ) and basic region leucine zipper transcription factor (TF) (Oyake et al., 1996). It is ubiquitously expressed in various human tissues and forms heterodimers with the protooncogene family member MafK, serving as transcription repressor or activator in mammalian cells. BACH1 nucleates generation of a multiprotein complex *in vitro*, binding to NF-E2 binding site and Maf recognition elements (MARE) which are present in the regulatory domains of various genes involved in heme metabolism such as oxidative stress-response heme oxygenase-1 (HO-1) gene (Ogawa et al., 2001). Gene targeting experiments in mice revealed that HO-1 became expressed constitutively at high levels in various tissues in the absence of BACH1 (Sun et al., 2002). In addition, Katsuoka et al. (2003) have reported that mafG::mafK — mutant mice display neuronal degeneration. The key biochemical deficiency in neurons of mutant is that Bach/small Maf heterodimers were significantly underrepresented and BACH proteins fail to accumulate normally in nuclei, thus HO-1 was markedly induced coincident with a clear reduction in the Bach/Maf complex in brain extracts. Furthermore, Kitamuro et al. (2003) have shown that BACH1 functions as a hypoxia inducible repressor for the HO-1 gene, suggesting that BACH1 could contribute to the fine-tuning of oxygen homeostasis in human cells through regulation of HO-1. Moreover, it has been proposed that chronic over expression of HO-1 could exacerbate oxidative stress in brain tissues by stimulating reactive oxygen species generation which is considered as pathogenetic principle for the development of neurodegenerative disorders including DS and AD (Schipper, 2000). In this study, we observed significantly decreased expression of BACH1 in adult brain of patients with DS, and a non-significant decrease in AD, suggesting that deteriorated BACH1 protein expression may result in the stoichiometric imbalance of the BACH1/Maf complex, thus lead to dysregulation of antioxidant systems in adult DS and probably in AD. Reduction of chromosome 21 encoded BACH1 in adult DS does not support the gene dosage hypothesis but may reflect or contribute to pathogenesis, i.e. neurodegeneration, in adult DS. BACH1 in fetal DS brain, however, was overexpressed during the early second trimester of gestation and may be therefore involved in the development of AD like neuropathology later in life (Ferrando-Miguel and Lubec, 2003).

ERG belongs to the *ets* gene family acting as a regulator of genes required for maintenance and/or differentiation of early hematopoietic and non-hematopoietic cells (Trojanowska, 2000). During development in the mouse, ERG expression has been detected in cells derived from mesoderm and in neuroectoderm by northern analysis (Maroulakaou and Bowe, 2000). ERG has two trans-activation and a DNA-binding domain which form homo- and hetero- dimers with other *ets* family members and differential interactions with several TF. Using the yeast two hybrid system, it has been demonstrated that ERG directly interacts with histone methyltransferase, ESET, forming a multi protein complex with histone deacetylase and transcription co-repressors to regulate transcription (Yang et al., 2002). Additionally, ERG builds a ternary complex with c-Fos/c-Jun heterodimer and ETS2, specific for the stromelysin-1 (MMP-3) promoter (Verger et al., 2001). ERG-3 forms dimers with itself and other isoforms and complexes with c-Fos/c-Jun heterodimer via the ETS domain (Basuyaux et al., 1997). The TF c-Fos responds to neuronal excitation enabling long-term potentiation, learning and memory, and initiates programmed cell death in neurons (Engidawork and Lubec, 2003). Previously our group reported increased mRNA levels of c-Fos in adult DS frontal, parietal and temporal cortices (Greber-Platzer et al., 1999). Recently it was observed that expression of ERG inhibited apoptosis of NIH3T3 cells induced by either serum deprivation or by treatment with calcium ionophore, suggesting that ERG may play a role in cell growth and differentiation through an apoptotic pathway (Yi et al., 1997). Based on the interaction between Fos and ERG systems along with the importance of Fos in plasticity and long-term potentiation for cognitive functions such as memory and learning, increased expression of ERG alone or along with aberrant c-Fos impaired plasticity, wiring of the brain and apoptosis in DS and AD brain. Increased ERG in DS and AD may point to common pathogenetic mechanisms and such a common mechanism has to be postulated as all individuals with DS develop AD like neuropathology from the fourth decade (Cairns, 1999). The specificity of ERG overexpression is supported by findings of comparable chromosome 21 encoded TF SIM2 and RUNX1.

The single minded (*sim*) gene encodes a transcriptional regulator that functions as a key determinant of CNS midline development in *Drosophila* (Moffett et al., 1996). The SIM protein contains a basic helix-loop-helix (bHLH) motif and localizes to the cell nucleus. bHLH transcription factors appear to play important roles in directing cell fate and controlling cell proliferation and differentiation during embryonic development. In addition to *Drosophila sim*, two murine homologs of SIM, SIM1 and SIM2, were identified and shown that its expression is developmentally regulated (Fan et al., 1996). SIM1 has been shown to be essential in hypothalmic development by targeted gene inactivation in the mouse. SIM2 was first identified by exon trapping of a region of the human chromosome 21 proposed to be associated with many of the pathological features of DS (Oyake et al., 1996). SIM2 is expressed early during development in many tissues affected in DS including the developing forebrain, ribs, vertebrae, and skeletal muscles of limbs and kiney (Vialard et al., 2000). Analysis of transgenic mice overexpessing the mouse SIM2 gene located on synthetic mouse chromosome 16 has demonstrated mild impairment of learning tasks such as fear conditioning and the spatial water maze test (Ema et al., 1999). Recently, it has been shown that artificial chromosome transgenic mice overexpressing one or two additional copies of mouse SIM2 expressed in the same spatial pattern as physiologically (Chrast et al., 2000) exhibit abnormal anxiety-related/reduced exploratory behaviour and sensitivity to pain, suggesting that overexpression of SIM2 contributes to some symptoms of the complex DS phenotype (Seidl et al., 2003). Herein, however, SIM2 levels were statistically comparable between DS, AD and controls although SIM2 levels were increased. Re-analysis with a higher number of brain samples may reveal a statistically proven increase, in this study we focused on a small but homogeneous cohort and the male sex exclusively. The inherent problems of postmortem specimen and vast variety of human expression of DS and AD may be solved by testing hundreds of specimen with multivariance analysis only.

The RUNX1/AML1 gene resides on the long arm of chromosome 21 (21q22) and is one out of three members in the *runt* domain gene family, AML1, AML2, AML3 (Westendorf et al., 1999). All AML proteins bind DNA as heterodimers formed with the β subunit of core-binding factor (CBFβ) (Bernardin and Friedman, 2002). Although RUNX1/AML1 shows relatively weak transcription activation itself, it becomes an effective transcription enhancer or repressor, when cooperating with other TF such as Myb, Ets, p300/CBP and mSin3A (Perry et al., 2002). These do not bind DNA directly but stimulate transcription by acetylating histones and/or recruiting

the RNA polymerase II transcription initiation complex. RUNX1/AML1 mRNA is expressed at high level in the thymus, bone marrow and hematopoietic system in embryo and adult mouse, but weakly detected in small diameter neurons during mouse embryogenesis (Levanon et al., 2001). Mice lacking the RUNX1/AML1 gene die *in utero*, because of hemorrhage in the central nervous system and lack of fetal liver hematopoiesis, demonsrating that RUNX1/AML1 is essential for terminal differentiation of blood cell lineages and normal hematopoiesis (Okuda et al., 1996). Recently an immunohistochemistry study has demonstrated that RUNX1/AML1 was observed in astrocytic tumors and metastatic brain melanomas consistent with the expression of its homologs in Drosophila neuroblasts (Perry et al., 2002), suggesting a yet-unrecognized role for this TF in the mammalian nervous system. RUNX1 as well as SIM2 were also comparable to controls in brain of the fetal period of DS.

Methodologically, the use of housekeeping proteins as e.g. actin or glyceraldehyde-3-phosphate-dehydrogenase (GAPDH) for the normalisation of protein levels has been challenged as these structures were shown to be deranged and actins are subjects to the apoptotic process and proteolytic degradation taking place in neurodegeneration (Guenal et al., 1997; LeBlanc et al., 1998; Sawa, 1999; Mazzola and Sirover, 2002). In our study results for DS were identical for DS frontal cortex but statistical significance disappeared for ERG in AD. Normalisation therefore has to be interpreted with caution.

In conclusion, this is the first study of chromosome 21-encoded TF in human brain showing aberrant expression of BACH1 and ERG in frontal cortex at the protein level. The major finding is that BACH1, involved in reactive oxygen species homeostasis, is statistically decreased in brain of patients with DS, a finding in agreement with the proposed involvement of reactive oxygen species in neurodegeneration. The fact that increased ERG is common to DS and AD brain suggests a common pathogenetic principle for the development of AD-like neuropathology in adult DS and AD per se and this may help to explain why all DS patients develop AD from the fourth decade. Last not least impairment of the concerted action of several TF and TF systems in DS and AD may lead to the multitude of brain deficits.

Acknowledgements

We thank Dr. N. J. Cairns (MRC London Brain Bank for Neurodegenerative Disease, Institute of Psychiatry, King's College, UK) for post-mortem human brain tissues and Dr. Y. Groner (Department of Molecular Genetics, The Weizmann Institue of Science, Israel) for the antibody against RUNX1. We are highly indebted to the Red Bull Company, Salzburg, for generous financial support.

References

Asou N (2003) The role of a Runt domain transcription factor AML1/RUNX1 in leukemogenesis and its clinical implications. Crit Rev Oncol Hematol 45: 129–150

Aziz-Aloya RB, Levanon D, Karn H, Kidron D, Goldenberg D, Lotem J, Polak-Chaklon S, Groner Y (1998) Expression of AML1-d, a short human AML1 isoform, in embryonic stem cells suppresses in vivo tumor growth and differentiation. Cell Death Differ 5: 765–773

Basuyaux JP, Ferreira E, Stehelin D, Buttice G (1997) The Ets transcription factors interact with each other and with the c-Fos/c-Jun complex via distinct protein domains in a DNA-dependent and -independent manner. J Biol Chem 272: 26188–26195

Bernardin F, Friedman AD (2002) AML1 stimulates G1 to S progression via its transactivation domain. Oncogene 21: 3247–3252

Blom N, Gammeltoft S, Brunak S (1999) Sequence and structure-based prediction of eukaryotic protein phosphorylation sites. J Mol Biol 294: 1351–1362

Cairns NJ (1999) Neuropathology. J Neural Transm [Suppl] 57: 61–74

Cheon MS, Bajo M, Kim SH, Claudio JO, Stewart AK, Patterson D, Kruger WD, Kondoh H, Lubec G (2003) Protein levels of genes encoded on chromosome 21 in fetal Down syndrome brain: challenging the gene dosage effect hypothesis, part II. Amino Acids 24: 119–125

Cheon MS, Kim SH, Ovod V, Kopitar Jerala N, Morgan JI, Hatefi Y, Ijuin T, Takenawa T, Lubec G (2003) Protein levels of genes encoded on chromosome 21 in fetal Down syndrome brain: challenging the gene dosage effect hypothesis, part III. Amino Acids 24: 127–134

Cheon MS, Kim SH, YaspoML, Blasi F, Aoki Y, Melen K, Lubec G (2003) Protein levels of genes encoded on chromosome 21 in fetal Down syndrome brain: challenging the gene dosage effect hypothesis, part I. Amino Acids 24: 111–117

Cheon MS, Shim KS, Kim SH, Hara A, Lubec G (2003) Protein levels of genes encoded on chromosome 21 in fetal Down syndrome brain: challenging the gene dosage effect hypothesis, part IV. Amino Acids 25: 41–47

Chrast R, Scott HS, Madani R, Huber L, Wolfer DP, Prinz M, Aguzzi A, Lipp HP, Antonarakis SE (2000) Mice trisomic for a bacterial artificial chromosome with the single-minded 2 gene (Sim2) show phenotypes similar to some of those present in the partial trisomy 16 mouse models of Down syndrome. Hum Mol Genet 9: 1853–1864

Ema M, Ikegami S, Hosoya T, Mimura J, Ohtani H, Nakao K, Inokuchi K, Katsuki M, Fujii-Kuriyama Y (1999) Mild impairment of learning and memory in mice overexpressing the mSim2 gene located on chromosome 16: an animal model of Down's syndrome. Hum Mol Genet 8: 1409–1415

Engidawork E, Lubec G (2003) Molecular changes in fetal Down syndrome brain. J Neurochem 84: 895–904

Engidawork E, Balic N, Fountoulakis M, Dierssen M, Greber-Platzer S, Lubec G (2001) Beta-amyloid precursor protein, ETS-2 and collagen alpha 1 (VI) chain precursor, encoded on chromosome 21, are not overexpressed in fetal Down syndrome: further evidence against gene dosage effect. J Neural Transm [Suppl] 61: 335–346

Epstein CJ (1995) Down syndrome. In: Scriver CR, Beaudet AL, Sly WS, Valle D (eds) The metabolic and molecular bases of inherited disease, 7th edn, vol I. McGraw Hill, New York, pp 749–794

Fan CM, Kuwana E, Bulfone A, Fletcher CF, Copeland NG, Jenkins NA, Crews S, Martinez S, Puelles L, Rubenstein LR, Tessier-Lavigne M (1996) Expression patterns of two murine homologs of Drosophila single-minded suggest possible roles in embryonic patterning and in the pathogenesis of Down syndrome. Mol Cell Neurosci 7: 1–16

Fang-Kircher SG, Labudova O, Kitzmueller E, Rink H, Cairns N, Lubec G (1999) Increased steady state mRNA levels of DNA-repair genes XRCC1, ERCC2 and ERCC3 in brain of patients with Down syndrome. Life Sci 64: 1689–1699

Ferrando-Miguel R, Lubec G (2003) Overexpression of transciption factor BACH1 in fetal Down syndrome brain. J Neural Transm [Suppl] 67 (this volume)

Freidl M, Gulesserian T, Lubec G, Fountoulakis M, Lubec B (2001) Deterioration of the transcriptional, splicing and elongation machinery in brain of fetal Down syndrome. J Neural Transm [Suppl] 61: 47–57

Greber-Platzer S, Balcz B, Cairns N, Lubec G (1999) c-fos expression in brains of patients with Down syndrome. J Neural Transm [Suppl] 57: 75–85
Guenal I, Risler Y, Mignotte B (1997) Down-regulation of actin genes precedes microfilament network disruption and actin cleavage during p53-mediated apoptosis. J Cell Sci 110: 489–495
Gulesserian T, Engidawork E, Fountoulakis M, Lubec G (2001) Antioxidant proteins in fetal brain: superoxide dismutase-1 (SOD-1) protein is not overexpressed in fetal Down syndrome. J Neural Transm [Suppl] 61: 71–84
He X, Rosenfeld MG (1991) Mechanisms of complex transcriptional regulation: implications for brain development. Neuron 7: 183–196
Hewett PW, Nishi K, Daft EL, Clifford Murray J (2001) Selective expression of erg isoforms in human endothelial cells. Int J Biochem Cell Biol 33: 347–355
Katsuoka F, Motohashi H, Tamagawa Y, Kure S, Igarashi K, Engel JD, Yamamoto M (2003) Small Maf compound mutants display central nervous system neuronal degeneration, aberrant transcription, and Bach protein mislocalization coincident with myoclonus and abnormal startle response. Mol Cell Biol 23: 1163–1174
Kitamuro T, Takahashi K, Ogawa K, Udono-Fujimori R, Takeda K, Furuyama K, Nakayama M, Sun J, Fujita H, Hida W, Hattori T, Shirato K, Igarashi K, Shibahara S (2003) Bach1 functions as a hypoxia-inducible repressor for the heme oxygenase-1 gene in human cells. J Biol Chem 278: 9125–9133
Labudova O, Krapfenbauer K, Moenkemann H, Rink H, Kitzmuller E, Cairns N, Lubec G (1998) Decreased transcription factor junD in brains of patients with Down syndrome. Neurosci Lett 252: 159–162
Labudova O, Kitzmueller E, Rink H, Cairns N, Lubec G (1999) Gene expression in fetal Down syndrome brain as revealed by subtractive hybridization. J Neural Transm [Suppl] 57: 125–136
LeBlanc A (1998) Detection of actin cleavage in Alzheimer's disease. Am J Pathol 152: 329–332
Levanon D, Brenner O, Negreanu V, Bettoun D, Woolf E, Eilam R, Lotem J, Gat U, Otto F, Speck N, Groner Y (2001) Spatial and temporal expression pattern of Runx3 (Aml2) and Runx1 (Aml1) indicates non-redundant functions during mouse embryogenesis. Mech Dev 109: 413–417
Lubec G, Engidawork E (2002) The brain in Down syndrome (TRISOMY 21). J Neurol 249: 1347–1356
Maroulakou IG, Bowe DB (2000) Expression and function of Ets transcription factors in mammalian development: a regulatory network. Oncogene 19: 6432–6442
Mazzola JL, Sirover MA (2002) Alteration of intracellular structure and function of glyceraldehyde-3-phosphate dehydrogenase: a common phenotype of neurodegenerative disorders. Neurotoxicology 23: 603–609
Mirra SS, Heyman A, McKeel D, Sumi SM, Crain BJ, Brownlee LM, Vogel FS, Hughes JP, van Belle G, Berg L (1991) The Consortium to Establish a Registry for Alzheimer's Disease (CERAD) Part 2 Standardization of the neuropathologic assessment of Alzheimer's disease. Neurology 41: 479–486
Moffett P, Dayo M, Reece M, McCormick MK, Pelletier J (1996) Characterization of msim, a murine homologue of the Drosophila sim transcription factor. Genomics 35: 144–155
Muenke M, Bone LJ, Mitchell HF, Hart I, Walton K, Hall-Johnson K, Ippel EF, Dietz-Band J, Kvaloy K, Fan CM (1995) Physical mapping of the holoprosencephaly critical region in 21q22.3, exclusion of SIM2 as a candidate gene for holoprosencephaly, and mapping of SIM2 to a region of chromosome 21 important for Down syndrome. Am J Hum Genet 57: 1074–1079
Nicham R, Weitzdörfer R, Hauser E, Freidl M, Schubert M, Wurst E, Lubec G, Seidl R (2003) Spectrum of cognitive, behavioural and emotional problems in children and young adults with Down syndrome. J Neural Transm [Suppl 67] (this volume)
Ogawa K, Sun J, Taketani S, Nakajima O, Nishitani C, Sassa S, Hayashi N, Yamamoto M, Shibahara S, Fujita H, Igarashi K (2001) Heme mediates derepression of Maf recog-

nition element through direct binding to transcription repressor Bach1. EMBO J 20: 2835–2843
Okuda T, van Deursen J, Hiebert JW, Grosveld G, Downing JR (1996) AML1, the target of multiple chromosomal translocations in human leukemia, is essential for normal fetal liver hematopoiesis. Cell 84: 321–330
Oyake T, Itoh K, Motohashi H, Hayashi N, Hoshino H, Nishizawa M, Yamamoto M, Igarashi K (1996) Bach proteins belong to a novel family of BTB-basic leucine zipper transcription factors that interact with MafK and regulate transcription through the NF-E2 site. Mol Cell Biol 16: 6083–6095
Perry C, Eldor A, Soreq H (2002) Runx1/AML1 in leukemia: disrupted association with diverse protein partners. Leuk Res 26: 221–228
Perry C, Sklan EH, Birikh K, Shapira M, Trejo L, Eldor A, Soreq H (2002) Complex regulation of acetylcholinesterase gene expression in human brain tumors. Oncogene 21: 8428–8441
Sawa A (1999) Neuronal cell death in Down's syndrome. J Neural Transm [Suppl] 57: 87–97
Schipper HM (2000) Heme oxygenase-1: role in brain aging and neurodegeneration. Exp Gerontol 35: 821–830
Sun J, Hoshino H, Takaku K, Nakajima O, Muto A, Suzuki H, Tashiro S, Takahashi S, Shibahara S, Alam J, Taketo MM, Yamamoto M, Igarashi K (2002) Hemoprotein Bach1 regulates enhancer availability of heme oxygenase-1 gene. EMBO J 21: 5216–5224
Tierney MC, Fisher RH, Lewis AJ, Zorzitto ML, Snow WG, Reid DW, Nieuwstraten P (1998) The NINCDS-ADRDA Work Group criteria for the clinical diagnosis of probable Alzheimer's disease: a clinicopathologic study of 57 cases. Neurology 38: 359–364
Trojanowska M (2000) Ets factors and regulation of the extracellular matrix. Oncogene 19: 6464–6471
Tsuji K, Noda M (2000) Identification and expression of a novel 3′-exon of mouse Runx1/Pebp2 alphaB/Cbfa2/AML1 gene. Biochem Biophys Res Commun 274: 171–176
Verger A, Buisine E, Carrere S, Wintjens R, Flourens A, Coll J, Stehelin D, Duterque-Coquillaud M (2000) Identification of amino acid residues in the ETS transcription factor Erg that mediate Erg-Jun/Fos-DNA ternary complex formation. J Biol Chem 276: 17181–17189
Vialard F, Toyama K, Vernoux S, Carlson EJ, Epstein CJ, Sinet PM, Rahmani Z (2000) Overexpression of mSim2 gene in the zona limitans of the diencephalon of segmental trisomy 16 Ts1Cje fetuses, a mouse model for trisomy 21: a novel whole-mount based RNA hybridization study. Brain Res Dev Brain Res 121: 73–78
Westendorf JJ, Hiebert SW (1999) Mammalian runt-domain proteins and their roles in hematopoiesis, osteogenesis, and leukaemia. J Cell Biochem [Suppl] 32: 51–58
Yang L, Xia L, Wu DY, Wang H, Chansky HA, Schubach WH, Hickstein DD, Zhang Y (2002) Molecular cloning of ESET, a novel histone H3-specific methyltransferase that interacts with ERG transcription factor. Oncogene 21: 148–152
Yeghiazaryan K, Turhani-Schatzmann D, Labudova O, Schuller E, Olson EN, Cairns N, Lubec G (1999) Downregulation of the transcription factor scleraxis in brain of patients with Down syndrome. J Neural Transm [Suppl] 57: 305–314
Yi H, Fujimura Y, Ouchida M, Prasad DD, Rao VN, Reddy ES (1997) Inhibition of apoptosis by normal and aberrant Fli-1 and erg proteins involved in human solid tumors and leukemias. Oncogene 14: 1259–1268

Authors' address: Prof. Dr. G. Lubec, CChem, FRSC (UK), Department of Pediatrics, University of Vienna, Währinger Güertel 18, A-1090 Vienna, Austria, e-mail: gert.lubec@akh-wien.ac.at

Cell cycle and cell size regulation in Down Syndrome cells

M. Rosner, A. Kowalska, A. Freilinger, A-R. Prusa, E. Marton, and M. Hengstschläger

Obstetrics and Gynecology, Prenatal Diagnosis and Therapy, University of Vienna, Vienna, Austria

Summary. Although the neuropathological features typical for Down Syndrome obviously result from deregulation of both, cell cycle control and differentiation processes, so far research focused on the latter. Considering the known similarities between the neuropathology of Down Syndrome and Alzheimer's disease and the knowledge, that in Alzheimer's disease neuronal degeneration is associated with the activation of mitogenic signals and cell cycle activation, it is tempting to investigate the consequences of an additional chromosome 21 on mammalian cell cycle regulation. We analysed the distribution of cells in different cell cycle phases on the flowcytometer and the cell size of human amniotic fluid cells with normal karyotypes and with trisomy 21. We could not detect any significant differences suggesting that the presence of an additional copy of the about 225 genes on human chromosome 21 does not trigger cell cycle effects in amniotic fluid cells. These data provide new insights into the cell biology of trisomy 21 cells.

Introduction

Down Syndrome is the most frequent genetic cause of mental retardation, affecting up to 1 in 700 pregnancies. 95% of the patients have complete trisomy of chromosome 21 due to nondisjunction during gamete formation, while in the remaining patients the syndrome is due to chromosomal translocation (Antonarakis, 1993; Yoon, 1996). This syndrome is characterized by a specific phenotype featured by the classic facial appearance, mental retardation and other major congenital malformations such as those of the heart and of the gastrointestinal tract (Hernandez, 1996).

Despite a widespread interest in Down Syndrome research the relationship between mental retardation and this aneuploidy remains elusive. The neurohistological features of this syndrome are complex and variable. Brains of Down Syndrome patients have been reported to exhibit two hallmark lesions, senile plaques and neurofibrillary tangles, and delayed myelination, fewer neurons, lower neuronal density and distribution, and abnormal synaptic density and length (reviewed in Cairns, 1999; Engidawork, 2003). Although these features obviously result from overlapping effects of both, deregulated differentiation and altered cell cycle control, research on Down Syndrome focuses on the investigation of modulated differentiation control. This is particularly surprising considering the known similarities between Down Syndrome and Alzheimer's disease. Down Syndrome is characterized by an Alzheimer's disease-like neuropathology (β-amyloid plaques, neurofibrillary tangels and neuronal loss) known to supervene later in life (recently reviewed in Engidawork, 2003). Degeneration in Alzheimer's disease is well known to be associated with the activation of mitogenic signaling and cell cycle activation (Arendt, 2002). Detailed investigations of the cell cycle regulation in cells harbouring an additional chromosome 21 are missing.

The somatic mammalian cell cycle is typically divided into four phases. The periods associated with replication (S phase) and mitosis (M phase) are separated by gaps of varying length, called G1 and G2. The major regulatory points of this cycle are controlled by a class of protein kinases known as cyclin-dependent kinases (Cdks), composed of a catalytic subunit and a regulatory subunit known as cyclin (Hengstschläger, 1999a). The processes of cell cycle progression (cell proliferation) and cell size regulation (cell growth) are coordinated since initiation of DNA replication does not occur until cells reach a minimum size. In yeast, this control acts at a point in G1 phase, called START, and in mammals commitment to DNA replication is also thought to occur at a point in G1, termed the restriction point (Nurse, 1975; Pardee, 1974; Zetterberg, 1995; Hengstschläger, 1999a).

In the here reported study we compared Down Syndrome cells with normal control cells by investigating the distribution of cells in the different cell cycle phases and by cell size analyses. We decided to perform this study using human amniotic fluid cells as the biological system because of the following reasons: 1) These are primary cells, which are neither immortalized nor transformed, making them at least to some extent more representative for *in vivo* situations. 2) The mixture of cells in human amniotic fluid is speculated to origin from all three germ layers, ectoderm, mesoderm and endoderm (Prusa, 2002). Accordingly, the drawn conclusions are not restricted to only one specific cell type. 3) These cells exhibit natural occuring proliferative potential and need not to be artificially stimulated to proliferate, such as e.g. lymphocytes via phytohemagglutinin.

Materials and methods

Amniotic fluid cell cultivation

Human amniotic fluid cell samples were obtained from amniocentesis performed for routine prenatal genetic diagnosis. This project has been reviewed and accepted by the ethics commission of the University of Vienna, General Hospital, Austria (project number: 036/2002) and each patient signed a written informed consent. Amniotic fluid cells were grown according to the standard *in vitro* culturing procedure in standard medium: Nutrient Mixture Ham's F10 (Gibco, Austria) supplemented with 10% fetal calf serum (Gibco, Austria), Ultroser G (BioSepra, France), Gentamycin (Biochrom, Germany), L-

Glutamin (Biochrom, Germany) in a 95% humidified, 5% CO_2 chamber at 37°C. Cytogenetic analysis was performed according to standard protocols and chromosome banding was produced by means of the Trypsin-Giemsa method. An up-to-date summary of standard cytogenetic protocols, which were used for this study, can be found in (Wegner, 1999).

Cell cycle analyses

Logarithmically growing amniotic fluid cells were harvested by trypsinization and fixed by rapid submersion in icecold 85% ethanol. After overnight fixation at −20°C, cells were pelleted and, depending on the cell amount, resuspended in an appropriate volume of DNA staining solution (0.25 mg/ml propidium iodide, 0.05 mg/ml RNase, 0.1% Triton X-100 in citrate buffer, pH 7.8). Determination of DNA distribution was performed on a Becton Dickinson FACScan (Becton Dickinson Austria, Wien Schwechat, Austria). Data acquisition for subsequent DNA analysis was performed by using the CELLQUEST acquisition and analysis software. At least 20000 cells were collected for each analysis. Data analyses of cells in different cell cycle phases were then performed using the MODFIT analysis software (compare also Soucek, 2001).

Cell size analyses

Control amniotic fluid cell samples and trisomy 21 samples were divided into triplets, plated in 6 well culture dishes and cultivated separately in standard amniotic fluid growth medium. Analyses of cell size, respectively cell volume, were performed on a CASY® Cell Counter + Analyser System, Modell TTC (Schärfe System, Reutlingen, Germany). Such measurements are based on the technology of pulse area analysis combined with the principles of resistance measurement. Vital, non-fixed cells are suspended in electrolyte buffer and aspirated through a measuring pore with defined size. While passing through, cells are scanned with high frequency in a low voltage field and are analysed for both cell count and cell volume. The pulse area of the resulting signal is strictly proportional to the volume of the particle generating the signal. Cell signals are then cumulated and assigned in a calibrated multi-channel analyser consisting of 512000 size channels. Size distribution of the so analysed sample is displayed graphically using 1024 size channels. Units for measurements of the absolut cell volume are femtoliter. For further detailed information on this method see www.CASY-Technology.com.

Logarithmically growing amniotic fluid cells were harvested by trypsinization and cell pellets were resuspended in cold phosphate-buffered saline. Depending on the amount of cells, about 10–200 μl of cell suspension were diluted in 10 ml of CASYton isotonic dilution buffer and 0.4 ml of total volume were then measured in triplets. Discrimination between vital and dead cells / cell debris was done by appropriate cursor setting, depending on the analysed cell type. The analysed parameter of cell volume was Vol_{max}, representing the average peak volume of the analysed cell population (compare Pusch, 1997). The remaining cell suspension was fixed by rapid submersion in icecold 85% ethanol for cytofluorometric analyses of DNA distribution (see above).

Results

To compare the distributions of cells in different cell cycle phases logarithmically growing human amniotic fluid control samples and cell samples with cytogenetically detected free trisomy 21 were cultivated. Logarithmically growing cell cultures were fixed and DNA was stained. Two representative examples for the performed cytofluorometric analyses are presented in Fig. 1A. Different control samples and different Down Syndrome samples were investigated, each in three independent determinations (Fig. 1B). The observed cell cycle distributions were comparable to distributions of relatively fast proliferating cells, what is represented by over 20% of cells in S phase, in which replication occurs. Comparison of control samples with Down Syndrome samples revealed no significant differences in the percentages of S phase cells. Whereas it appeared that in Down Syndrome samples the percentage of G1 cells was slightly higher, these results were only on the borderline to statistical significance (Fig. 1B). In conclusion, in this study no significant differences in cell cycle distributions between Down Syndrome cells and control amniotic fluid cell samples could be detected.

DNA degradation as one typical feature of apoptotic processes can be visualized by the appearance of a so called sub-G1 peak in flowcytometric analyses of DNA distributions (Hengstschläger, 1999b). In the here performed cytofluorometric DNA analyses of amniotic fluid cell samples no sub-G1 peaks could be detected (Fig. 1A). These data demonstrate that viable, proliferating, non-apoptotic cells were analysed in this study. In addition, these findings provide evidence that under the conditions of the here performed experiments an additional chromosome 21 does not induce higher rates of apoptosis in human amniotic fluid cells.

Since in mammalian cells DNA replication does not occur until cells reach a minimum size, cell size and cell cycle regulations are connected (Nurse, 1975; Pardee, 1974; Zetterberg, 1995; Hengstschläger, 1999a). This prompted us to include cell size investigations in this study. Logarithmically growing amniotic fluid cell samples with normal karyotype and trisomy 21 samples were analysed for cell size on a CASY® Cell Counter + Analyser System. The presented representative analyses demonstrate that the observed cell size distributions reflect the typical distribution of a cycling cell population, in which cells double their size during passage from G1 to mitosis (Fig. 2A). The investigation of cell size in femtoliter of different amniotic fluid cell samples further revealed that no significant differences could be detected comparing normal cells with aneuploid cells (Fig. 2B).

Discussion

A variety of arguments make it tempting to speculate that the presence of an additional copy of the about 225 genes on human chromosome 21 would trigger effects on the control of mammalian cell cycle regulation. Within this set of genes on chromosome 21 several candidates might exhibit direct functions in cell cycle control. For example, although the function for the amyloid precursor protein (the APP gene maps to the chromosome region 21q21.3–22.05) is not precisely known, it has also been implicated in the control of cell proliferation (Capone, 2001). The S100β protein (the S100β gene maps to the chromosome region 21q22.2–22.3) is known to be involved in signal

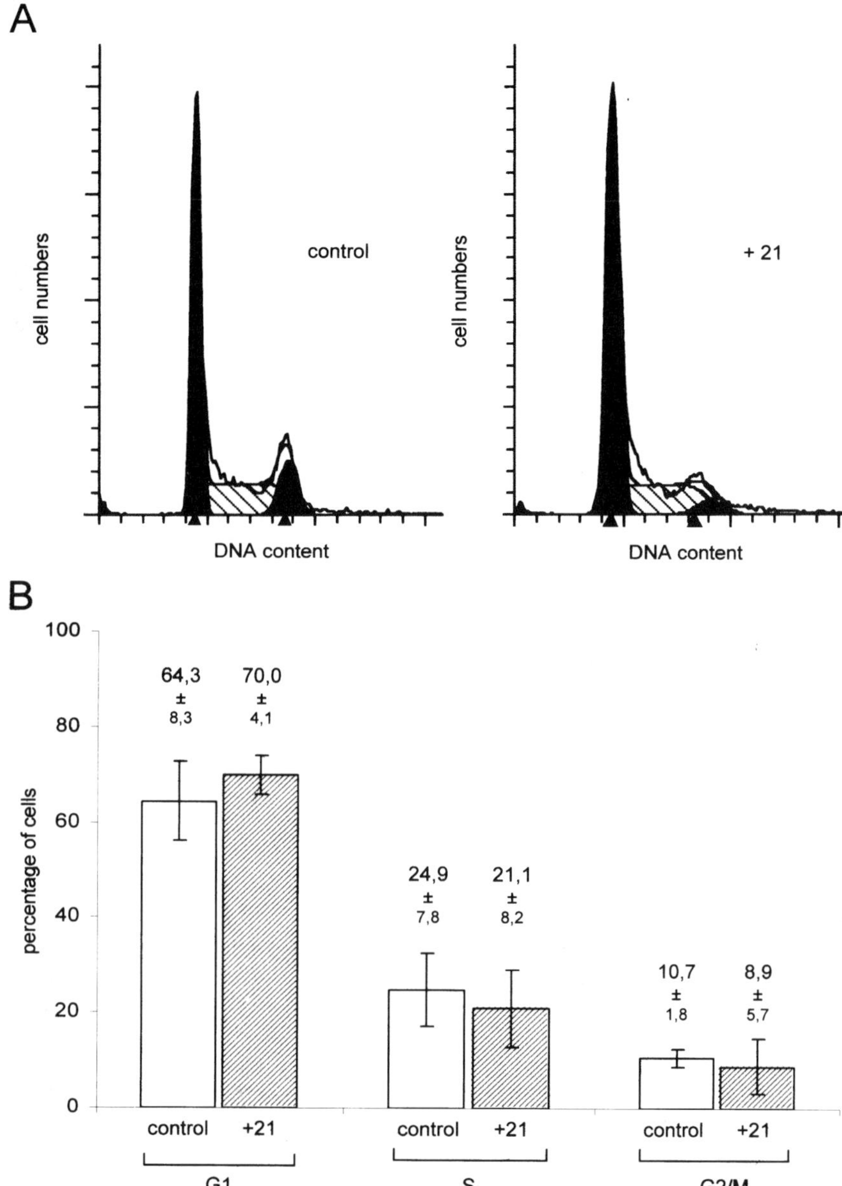

Fig. 1. Flowcytometric analyses of the DNA distribution of cells (in different phases of the cell cycle) in Down Syndrome amniotic fluid cell samples and control amniotic fluid cell samples. **A** Flowcytometric analyses of a representative control sample (control) and a Down Syndrome amniotic fluid cell sample (+21). **B** Statistical analyses of the percentages of cells in the different cell cycle phases (G1, S and G2/M phase). The results of three independent determinations of four control samples (control) and of three Down Syndrome amniotic fluid cell samples (+21) are shown. The numbers given above represent mean values ± standard deviations

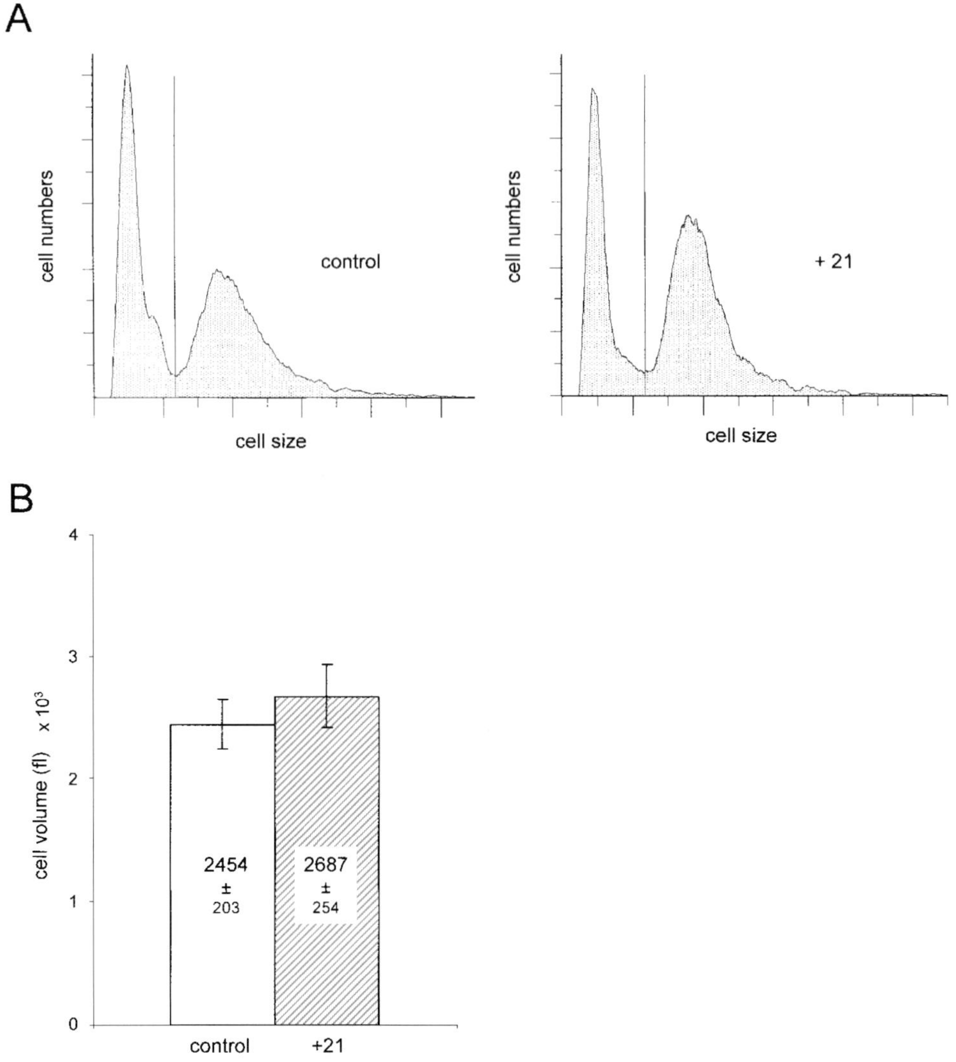

Fig. 2. Cell size analyses of Down Syndrome amniotic fluid cell samples and control amniotic fluid cell samples. **A** Analyses of cell size on a CASY® Cell Counter + Analyser System of a representative control sample (control) and a representative Down Syndrome amniotic fluid cell sample (+21). **B** Results of cell size analyses given in femtoliter. The results of three independent determinations of three control samples (control) and of three Down Syndrome amniotic fluid cell samples (+21) are shown. The numbers represent mean values in femtoliter ± standard deviations

transduction pathways, which regulate the cell cycle and neuronal differentiation (Kligman, 1988). Recently, the Down Syndrome critical region gene 2 (located between the STS markers D21S343 and D21S268 on chromosome 21) has been suggested to play a role in the control of mammalian cell proliferation (Vidal-Taboada, 2000).

The consequences of increased dosage of specific candidates within the 225 genes located on chromosome 21 depend, in part, on the biological function of the gene products themselves. In addition, these genes might regulate the expression of other genes located on other chromosomes, which then also might contribute to the known developmental consequences of Down Syndrome. This is particularly easy to imagine, since it is known that this spectrum of 225 genes also contains transcription factors (Capone, 2001; Lubec, 2002). In addition to such transcriptional regulations a variety of other posttranscriptional mechanisms must be taken in account, since they are able to affect the activity of a wide spectrum of protein products. For example, in connection to cell cycle regulation, it has been demonstrated that protein levels and activity of the Cdk p34cdc2 kinase are decreased in brain of patients with Down Syndrome (Bernert, 1996). Here it is important to mention that many Cdks are known to be also involved in the regulation of biological processes unrelated to cell cycle control (Hengstschläger, 1999a).

Taken together, these data prompted us to investigate the two connected regulations, cell cycle control and cell size control, in human amniotic fluid cells with and without an additional chromosome 21. We did not find significant differences, neither in the distribution of cells in different cell cycle phases nor in cell size given in femtoliter. From the here reported data we cannot exclude that in another cellular background, for example in neurons, the effects of an additional chromosome 21 on cell cycle regulation would be different. However, our findings allow the conclusion that cell cycle deregulation cannot be a ubiquitous phenomenon in mammalian trisomy 21 cells and therefore cannot explain the entire phenotypical spectrum of Down Syndrome. This is an important and useful prerequisite for future work to further elucidate the molecular mechanisms responsible for this phenotype.

References

Antonarakis SE, Avramopoulos D, Blouin JL, Talbot CC, Schinzel AA (1993) Mitotic errors in somatic cells cause trisomy 21 in about 4.5% of cases and are not associated with maternal age. Nat Genet 3: 146–149

Arendt T (2002) Dysregulation of neuronal differentiation and cell cycle control in Alzheimer's disease. J Neural Transm [Suppl] 62: 77–85

Bernert G, Nemethova M, Herrera-Marschitz M, Cairns N, Lubec G (1996) Decreased cyclin dependent kinase in brain of patients with Down Syndrome. Neurosci Lett 216: 68–70

Cairns NJ (1999) Neuropathology. J Neural Transm [Suppl] 57: 61–74

Capone GT (2001) Down syndrome: advances in molecular biology and the neurosciences. Dev Behav Ped 22: 40–59

Engidawork E, Lubec G (2003) Molecular changes in fetal Down syndrome brain. J Neurochem 84: 895–904

Hengstschläger M, Braun K, Soucek T, Miloloza A, Hengstschläger-Ottnad E (1999a) Cyclin-dependent kinases at the G1-S transition of the mammalian cell cycle. Rev Mutat Res 436: 1–9

Hengstschläger M, Hölzl G, Hengstschläger-Ottnad E (1999b) Different regulation of c-Myc-and E2F-1-induced apoptosis during the ongoing cell cycle. Oncogene 18: 843–848

Hernandez D, Fisher EMC (1996) Down syndrome genetics: unravelling a multifactorial disorder. Hum Mol Genet 5: 1411–1416

Kligman D, Hilt CD (1988) The S100 protein family. Trends Biochem Sci 13: 437–443
Lubec G, Engidawork E (2002) The brain in Down syndrome (TRISOMY 21). J Neurol 249: 1347–1356
Nurse P (1975) Genetic control of cell size at cell division in yeast. Nature 256: 547–551
Pardee AB (1974) A restriction point for control of normal animal proliferation. Proc Natl Acad Sci USA 71: 1286–1290
Prusa A-R, Hengstschläger M (2002) Amniotic fluid cells and human stem cell research — a new connection. Med Sci Monit 8: 253–257
Pusch O, Bernaschek G, Eilers M, Hengstschläger M (1997) Activation of c-Myc uncouples DNA replication from activation of G1-cyclin-dependent kinases. Oncogene 15: 649–656
Soucek T, Rosner M, Miloloza A, Kubista M, Cheadle J, Sampson J, Hengstschläger M (2001) Tuberous sclerosis causing mutants of the TSC2 gene product affect proliferation and p27 expression. Oncogene 20: 4904–4909
Vidal-Taboada J, Lu A, Pique M, Pons G, Gil J, Oliva R (2000) Down Syndrome critical region gene 2: expression during mouse development and in human cell lines indicates a function related to cell proliferation. Biochem Biophys Res Com 272: 156–163
Wegner R-D (1999) Diagnostic cytogenetics. Springer, Berlin Heidelberg New York Tokyo
Yoon PW, Freeman SB, Sherman SL, Taft LF, Gu Y, Pettay D, Flanders WD, Khoury MJ, Hassold TJ (1996) Advanced maternal age and the risk of Down syndrome characterized by the meiotic stage of the chromosomal error: a population-based study. Am J Hum Genet 58: 628–633
Zetterberg A, Larsson O, Wiman KG (1995) What is the restriction point? Curr Opin Cell Biol 7: 835–842

Authors' address: M. Hengstschläger, PhD, Prof., Obstetrics and Gynecology, Prenatal Diagnosis and Therapy, University of Vienna, Währinger Gürtel 18-20, A-1090 Vienna, Austria, e-mail: markus.hengstschlaeger@akh-wien.ac.at

Transcription factor REST dependent proteins are comparable between Down Syndrome and control brains: challenging a hypothesis

S. Y. Sohn[1], **R. Weitzdoerfer**[2], **N. Mori**[3], and **G. Lubec**[1]

[1] Department of Pediatrics, and
[2] Department of Neonatology University of Vienna, Vienna, Austria
[3] Department of Molecular Genetic Research, National Institute for Longevity Sciences, Oobu, Aichi, Japan

Summary. Impairment of the RE-1-silencing transcription factor (REST) and REST — dependent genes in Down Syndrome (DS) neuronal progenitor cells and neurospheres has been published recently. As dysregulation of this system has been shown at the RNA level and considering the long and unpredictable way from RNA to proteins, and as it is the proteins that do the function in brain, we decided to test this hypothesis at the protein level.

Cortex of brains of patients with Down Syndrome at the early second trimester were used. REST — dependent structures as synapsin I, brain derived neurotrophic factor BDNF and neuronal growth-associated protein SCG10 were determined at the protein level using immunoblotting.

Proteins were comparably expressed in fetal Down syndrome and control brains. Even when normalized versus housekeeping genes (glyceraldehyde-6-phosphate-dehydrogenease) and a marker for neuronal density (neuron — specific enolase) DS results were resembling controls. Therefore, we cannot confirm the REST-hypothesis by our studies in the 18/19th week of gestation at the protein level in brain and taking into account that the hypothesis was based upon studies in progenitor cells.

Introduction

Down syndrome (DS), the most common genetic abnormality associated with early mental retardation and neurological abnormalities, results from triplication of the whole or distal part of human chromosome 21. Neuropathologically DS is characterized by reduced number and density of cortical neurons, malformed dendritic spines, defective lamination of the cortex, and abnormal synapses from the 22nd week of gestation (Becker et al., 1986, 1991; Epstein, 1995; Wisniewski and Kida, 1994). In addition to developmental brain abnormalities, almost all DS brains over 40 years old manifest neuropathology similar to Alzheimer's disease (AD) (Wisniewski et al.,

1985). Although there is a series of candidate RNAs and proteins explaining the brain deficit in DS, studies in early human life are limited (Engidawork et al., 2001, 2002).

Transcription factors, representing positive or negative regulators of gene transcription, are essential for development, regulating the acquisition of a specific cellular phenotype and the final cell differentiation, a DNA element was identified by two different research groups and named restrictive element-1 (RE-1) (Kraner et al., 1992) or neuron restrictive silencer element (NRSE), to which RE-1-silencing transcription factor (REST) (Chong et al., 1995), also known as neuron-restrictive silencer factor (NRSF) (Schoenherr and Anderson, 1995), is binding. Furthermore, it was shown that REST is recruiting co-repressors such as mSin3 and histone deacetylase to repress gene transcription (Naruse et al., 1999).

REST is a transcriptional factor that during early embryogenesis is required to repress a subset of neuron-specific genes in non-neural tissues and undifferentiated neural precursors. Though originally thought to be exclusively expressed in non-neuronal cell types, various isoforms of REST were shown to be generated by alternative splicing of its mRNA, with tissue specific expression of each isoform, including at least one neuron-specific form (Palm et al., 1998). Overexpression of REST in spinal cord neurons of the developing chicken resulted in impaired axon guidance and repression of neuron-specific gene expression (Paquette et al., 2000).

Recently, reduced expression of several developmental genes regulated by the REST transcription factor in DS stem-cell and progenitor-cell pool were reported to correlate with changes in neuron morphology after differentiation (Bahn et al., 2002).

To test the hypothesis of transcription factor REST involvement in pathomechanisms leading to the brain deficit in DS, we evaluated protein levels of several REST-regulated proteins in brain regions of fetuses with DS compared to controls.

Materials and methods

Brain samples

Fetal brain tissue (cerebral cortex) of controls (fetuses with no obvious abnormalities (n = 4; all female, with gestational age of 18–19 weeks) and DS (n = 4; all females with gestational age of 18–19 weeks) was used for the experiment. Fetuses were karyotyped and obtained from Dr. M. Dierssen and Dr. J. C. Ferreres, Department of Pathology UDIAT-CD, Corporacis Saniria Parc Tauli, Sabadell, Barcelona, Spain.

Western blotting

Brain samples were ground with mortar under liquid nitrogen, directly suspended and homogenized with homogenization buffer containing 10 mM Tris-HCl (pH 7.5), 150 mM NaCl, 0.05% (v/v) Tween 20, 1 mM PMSF (phenylmethylsulfonyl fluoride,

Sigma, Austria) and 1 tablet of protease inhibitor cocktail (Roche, Austria). The homogenized samples were centrifuged at 8,000 3 g for 10 minutes at 4°C. The BCA protein assay kit (Pierce, USA) was used to determine protein concentration in the supernatant. Samples (10 ug) were mixed with the same volume of sample buffer (60 mM Tris-HCI, 2% SDS, 0.1% bromophenol blue, 25% glycerol, 14.4 mM 2-mercaptoethanol, pH 6.8), incubated at 95°C for 15 minutes and loaded onto a 12.5% ExcelGel SDS homogeneous gel (Amersham Pharmacia Biotech, Sweden).

Electrophoresis was performed with Multiphor II Electrophoresis System (Amersham Pharmacia Biotech, Sweden). Proteins separated on the gel were transferred onto PVDF membrane (Millipore, Bedford, USA) and the membrane was blocked in blocking buffer (10 mM Tris-HCI, pH 7.5, 150 mM NaCI, 2% non-fat dried milk and 0.1% Tween 20). The membrane was incubated for 2 hours at room temperature with diluted primay antibodies (Monoclonal anti-SCG10 was kindly provided by Dr. N. Mori (Department of Molecular Genetic Research, National Institute for Longevity Sciences, Japan). Synapsin I (s193, Sigma), BDNF (sc-546, Santa Cruz), GAPDH (MAB374, Chemicon) and NSE (AB951, Chemicon) were purchased). After 4 times washing for 10 min with blocking buffer, membranes were probed with secondary antibody coupled to horseradish peroxidase (Southern Biotechnology Associates, Inc., Alabama, USA) for 1 hour. Subsequent washing 4 times for 10 minutes and developing with the Western blot Chemiluminescene reagent (NEN™ Life Science Products, Inc., USA) followed.

Analysis and statistics

The density of detected bands was measured using RFLP scan version 2.1 software program (Scanalytics, USA). Between group differences were calculated by non-parametric Mann-Whitney U test. The statistical analysis was performed by GraphPad Instat 2 software version 2.05 and the level of significance was set at $P < 0.05$.

Results

Protein levels of Synapsin I were comparable between control and DS samples. The densities of detected bands representing BDNF were comparable between groups. SCG10 levels were comparable between controls and DS fetuses. Results are shown in Fig. 1.

Even when proteins were normalized versus a neuronal marker (neuron-specific enolase, NSE) or a housekeeping protein, glyceraldehyde — 3 — phosphate dehydrogenase (GAPDH), ratios were comparable between groups thus ruling out neuronal loss or cell loss per se.

Discussion

A mouse knock-out model of REST, by targeted disruption of the Rest gene assigned to 4q12 (Cowan et al., 1996), was generated to elucidate the functional role in the living animal. Gene disruption was leading to precocious neuronal differentiation with ectopic expression of some target genes (Chen et al., 1998). Large input was expected from the analysis of REST target genes in these knock-out mice, however these mice die by embryonic day 10 (E10),

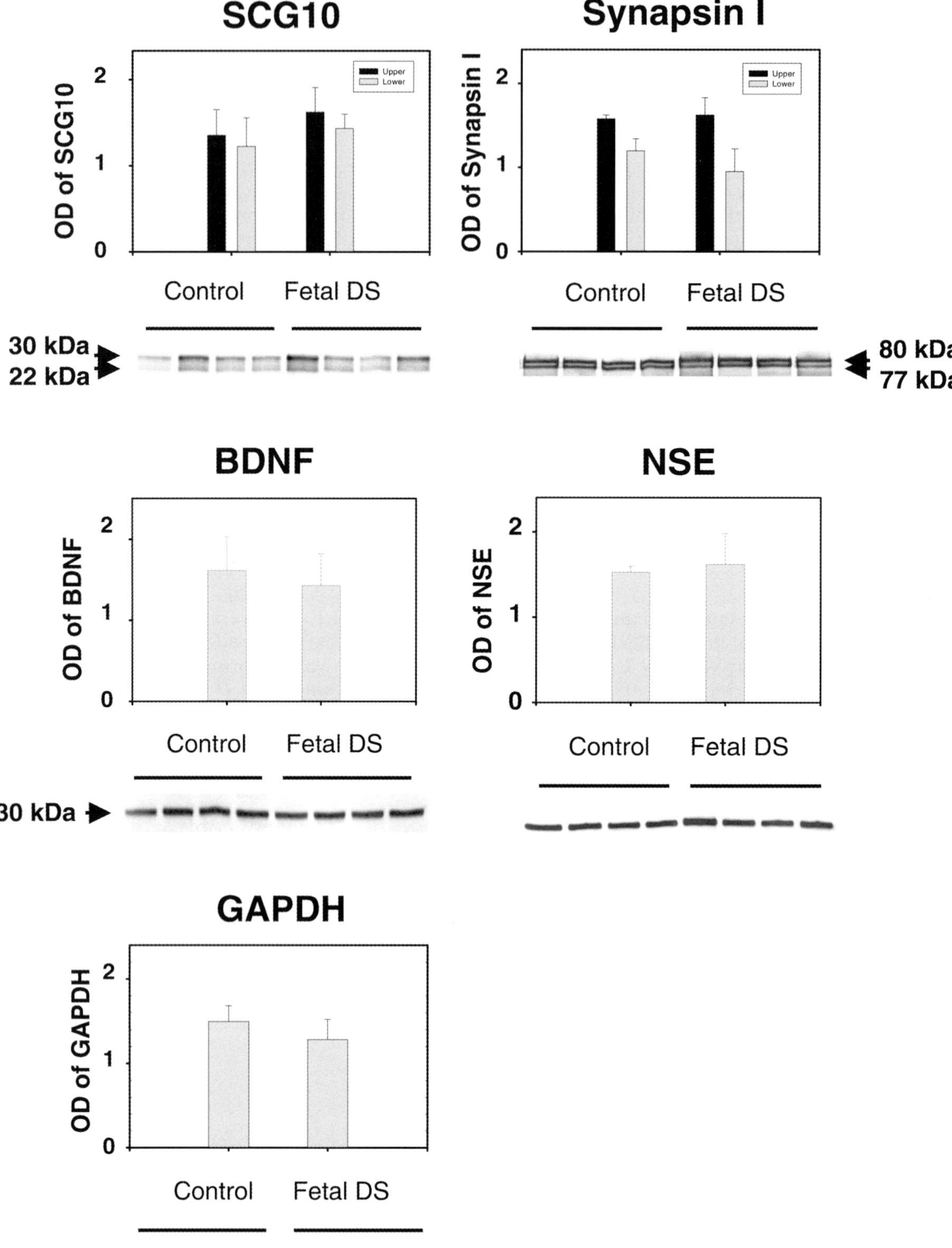

Fig. 1. Western blot analysis of the expression levels of SCG10, Synapsin I, BDNF, NSE and GAPDH in cerebral cortex of fetuses with DS and controls. No differences between expression levels were found between control and DS samples

a time point at which expression of neural specific genes is incomplete. Nevertheless, a large number of target genes of REST were identified and were proposed to be of great impact in neurogenesis and neural function (Schoenherr et al., 1996). They include, among others, synapsin I (Li et al., 1993; Schoch et al., 1996), brain-derived neurotrophic factor (BDNF) (Timmusk et al., 1999), NMDAR1 (Bai et al., 1998), corticotropin-releasing hormone (CRH) (Seth and Majzoub, 2001), SCG10 (Mori et al., 1992), choline acetyltransferase (ChAT), the m4 muscarinic acetylcholine receptor (Mieda et al., 1996; Wood et al., 1996), type II sodium channel (Kraner et al., 1992) and the adhesion proteins L1 and NgCAM (Thiel et al., 1999). There is evidence that REST/RE-1 may serve as a repressor as well as an activator depending on its spatial and temporal expression (Bessis et al., 1997; Kallunki et al., 1998; Seth and Majzoub, 2001).

SCG10, one of the neuronal growth-associated proteins, may be involved in the regulation of differentiation, activities, and plasticity of the nervous system as it is highly enriched in growth cones during development (Di Paolo et al., 1997; Lutjens et al., 2000) playing an important role in cytoskeletal response towards guidance and growth signals (Antonsson et al., 1998). SCG10 expression is developmentally regulated and expressed predominantly in the neonatal stages, and reduced drastically during postnatal maturation (Riederer et al., 1997; Sugiura and Mori, 1995). Additionally, up-regulation of the SCG10 mRNA has been described during regeneration in the central nervous system and has been proposed to be determining regenerative ability (Pellier-Monnin et al., 2001; Mason et al., 2002). Using western blot technique unchanged levels of SCG10 were found in brains of Alzheimer's disease patients, however a positive correlation to tangle number was described (Okazaki et al., 1995).

Brain-derived neurotrophic factor (BDNF) is a member of the neurotrophin family and abundantly and widely expressed in fetal and adult mammalian brain. BDNF exerts multiple effects on the development and maintenance of the nervous system, including regulating synaptic plasticity and promoting survival of neurons (for review: Schuman, 1999).

The synapsin I gene, encoding the synaptic vesicle protein synapsin I, is regulated by REST as well. Synapsins are a family of neuron-specific, synaptic vesicle-associated phosphoproteins. In addition to regulating neurotransmitter release at fully functional synapses, the synapsins may participate in the functional maturation of synapses during development and possibly even synaptogenesis in the adult nervous system. The introduction of synapsin I or synapsin II into several neuronal preparations has been shown to promote the maturation of synapses (for review: Ferreira and Rapoport, 2002). In fact, axonal development and synaptogenesis are impaired in mice deficient in synapsin I (Chin et al., 1995).

Another REST — dependent structure, ChAT, has been already described to be comparable between fetal DS and controls (Lubec et al., 2001a) and this has been confirmed at the acetylcholine neurotransmitter level from fetal until early postnatal life indicating that DS patients start life with normal cholinergic innervation (Kish et al., 1989; Brooksbank et al., 1989).

Our observation is in agreement with morphological findings showing normal histoarchitecture until the 22nd week of gestation when neuritic / dendritic outgrowth starts and only from then first subtle morphological changes of the dendritic tree and arborization are observed by silver staining.

The very first abnormalities in protein expression are found from the 19th week of gestation and are represented by reduction of cytoskeleton proteins including actin — related protein 2/3 (Weitzdoerfer et al., 2002), centractin and capping proteins (Gulesserian et al., 2002), moesin (Lubec et al., 2001b), and synaptojanin, a synaptosomal associated protein encoded on chromosome 21, is overexpressed at that time point of development (Arai et al., 2002, Cheon et al., submitted). It is therefore highly unlikely, that REST-dependent structures are involved in the abnormal development of DS brain, when at a much later developmental stage the brain of patients with DS is still morphologically intact and as shown, herein, does not present with decreased or dysregulated synapsin I, SCG10 and BDNF.

Acknowledgements

We are highly indebted to Dr. G. Mandel, Howard Hughes Medical Institute, State University of New York, Stony Brook, 11794, USA and Dr. G. Thiel, Department of Medical Biochemistry and Molecular Biology, University of Saarland Medical Center, D-66421 Homburg, Germany, for supplying antibodies in context with the study but were not used.

References

Antonsson B, Kassel DB, Di Paolo G, Lutjens R, Riederer BM, Grenningloh G (1998) Identification of in vitro phosphorylation sites in the growth cone protein SCG10. Effect of phosphorylation site mutants on microtubule-destabilizing activity. J Biol Chem 273: 8439–8446

Arai Y, Ijuin T, Takenawa T, Becker LE, Takashima S (2002) Excessive expression of synaptojanin in brains with Down syndrome. Brain Dev 24: 67–72

Bahn S, Mimmack M, Ryan M, Caldwell M, Jauniaux E, Starkey M, Svendsen C, Emson P (2002) Neuronal target genes of the neuron-restrictive silencer factor in neurospheres derived from fetuses with Down's syndrome: a gene expression study. Lancet 359: 310–315

Bai G, Norton D, Prenger M, Kusiak J (1998) Single-stranded DNA-binding proteins and neuron-restrictive silencer factor participate in cell-specific transcriptional control of the NMDAR1 gene. J Biol Chem 273: 1086–1091

Becker L, Armstrong D, Chan F (1986) Dendritic atrophy in children with Down Syndrome. Ann Neurol 20: 520–526

Becker L, Mito T, Takashima S, Onodera K (1991) Growth and development of the brain in Down syndrome. Prog Clin Biol Res 373: 133–152

Bessis A, Champtiaux N, Chatelin L, Changeux JP (1997) The neuron-restrictive silencer element: a dual enhancer/silencer crucial for patterned expression of a nicotinic receptor gene in the brain. Proc Natl Acad Sci USA 94: 5906–5911

Brooksbank B, Walker D, Balasz R, Jorgensen OS (1989) Neuronal maturation in the foetal brain in Down's syndrome. Early Hum Dev 18: 237–246

Chen ZF, Paquette A, Anderson D (1998) NRSF/REST is required in vivo for repression of multiple neuronal target genes during embryogenesis. Nat Genet 20: 136–142

Cheon MS, Shim KS, Kim SH, Hara A, Lubec G (2003) Protein levels of genes encoded on chromosome 21 in fetal Down syndrome brain: challenging the gene dosage effect hypothesis, part IV. Amino Acids 25: 41–47

Chin LS, Li L, Ferreira A, Kosik K, Greengard P (1995) Impairment of axonal development and of synaptogenesis in hippocampal neurons of synapsin I-deficient mice. Proc Natl Acad Sci USA 92: 9230–9234

Chong JA, Tapia-Ramirez J, Kim S, Toledo-Aral J, Zheng Y, Boutros M, Altshuller Y, Frohman M, Kraner S, Mandel G (1995) REST: a mammalian silencer protein that restricts sodium channel gene expression to neurons. Cell 80: 949–957

Cowan J, Powers J, Tischler A (1996) Assignment of the REST gene to 4q12 by fluorescence in situ hybridization. Genomics 34: 260–262

Di Paolo G, Lutjens R, Osen-Sand A, Sobel A, Catsicas S, Grenningloh G (1997) Differential distribution of stathmin and SCG10 in developing neurons in culture. J Neurosci Res 50: 1000–1009

Engidawork E, Lubec G (2001) Protein expression in Down syndrome brain. Amino Acids 21: 331–361

Engidawork E, Lubec G (2003) Molecular changes in fetal Down Syndrome brain. J Neurochem 84: 895–904

Epstein C (1995) The metabolic and molecular bases of inherited disease. In: Scriver SR, Beaudet AL, Sly WS, Valle D (eds) Down Syndrome (Trisomy 21). McGraw Hill, New York, pp 749–794

Ferreira A, Rapoport M (2002) The synapsins: beyond the regulation of neurotransmitter release. Cell Mol Life Sci 59: 589–595

Gulesserian T, Kim SH, Fountoulakis M, Lubec G (2002) Aberrant expression of centractin and capping proteins, integral constituents of the dynactin complex, in fetal Down syndrome brain. Biochem Biophys Res Commun 291: 62–67

Kallunki P, Edelman G, Jones F (1998) The neural restrictive silencer element can act as both a repressor and enhancer of L1 cell adhesion molecule gene expression during postnatal development. Proc Natl Acad Sci USA 95: 3233–3238

Kish S, Karlinsky H, Becker L, Gilbert J, Rebbetoy M, Chang LJ, DiStefano L, Hornykiewicz O (1989) Down's syndrome individuals begin life with normal levels of brain cholinergic markers. J Neurochem 52: 1183–1187

Kraner SD, Chong JA, Tsay HJ, Mandel G (1992) Silencing the type II sodium channel gene: a model for neural-specific gene regulation. Neuron 9: 37–44

Li L, Suzuki T, Mori N, Greengard P (1993) Identification of a functional silencer element involved in neuron-specific expression of the synapsin I gene. Proc Natl Acad Sci USA 90: 1460–1464

Lubec B, Yoo BC, Dierssen M, Balic N, Lubec G (2001a) Down syndrome patients start early prenatal life with normal cholinergic, monoaminergic and serotoninergic innervation. J Neural Transm [Suppl] 61: 303–310

Lubec B, Weitzdoerfer R, Fountoulakis M (2001b) Manifold reduction of moesin in fetal Down syndrome brain. Biochem Biophys Res Commun 286: 1191–1194

Lutjens R, Igarashi M, Pellier V, Blasey H, Di Paolo G, Ruchti E, Pfulg C, Staple JK, Catsicas S, Grenningloh G (2000) Localization and targeting of SCG10 to the trans-Golgi apparatus and growth cone vesicles. Eur J Neurosci 12: 2224–2234

Mason M, Lieberman A, Grenningloh G, Anderson P (2002) Transcriptional upregulation of SCG10 and CAP-23 is correlated with regeneration of the axons of peripheral and central neurons in vivo. Mol Cell Neurosci 20: 595

Mieda M, Haga T, Saffen D (1997) Expression of the rat m4 muscarinic acetylcholine receptor gene is regulated by the neuron-restrictive silencer element/repressor element 1. J Biol Chem 272: 5854–5860

Mori N, Schoenherr C, Vandenbergh D, Anderson D (1992) A common silencer element in the SCG10 and type II Na+ channel genes binds a factor present in nonneuronal cells but not in neuronal cells. Neuron 9: 45–54

Naruse Y, Aoki T, Kojima T, Mori N (1999) Neural restrictive silencer factor recruits mSin3 and histone deacetylase complex to repress neuron-specific target genes. Proc Natl Acad Sci USA 96: 13691–13696

Okazaki T, Wang H, Masliah E, Cao M, Johnson S, Sundsmo M, Saitoh T, Mori N (1995) SCG10, a neuron-specific growth-associated protein in Alzheimer's disease. Neurobiol Aging 16: 883–894

Palm K, Belluardo N, Metsis M, Timmusk T (1998) Neuronal expression of zinc finger transcription factor REST/NRSF/XBR gene. J Neurosci 18: 1280–1296

Paquette AJ, Perez SE, Anderson DJ (2000) Constitutive expression of the neuron-restrictive silencer factor (NRSF)/REST in differentiating neurons disrupts neuronal gene expression and causes axon pathfinding errors in vivo. Proc Natl Acad Sci USA 97: 12318–12323

Pellier-Monnin V, Astic L, Bichet S, Riederer BM, Grenningloh G (2001) Expression of SCG10 and stathmin proteins in the rat olfactory system during development and axonal regeneration. J Comp Neurol 433: 239–254

Riederer BM, Pellier V, Antonsson B, Di Paolo G, Stimpson SA, Lutjens R, Catsicas S, Grenningloh G (1997) Regulation of microtubule dynamics by the neuronal growth-associated protein SCG10. Proc Natl Acad Sci USA 94: 741–745

Schoenherr CJ, Anderson DJ (1995) The neuron-restrictive silencer factor (NRSF): a coordinate repressor of multiple neuron-specific genes. Science 267: 1360–1363

Schoenherr CJ, Paquette AJ, Anderson DJ (1996) Identification of potential target genes for the neuron-restrictive silencer factor. Proc Natl Acad Sci USA 93: 9881–9886

Schoch S, Cibelli G, Thiel G (1996) Neuron-specific gene expression of synapsin I. Major role of a negative regulatory mechanism. J Biol Chem 271: 3317–3323

Schuman EM (1999) Neurotrophin regulation of synaptic transmission. Curr Opin Neurobiol 9: 105–109

Seth KA, Majzoub JA (2001) Repressor element silencing transcription factor/neuron-restrictive silencing factor (REST/NRSF) can act as an enhancer as well as a repressor of corticotropin-releasing hormone gene transcription. J Biol Chem 276: 13917–13923

Sugiura Y, Mori N (1995) SCG10 expresses growth-associated manner in developing rat brain, but shows a different pattern to p19/stathmin or GAP-43. Brain Res Dev Brain Res 90: 73–91

Thiel G, Lietz M, Leichter M (1999) Regulation of neuronal gene expression. Naturwissenschaften 86: 1–7

Timmusk T, Palm K, Lendahl U, Metsis M (1999) Brain-derived neurotrophic factor expression in vivo is under the control of neuron-restrictive silencer element. J Biol Chem 274: 1078–1084

Weitzdoerfer R, Fountoulakis M, Lubec G (2002) Reduction of actin-related protein complex 2/3 in fetal Down syndrome brain. Biochem Biophys Res Commun 293: 836–841

Wisniewski KE, Wisniewski HM, Wen GY (1985) Occurrence of neuropathological changes and dementia of Alzheimer's disease in Down's syndrome. Ann Neurol 17: 278–282

Wisniewski K, Kida E (1994) Abnormal neurogenesis and synaptogenesis in Down Syndrome brain. Dev Brain Dysfunct 7: 289–301

Wood IC, Roopra A, Buckley NJ (1996) Neural specific expression of the m4 muscarinic acetylcholine receptor gene is mediated by a RE1/NRSE-type silencing element. J Biol Chem 271: 14221–14225

Authors' address: Prof. Dr. G. Lubec, Department of Pediatrics, University of Vienna, Währinger Gürtel 18, A-1090 Vienna, Austria, e-mail: gert.lubec@akh-wien.ac.at

An altered antioxidant balance occurs in Down syndrome fetal organs: implications for the "gene dosage effect" hypothesis

J. B. de Haan[1], **B. Susil**[2], **M. Pritchard**[1], and **I. Kola**[3]

[1] Monash Institute of Reproduction and Development, Centre for Functional Genomics and Human Disease, Monash University, and
[2] Department of Pathology, Monash Medical Centre, Clayton, Victoria, Australia
[3] Merck Research Laboratories, Merck & Co., Inc, Rahway, NJ, U.S.A.

Summary. Down syndrome (DS) is the congenital birth defect responsible for the greatest number of individuals with mental retardation. It arises due to trisomy of human chromosome 21 (HSA21) or part thereof. To date there have been limited studies of HSA21 gene expression in trisomy 21 conceptuses. In this study we investigate the expression of the HSA21 antioxidant gene, Cu/Zn-superoxide dismutase-1 (SOD1) in various organs of control and DS aborted conceptuses. We show that SOD1 mRNA levels are elevated in DS brain, lung, heart and thymus. DS livers show decreased SOD1 mRNA expression compared with controls. Since non-HSA21 antioxidant genes are reported to be concomitantly upregulated in certain DS tissues, we examined the expression of glutathione peroxidase-1 (GPX1) in control and DS fetal organs. Interestingly, GPX1 expression was unchanged in the majority of DS organs and decreased in DS livers. We examined the SOD1 to GPX1 mRNA ratio in individual organs, as both enzymes form part of the body's defense against oxidative stress, and because a disproportionate increase of SOD1 to GPX1 results in noxious hydroxyl radical damage. All organs investigated show an approximately 2-fold increase in the SOD1 to GPX1 mRNA ratio. We propose that it is the altered antioxidant ratio that contributes to certain aspects of the DS phenotype.

Abbreviations

SOD1: human Cu/Zn-superoxide dismutase; GPX1: human selenium-dependent glutathione peroxidase; DS: Down syndrome; HSA21: human chromosome 21

Introduction

Down syndrome (DS) occurs at a frequency of 1 in 700 to 1000 live births within the general population, and of all the cytogenetic abnormalities it is the one that most frequently comes to term. DS is responsible for the genesis of the greatest number of individuals with mental retardation (Patterson, 1987). Furthermore, individuals with DS suffer from a wide range of other abnormalities that may include: congenital heart defects (Rehder, 1981); in utero growth retardation resulting in reduced birth weight of approximately 10%; increased susceptibility to infections (Chaushu et al., 2002); a 20 to 50 fold higher incidence of leukaemia (Fong and Brodeur, 1987); eye lens defects (Patterson, 1987); and premature aging (Tam and Walford, 1980) with Alzheimer-type neuronal pathology (Wisniewski et al., 1985; Mann and Esiri, 1989). Indeed, individuals with DS show pathology in every major organ system in the body, although the penetrance and severity of the pathology varies between DS individuals.

Approximately 95% of DS individuals present with an extra copy of the entire human chromosome 21 (HSA21) and this has focussed research on gaining a better understanding of how extra chromosomal material may contribute to the syndrome. The "gene dosage effect" hypothesis, which proposes that increased expression of HSA21 genes in accordance with gene dosage directly contributes to the syndrome, has largely been accepted as the modus operandi for these defects associated with the DS phenotype (see review by Pritchard and Kola, 1999). However, some proponents of the "developmental instability" or "quantitative" hypothesis believe that it is the amount of additional chromosomal material that causes the DS phenotype (Hall, 1965; Shapiro, 1994). Most of the available data (the analyses of HSA21 gene products in DS tissues (Brooksbank and Balazs 1984; Anneren and Epstein, 1987; Lemieux et al., 1993; Pallister et al., 1997; Fuentes et al., 2000) and the genotype-phenotype relationship of transgenic mice overexpressing HSA21-specific genes (Sumarsono et al., 1996; Ceballos et al., 1991) support the "gene dosage effect" hypothesis. However not all genes are upregulated in DS tissues (Stefani et al., 1988), and some authors have interpreted a lack of upregulation of HSA21 genes in DS to argue against the "gene dosage effect" hypothesis (Greber-Platzer et al., 1999a,b; Cheon et al., 2003a,b,c).

In this study, we have investigated the expression of the antioxidant gene Cu/Zn-superoxide dismutase-1 (SOD1) which is located on HSA21 at 21q22.1 (Tan et al., 1973), in various tissues of control and DS aborted conceptuses. The purpose of our study was two-fold. First we wished to establish the expression profile of SOD1 in a broader range of DS fetal tissue since expression of this gene has only been investigated in fetal brain (Brooksbank and Balazs, 1984; Gulesserian et al., 2001a). Other studies have focused on a number of cell types primarily isolated from adult DS blood (Sinet et al., 1975a; Frischer et al., 1981; Neve et al., 1983; Pastor et al., 1998) and adult brain (Gulessarian et al., 2001b). Given the congenital nature of Down syndrome, an understanding of aberrant antioxidant expression during development in a range of organs would be of extreme significance. Furthermore, it would be incorrect to assume that SOD1 expression is elevated 1.5 fold in all DS fetal organs since not all genes investigated in DS tissues are expressed at the predicted gene dosage increase of 1.5 fold (Stefani et al., 1988). Second we wished to investigate the ratio of SOD1 to GPX1, the latter being an antioxidant gene not localized on HSA21 (Chada et al., 1990), since some controversy exists with respect to the level of GPX1 expression in DS organs and tissues (Frisher et al., 1981, Neve et al., 1983; Pastor et al., 1998; Sinet et al., 1975b). The rationale for the latter investigation is derived from the following. The SOD1 enzyme is a key player in the conversion of superoxide radicals ($O_2^{-\bullet}$) to hydrogen peroxide (H_2O_2) which constitutes the first step of the cell's natural defense against oxidative stress. A build-up of noxious levels of H_2O_2 is prevented by a second step in which two further antioxidant enzymes, namely glutathione peroxidase and catalase, neutralize this to water. Thus a delicate balance exists in cells and perturbations of this balance (as may be predicted from a gene-dosage increase in SOD1 with respect to second step antioxidant enzymes in DS) give rise to noxious hydroxyl radicals ($^{\bullet}OH$) through Fenton-type reactions of H_2O_2 with transition metals. It is these highly reactive $^{\bullet}OH$ species that damage DNA (Imlay et al., 1988), protein (Davies, 1987) and lipid molecules (Fridovich, 1978) and initiate many rounds of peroxidative damage to biologically important membranes (Sies and de Groot, 1992). For these reasons, it becomes important to investigate the ratio of first to second-step antioxidant enzymes as a determinant of cellular and organ damage in DS. Indeed, limited data exist to suggest that a compensatory rise in GPX1 levels occur in some but not all DS tissue (Neve et al., 1983; Brooksbank and Balazs, 1984), implying that some DS organs or tissues may be more at risk of peroxidative damage than others. Given the pathophysiological implications of an altered antioxidant balance in DS, we investigated this parameter in a range of organs from control and DS conceptuses.

Materials and methods

Human aborted conceptuses

Ethical approval for the study was obtained from the Monash Medical Centre Ethics Committee, Melbourne, Australia and from patients participating in the study. Most of the DS conceptuses were obtained from patients who were screened because of advanced maternal age (i.e. 37 yrs and older). Control conceptuses were obtained from patients who had other chromosomal (i.e. non-DS) and/or morphological abnormalities. Chromosomal analyses were done on all conceptuses, and one DS conceptus that was a mosaic was excluded from the analysis. All fetuses analysed were between 18 and 20 weeks of gestation. Abortions were performed using prostaglandins (Wadhera and Miller, 1994), thus ensuring that maceration of the tissue did not occur. Organs were immediately frozen in liquid nitrogen and stored at −70°C until RNA was extracted.

Probe preparation and RNA extraction

The 650 bp human SOD1 cDNA probe was released from pS61–10 with *Pst*1 (de Haan et al., 1992) and used to hybridize to 30 μg of total RNA. The GPX1 and β-actin probes have been described previously (de Haan et al., 1992, 1994). Labelling of the SOD1, GPX1 and β-actin probes and the extraction of total cellular RNA was performed as described previously (de Haan et al., 1992). Particular attention was paid to the integrity of the

RNA due to unavoidable variations in time between obtaining aborted material and being frozen in liquid nitrogen. RNA was found to be intact for most samples based on visualization of 18S and 28S ribosomal bands on ethidium bromide stained gels. Samples that appeared degraded were not included in the analysis.

Northern blotting

Northern blots were first hybridized with SOD1, the probe removed and then sequentially hybridized with GPX1 and β-actin as detailed before (de Haan et al., 1992, 1994). Washing conditions were stringent: 0.1 × SSPE, 0.1% SDS for 20 min at 65°C. Prior to rehybridization of filters, autoradiography was carried out to confirm that no residual signal remained. Autoradiography was performed using intensifying screens at −70°C and Fuji X-ray RX Medical film.

Preparation of slot blots

15-, 7.5-, 3.7-, 1.8- and 0.9-μg aliquots of total RNA were lyophilized and resuspended in 12 × SSC and 6% formaldehyde. The solution was then heated to 60°C for 15 min and rapidly cooled on ice. Bromophenol blue (0.01% final conc.) was added to each sample, which was then blotted under vacuum onto Hybond-N nylon membranes (Amersham) using a Bio-rad slot-blotting apparatus. Each well was rinsed twice with 20X SSC, the filters air-dried and the RNA bound to the membranes by UV crosslinking for 7 min and oven-baking at 80°C for 4 h.

Quantitation of Northern and slot blots

Northern and slot blot autoradiograms were densitometrically scanned for quantitation using an LKB densitometer. For Northern blot analysis, the path used by the laser densitometer was set in such a way as to always dissect vertically through the middle of a signal. Furthermore, the position at which the laser densitometer began its analysis and the position where this was terminated, was carefully defined prior to the analysis and was always the same across a filter. Care was taken to ensure that all autoradiographic results fell within the linear range of signal density versus radioactivity. SOD1 and GPX1 expression were corrected by expressing the data relative to β-actin. For slot blot analysis, the laser densitometer was set in such a way so as to dissect each band perpendicularly. Each sample was analysed three times at a different location within the slot blot band. This rigorous treatment of slot blots ensured that an accurate reading was obtained across the signal. The three integrated areas generated for each dilution of RNA were used in linear regression analysis to determine the amount of specific probe hybridized to the filter. This value was determined from the slope of the linear regression for each probe and the values corrected for β-actin mRNA levels. Samples were only included in the analysis when all three genes (SOD1, GPX1, and β-actin) gave statistically significant ($p < 0.001$) linear regression values.

Statistical analysis

Unpaired non-parametric statistics (Mann-Whitney U test) was applied to both control and DS groups. Non-parametric statistics makes no assumptions about the data distribution and is therefore more rigorous (Siegel, 1956). A p value of <0.05 was taken as significant. Median values were determined as this is consistent with the use of non-parametric statistics.

Results

Northern blot analysis of SOD1 expression in control and DS organs

Northern blot analysis was performed on all control and DS samples since a bank of RNA was prepared from the tissues obtained. This facilitated the analysis of a greater range of genes than would have been possible from enzymatic studies. Given the limited availability of such a resource, and the fact that enzymatic levels for both SOD1 and GPX1 correlate well with mRNA levels for these enzymes (Sherman et al., 1984; Delabar et al., 1987; de Haan et al., 1992) we felt it appropriate to proceed in this manner.

Hybridization with the SOD1 probe to RNA from aborted DS and control conceptuses detects 2 mRNA species of approximately 0.7 kb and 0.9 kb in all the human tissues studied consistent with published results (Sherman et al., 1984). Since both transcripts code for functionally active SOD1 protein (Sherman et al., 1984), both SOD1 mRNA bands were quantified and the values added together. SOD1 expression therefore represents total SOD1 message expressed by the organ. A representative Northern blot has been included to show the integrity of the two SOD1 mRNA species since in most instances there was a delay in obtaining the tissue before it could be frozen (see Fig. 1). Fortunately, this did not appear to affect the integrity of the RNA.

Northern blot analysis of brain, heart, lung and thymus showed an increase in the median SOD1 mRNA level in DS organs compared with controls (Fig. 2a). In three of the four organs (brain, heart and thymus) this increase was statistically significant ($p < 0.01$). Although median values showed an increase in SOD1 expression in DS lungs, significance was not reached ($p > 0.05$). Analysis of a larger sample size would help determine whether the increase in SOD1 levels in DS lungs is statistically significant. Unfortunately this was not possible due to the difficulty of obtaining more samples for analysis. The increase in SOD1 expression is approximately 1.5 fold in DS brain and lungs, and 2.3 fold in the DS heart compared with controls. A greater than dose-dependent increase in SOD1 levels was observed in DS fetal thymus (3.0 fold increase).

Of the five organs studied, DS liver was the only organ in which the median SOD1 mRNA level did not increase compared with the median value of control livers (Fig. 2b). Previously, in a preliminary study, we reported that DS liver showed elevated SOD1 expression levels compared with control liver (Kola et al., 1993). However in that study SOD1 expression was investigated in 4 DS livers and compared with one control liver. In this study we have expanded the number of samples analyzed to 6 controls and 5 DS livers, thus enabling statistical analysis of the data. Median values now show that DS

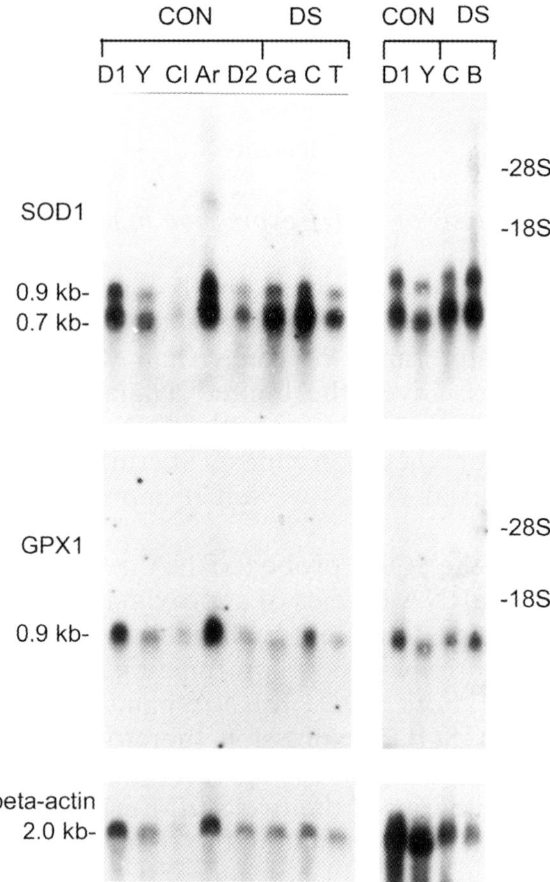

Fig. 1. Northern blot analysis of SOD1 and GPX1 expression in control and DS brains. Northern blots were sequentially hybridized with the SOD1, GPX1 and β-actin probes. Hybridization with the SOD1 probe detects two mRNA species of 0.7 and 0.9 kb. In most instances mRNA integrity was unaffected and enabled quantitation of total SOD1 mRNA in individual tissues. The position of 18 and 28S ribosomal bands for both hybridizations with SOD1 and GPX1 are indicated. To include the data of "B" a second filter was prepared that included total RNA of two controls (D1 and Y) and one DS sample (C). Comparison across filters was facilitated by normalizing the data relative to one of the controls (D1) and averaging the results of "Y" and "C". D1, Y, Cl, Ar and D2 are controls; Ca, C, T and B are Down syndrome samples

livers have an approximately 1.8 fold decrease in SOD1 mRNA compared with control livers ($p < 0.05$). To verify the decrease in SOD1 expression observed by Northern blot analysis, slot blot analysis (White and Bancroft, 1982) was performed on mRNA of DS and control liver samples. When the data were corrected for β-actin expression (Fig. 3a), a statistically significant decrease was noted again in DS livers when compared with controls ($p < 0.05$).

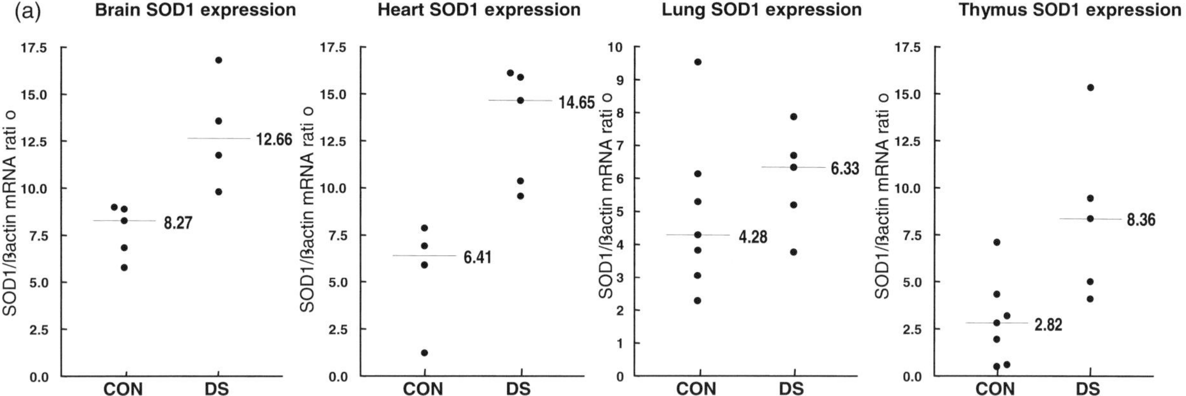

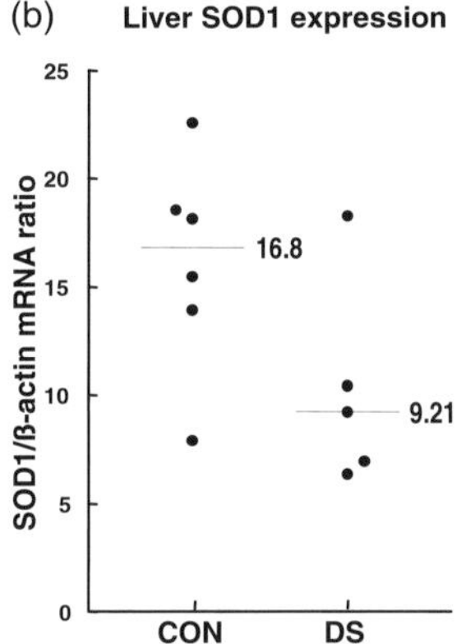

Fig. 2. Analysis of SOD1 mRNA levels in control and DS organs. **a** Northern blots were densitometrically scanned and the results expressed relative to β-actin. A comparison of median values (horizontal bars) shows that SOD1 mRNA levels are significantly elevated in DS brain, heart and thymus compared with the control median value for each tissue investigated ($p < 0.05$). Although the DS lung median is increased compared with controls, significance was not reached ($p > 0.05$). **b** Control and DS liver SOD1 mRNA levels expressed relative to β-actin. This is the only DS fetal organ to show a statistically significant decrease in the SOD1 mRNA median level compared with the control median level

Northern blot analysis of GPX1 expression in control and DS organs and an analysis of the SOD1 to GPX1 ratio in these organs

We investigated the expression of a non-HSA21 antioxidant gene, GPX1, in control and DS tissues since a number of studies have suggested that this enzyme is upregulated in certain DS tissues in an adaptive response to the increased levels of SOD1 (Frischer et al., 1981; Neve et al., 1983). Furthermore, an investigation of the GPX1 mRNA levels in these DS tissues allowed us to focus on the SOD1 to GPX1 mRNA ratio as this may be the more important determinant of cellular damage.

In the majority of organs studied (namely the brain, heart, lung and thymus), the median DS GPX1 mRNA level was not significantly different

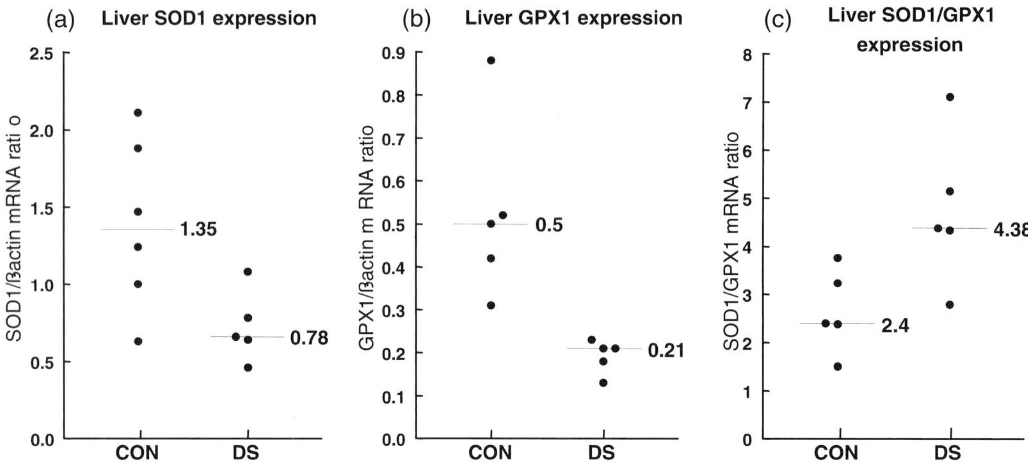

Fig. 3. Slot blot analysis of total RNA extracted from control and DS livers. Serial dilutions of total RNA (ranging from 15 to 0.9μg) were blotted onto Hybond N and sequentially hybridized with the SOD1, GPX1 and β-actin probes (data not shown). The slot blots were densitometrically scanned and the results corrected relative to β-actin. Median values (horizontal bars) are indicated for both control and DS groups. **a** SOD1 mRNA levels. The DS SOD1 mRNA median is significantly reduced (1.3-fold) compared with the control liver median; **b** GPX1 mRNA levels. The median values indicate a 2.4-fold decrease in DS livers compared with controls; **c** SOD1/GPX1 mRNA levels. The median SOD1/GPX1 level is elevated (1.4 fold) in DS livers compared with controls

from that of controls (Fig. 4a; $p > 0.05$). Only the DS liver showed a significant 4.5 fold decrease in the median GPX1 mRNA level (Fig. 4b; $p < 0.005$). The median SOD1 to GPX1 mRNA ratio (SOD1 mRNA levels are expressed relative to GPX1 mRNA levels within individual tissues) was significantly increased (1.5–2 fold) in all DS organs compared with controls (Fig. 5a; $p < 0.05$). Even DS fetal livers showed an increased median SOD1 to GPX1 mRNA ratio of approximately 2.0 fold since the decrease in GPX1 was far greater than the decrease in SOD1 mRNA levels. The latter data were obtained by both Northern (Fig. 5b) and slot blot (Fig. 3c) analysis of liver samples.

Discussion

This study addresses a number of issues relating to the expression of antioxidant genes in various tissues of control and DS aborted conceptuses. Prior to this study limited information was available on the expression of antioxidant genes such as SOD1 and GPX1 in DS tissues. The data already published mainly emanated from an analysis of blood and fibroblasts from individuals with DS (permanent fibroblast cell cultures established from DS individuals) (Anneren and Epstein, 1987) and adult brain (Gulesserian et al., 2001b). Only two fetal studies have been reported and both have been restricted to fetal

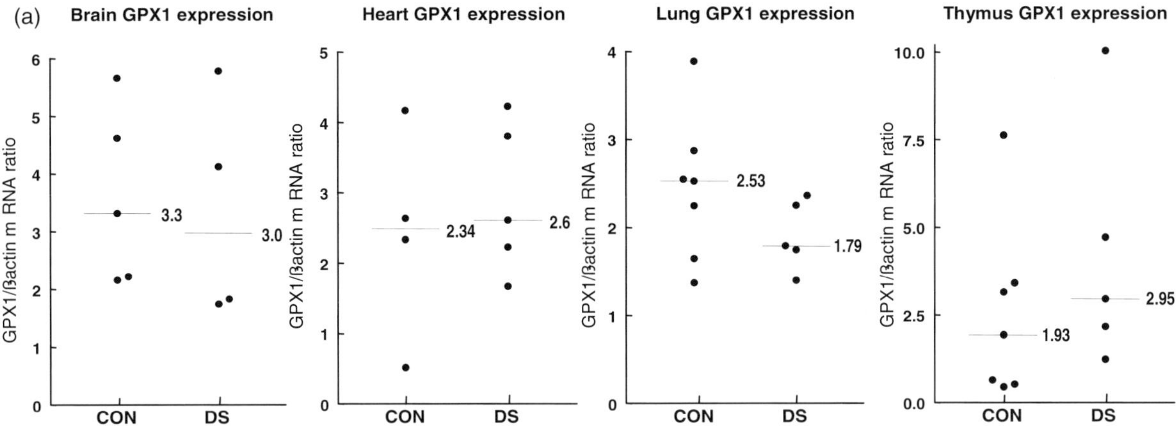

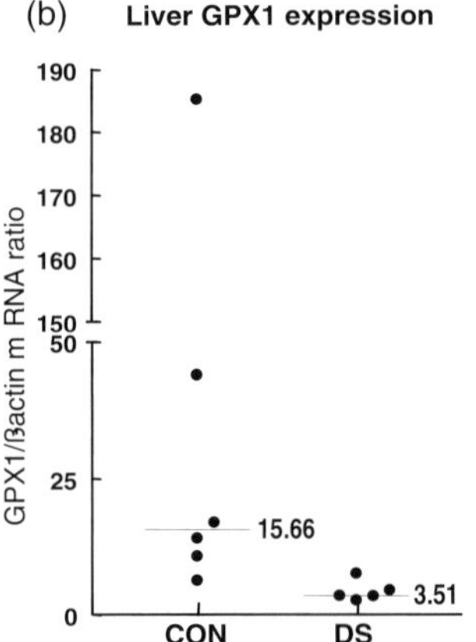

Fig. 4. Analysis of the GPX1 mRNA levels in control and DS organs. **a** Northern blots were densitometrically scanned and the results expressed relative to β-actin. Median values (horizontal bars) are indicated for both control and DS groups. No significant difference in the median GPX1 mRNA level was seen between DS and control organs ($p > 0.05$ for all four organs). **b** Control and DS liver GPX1 mRNA levels expressed relative to β-actin. The DS fetal liver is the only organ to show a statistically significant decrease in the median GPX1 expression level

brain (Brooksbank and Balazs, 1984; Gulessarian et al., 2001a). We felt that a study of a greater range of DS fetal tissue was required since during development, potential oxidative damage of vital organs may have severe consequences on the developing fetus.

The data of this study demonstrate that the expression of the HSA21 gene, SOD1, is elevated in four of the five fetal organs investigated (namely, brain, heart, thymus and lung). Furthermore, the level of elevation is in agreement with the 1.5 fold increase in gene dosage for three of the four organs (namely brain, heart and lung). Our data are in agreement with previous studies that show gene-dosage elevated SOD1 in red blood cells (Gilles et al., 1976; De La Torre et al., 1996; Pastor et al., 1998), platelets (Sinet et al., 1975a), leucocytes and fibroblasts (Feaster et al., 1977) of individuals with complete trisomy of

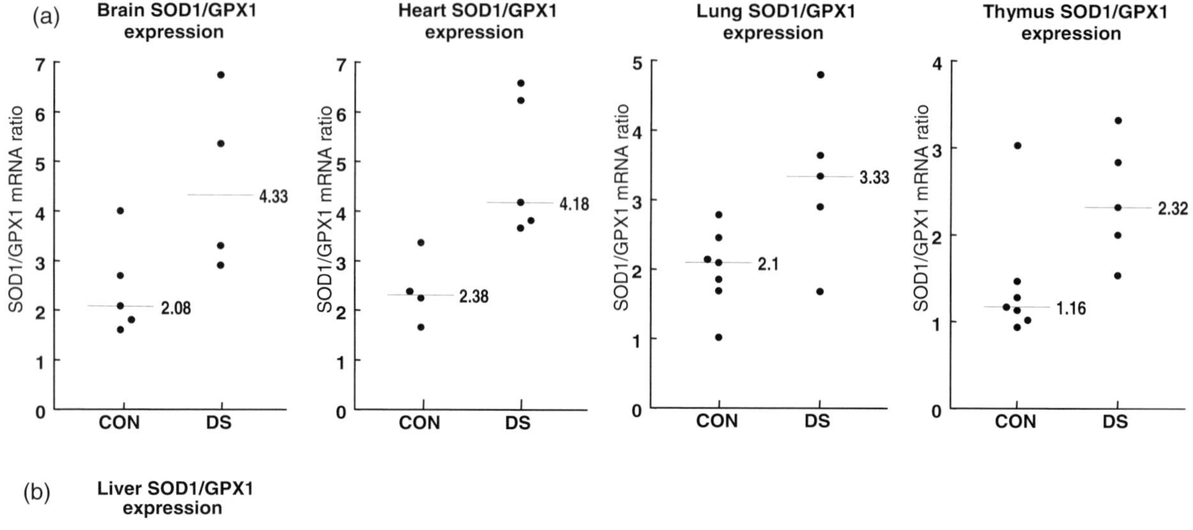

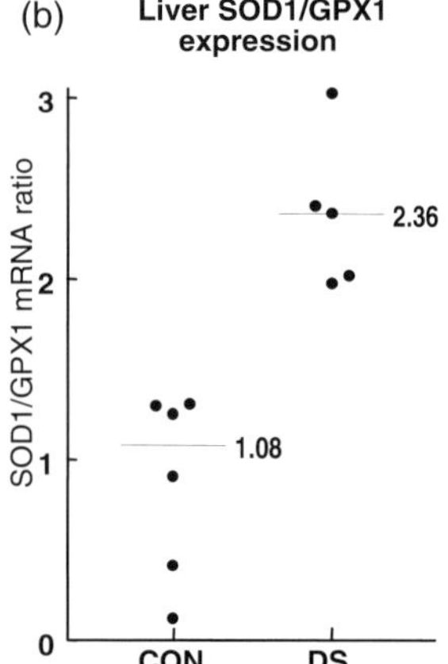

Fig. 5. Northern blot analysis of the SOD1 to GPX1 ratio in control and DS organs. In both **a** and **b** SOD1 mRNA levels are expressed relative to the level of GPX1 expression for each individual sample analyzed. An analysis of median values shows that the DS median is higher than the median of the control group for each tissue-type investigated. Use of non-parametric statistics (Mann-Whitney U test) shows that this increase is significant for all five tissues investigated

HSA21. Our data are also consistent with the gene dosage elevation in SOD1 activity seen in DS fetal brain (Brooksbank and Balazs, 1984). However, our data and that of Brooksbank and Balazs (1984) are not in agreement with recent data of Gulesserian et al. (2001a) who show decreased SOD1 protein levels in DS fetal brains compared with controls. At this time, it is unclear why this discrepancy has arisen. Future studies should evaluate expression and activity data within the same DS sample to eliminate such discrepancies, however, given the limitation in obtaining sufficient human material, this is not always possible.

Our study of the DS thymus showed a greater than expected increase in SOD1 mRNA levels (3 fold) compared with controls. Elevation of SOD1 expression above gene-dosage is not unprecedented, since previous reports support a departure from strict gene dosage in DS organs. Neve et al. (1988) show a 4-fold elevation in the mRNA for APP (amyloid precursor protein; the gene for APP is localized to HSA21) in the cortex of three fetal DS brains when compared with age-matched controls, while Stefani et al. (1988) report a 5 fold increase in two HSA21-specific mRNAs in DS brain compared with control brains. Furthermore, Holtzman et al. (1992) show that dysregulation of the APP gene in trisomy 16 fetuses (a mouse model for DS) results in a 2–3 fold elevation in APP mRNA in most organs investigated. Thus over-expression of genes residing on HSA21 may not always be in strict adherence with gene dosage in DS.

Although the majority of organs showed increased expression of SOD1 mRNA levels in this study, the DS liver showed decreased expression compared with control livers. This lack of elevation in the DS liver is consistent with other evidence in the literature for a number of HSA21 genes in this organ. Stefani et al. (1988) report decreased expression of one of four HSA21-specific sequences in DS fetal liver, while three others are expressed at the same level in control and DS livers. Interestingly, all four sequences were elevated in DS brains compared with controls. Taken together with the data of this study, these data suggest that the lack of increased expression in trisomy 21 is perhaps related to the organ per se. Consistent with this idea, Ceballos et al. (1991) have shown that mice transgenic for human SOD1 under the regulation of the SOD1 homologous promoter, express the transgene in a tissue-specific fashion in a range of murine organs, with higher amounts of SOD1 mRNA in brain, thymus and heart than liver. Similarly, Epstein et al. (1987) also demonstrate in a different SOD1 transgenic mouse line (again under homologous SOD1 promoter regulation), that the transgene is expressed in a tissue-specific manner, with expression being low in the livers of all transgenic lines generated. This poses an interesting question as to the reason for the decreased or unchanged expression of HSA21 genes (and perhaps non-HSA21 related genes since we find that GPX1 mRNA levels are also lower in this study) in DS livers and transgenic mice. Indeed, altered gene expression in DS livers may be etiological in some of the pathologies associated with the DS liver e.g. the increased incidence of liver fibrosis (Schwab et al., 1998) and increased serum cholesterol levels as a result of an intrinsic liver lipid metabolism abnormality (Diomede et al., 1999).

A lack of increase in the level of HSA21 gene expression in DS tissues has previously been interpreted as evidence against the "gene dosage effect" hypothesis. Greber-Platzer and colleagues measured the levels of ETS-2 message in DS hearts (Greber-Platzer et al., 1999a) and brains (Greber-Platzer et al., 1999b) and found unchanged expression in DS hearts compared with controls, and reduced expression in two brain regions, namely the frontal and temporal lobes, while three other brain regions showed unaltered ETS-2 expression. Recently the same group (Cheon et al., 2003 a,b,c) have investigated a vast range of HSA21 gene products by Western blot analysis and in most instances the protein levels were unaltered in DS fetal brains. These authors concluded that this provides evidence against the "gene dosage effect" hypothesis, since according to gene dosage, gene expression and gene products arising from HSA21 should be elevated by 1.5 fold in all DS tissues. The results of this study are in agreement with the notion that not all gene products of HSA21 are elevated 1.5 fold. However, we feel that departure from a 1.5 fold increase in gene expression should not be interpreted as evidence against the "gene dosage effect" hypothesis. In our opinion, the crux of the "gene dosage effect" hypothesis is that it is the qualitative increase of specific HSA21 gene products and not the quantity of extra chromosomal material that causes DS. The scientific community has no problem with the notion that certain gene products influence the expression of other genes. For instance, elevated levels of particular HSA21 gene products may cause an elevation or decrease in the levels of other gene products, which in turn may induce or repress other HSA21 genes in a common pathway. Thus relative abundances and gene interactions between HSA21 and non-HSA21 gene products can influence the expression of any particular gene in an organism. Furthermore, these influences may be context-dependent, accounting for the apparent variable levels of the same gene product observed in different organs. Thus what matters is not the absolute level of a HSA21 gene product, but the relative level of interacting products, whether or not they are HSA21-derived. Examination of the genomic and proteomic pathways involved will yield clearer understanding and lead to more meaningful analysis of the Down syndrome phenotype. However, it should be emphasized that analysis of HSA21 gene expression and activity of its gene products is still totally relevant and important in the understanding of the DS phenotype.

In light of our interpretation of the importance of interacting gene products in Down syndrome, we have investigated the expression of the non-HSA21 second-step antioxidant enzyme, GPX1 in the same DS fetal tissues. Prior to this study, the issue of whether a compensatory increase in the level of GPX1 occurs in all DS fetal organs remained unanswered, since a number of reports have detailed a rise in GPX1 activity in DS erythrocytes and lymphoid cells of approximately 40% (Frischer et al., 1981; Neve et al., 1983; Pastor et al., 1998; Sinet et al., 1975b), while another study showed no change in the DS fetal brain (Brooksbank and Balazs, 1984). This issue becomes important given that an upregulation of GPX1 in response to elevated SOD1 levels restores the balance in antioxidant activity which could limit Fenton-type reactions, thus reducing the potential for oxidative damage to biologically important molecules. Conversely a lack of upregulation of GPX1 could result in large-scale macromolecular damage through Fenton-type •OH radical-mediated reactions.

The results of this study show no adaptive rise in GPX1 mRNA in four of the five organs studied (namely brain, heart, lung and thymus). Our results are in agreement with Brooksbank and Balasz (1984) who failed to demonstrate a compensatory increase in GPX1 activity in the DS fetal brain, in response to a 50% increase in SOD1 activity. Our study has now extended the lack of compensation by GPX1 to the DS fetal heart, lung and thymus. Therefore the results of this study show that the brain is not unique in its inability to upregulate GPX1 expression in the developing DS fetus. One reason for the lack of adaptation by fetal tissues to the increased SOD1 activity may be that oxygen tension in these fetal organs is not high enough to trigger the adaptive response (Rudolf, 1984). If this is correct, then GPX1 adaptation to increased SOD1 expression may only occur in postnatal, childhood and adult DS organs. However, this has not been investigated due to obvious limitations of the availability of human material (other than blood and lymphoid tissue where indeed, GPX1 activity has been shown to be elevated). However, the possibility exists that some postnatal, childhood and adult organs fail to upregulate GPX1 expression in response to the elevated SOD1 levels, and in this way may contribute to the DS pathology of that particular organ. This issue needs further investigation.

The lack of adaptation by GPX1 to the elevated SOD1 expression in DS brain, lung, heart and thymus results in an altered SOD1 to GPX1 mRNA ratio in these organs. Even the liver, which did not show an increase in SOD1 mRNA levels, shows an altered SOD1 to GPX1 mRNA ratio as a consequence of a greater decline in GPX1 mRNA levels compared with SOD1 mRNA levels. Therefore, all DS fetal organs examined in this study display a disproportionate elevation in the ratio of SOD1 to GPX1 mRNA levels. Indeed, the altered SOD1 to GPX1 ratio may contribute to the oxidative-stress induced damage of DS fetal organs, since we (de Haan et al., 1996) and others (Brooksbank and Balazs, 1984; Kelner and Bagnell, 1990; Kelner et al., 1995) have shown that a disproportionate increase in SOD1 to GPX1 results in lipid damage and alters gene expression of non-HSA21 genes. In agreement with this, one study has shown extensive glycoxidation in DS fetal brains (Odetti et al., 1998). Further conclusions can be drawn from the extensive analysis done on mice transgenic for SOD1, where a disporportionate increase in SOD1 relative to GPX1 occurs (Ceballos-Picot et al., 1992). These mice show (a) enhanced lipid peroxidation of their brains; (b) impairment of their immune system with reduced function of macrophages (Mirochnitchenko and Inouye, 1996) and thymic involution (Nabarra et al., 1996); (c) premature aging changes, such as neuromuscular junction degeneration of tongues and hind limbs (Avraham et al., 1991); and (d) increased susceptibility of neurons to kainic acid-induced apoptosis (Bar-Peled et al., 1996). Indeed, many of these features are also seen in the DS phenotype.

Based on the results of the study, it is our contention that the imbalance in the antioxidant enzymatic pathway places the developing DS fetus in an environment of increased oxidative stress. Indeed, the interactions of reactive oxygen species have been shown to affect numerous biochemical pathways (Minc-Golomb et al., 1991) and the binding of transcription factors such as NF-$_{\kappa}$B and AP-1 to DNA (Meyer et al., 1993). It is therefore highly likely that this occurs in developing DS fetal organs, and in this manner may contribute to certain aspects of the DS phenotype. This study has shown that an altered ratio of first to second-step antioxidant gene expression exists in all major DS fetal organs. The radical-induced changes that may arise as a consequence of this may begin in utero and continue during the life-span of the individual with DS. This may in part explain the earlier onset of pathologies such as the premature aging and neurodegenerative disorders seen in DS individuals where damage by reactive oxygen species has been implicated.

Acknowledgements

We would like to thank P. Hertzog for critical reading of the manuscript. This work was funded in part by SmithKline Beecham Pharmaceuticals, and the National Health and Research Council of Australia (NH & MRC).

References

Anneren KG, Epstein CJ (1987) Lipid peroxidation and superoxide dismutase-1 and glutathione peroxidase activities in trisomy 16 fetal mice and human trisomy 21 fibroblasts. Pediatr Res 21: 88–92

Avraham KB, Sugarman H, Rotshenker S, Groner Y (1991) Down's syndrome: morphological remodelling and increased complexity in the neuromuscular junction of transgenic CuZn-superoxide dismutase mice. J Neurocytol 20: 208–215

Bar-Peled O, Korkotian E, Segal M, Groner Y (1996) Constitutive overexpression of Cu/Zn superoxide dismutase exacerbates kainic acid-induced apoptosis of transgenic-Cu/Zn superoxide dismutase neurons. Proc Natl Acad Sci 93: 8530–8535

Brooksbank BWL, Balazs R (1984) Superoxide dismutase, glutathione peroxidase and lipoperoxidation in Down's Syndrome fetal brain. Dev Brain Res 16: 37–44

Ceballos I, Nicole A, Briand P, Grimber G, Delacourte A, Flament S, Blouin JL, Thevenin M, Kamoun P, Sinet M (1991) Expression of human Cu-Zn superoxide dismutase gene in transgenic mice: model for gene dosage effect in Down syndrome. Free Rad Res Commun 12–13: 581–589

Ceballos-Picot I, Nicole A, Clement M, Bourre JM, Sinet PM (1992) Age-related changes in antioxidant enzymes and lipid peroxidation in brains of control and transgenic mice overexpressing copper-zinc superoxide dismutase. Mutat Res 275: 281–293

Chada S, Le Beau MM, Casey L, Newburger PE (1990) Isolation and chromosomal localization of the human glutathione peroxidase gene. Genomics 6: 268–271

Chaushu S, Yefenof E, Becker A, Shapira J, Chaushu G (2002) Severe impairment of secretory Ig production in parotid saliva of Down syndrome individuals. J Dent Res 81: 308–312

Cheon MS, Kim SH, Yaspo ML, Blasi F, Aoki Y, Melen K, Lubec G (2003a) Protein levels of genes encoded on chromosome 21 in fetal Down syndrome brain. Challenging the gene dosage effect hypothesis, part I. Amino Acids 24: 111–117

Cheon MS, Bajo M, Kim SH, Claudio JO, Stewart AK, Patterson D, Kruger WD, Kondoh H, Lubec G (2003b) Protein levels of genes encoded on chromosome 21 in fetal Down syndrome brain. Challenging the gene dosage effect hypothesis, part II. Amino Acids 24: 119–125

Cheon MS, Kim SH, Ovod V, Kopitas Jerala N, Morgan JI, Hatefi Y, Ijuin T, Takenawa Y, Lubec G (2003c) Protein levels of genes encoded on chromosome 21 in fetal Down syndrome brain. Challenging the gene dosage effect hypothesis, part III. Amino Acids 24: 127–134

Davies KJA (1987) Protein damage and degradation by oxygen radicals. J Biol Chem 262: 9895–9901

de Haan JB, Newman JD, Kola I (1992) Cu/Zn superoxide dismutase mRNA and enzyme activity, and susceptibility to lipid peroxidation, increases with aging in murine brains. Mol Brain Res 13: 179–186

de Haan JB, Tymms MJ, Cristiano F, Kola I (1994) Expression of copper/zinc superoxide dismutase and glutathione peroxidase in organs of developing mouse embryos, fetuses and neonates. Pediatr Res 35: 188–196

de Haan JB, Cristino C, Iannello R, Bladier C, Kelner MJ, Kola I (1996) Elevation in the ratio of Cu/Zn-superoxide dismutase to glutathione peroxidase activity induces features of cellular senescence and this effect is mediated by hydrogen peroxide. Hum Mol Genet 5: 283–292

Delabar JM, Nicole A, D'Auriol L, Jacob Y, Meunier-Rotival M, Galibert F, Sinet PM, Jerome H (1987) Cloning and sequencing of a rat CuZn superoxide dismutase cDNA: correlation between CuZn superoxide dismutase mRNA levels and enzyme activity in rat and mouse tissues. Eur J Biochem 166: 181–187

De La Torre R, Casado A, Lopez-Fernandez E, Carrascosa D, Ramirez V, Saez J (1996) Overexpression of copper-zinc superoxide dismutase in trisomy 21. Experientia 52: 871–873

Diomede L, Salmona M, Albani D, Bianchi M, Bruno A, Salmona S, Nicolini U (1999) Alteration of SREBP activation in liver of trisomy 21 fetuses. Biochem Biophys Res Commun 260: 499–503

Epstein CJ, Avraham KB, Lovett M, Smith S, Elroy-Stein O, Rotman G, Bry C, Groner Y (1987) Transgenic mice with increased Cu/Zn-superoxide dismutase activity: animal model of dosage effects in Down syndrome. Proc Natl Acad Sci USA 84: 8044–8048

Feaster WW, Kwok LW, Epstein C (1977) Dosage effects for superoxide dismutase-1 in nucleated cells aneuploid for chromosome 21. Am J Hum Genet 29: 563–570

Fong C, Brodeur GM (1987) Down's syndrome and leukemia: epidemiology, genetics, cytogenetics and mechanisms of leukemogenesis. Cancer Genet Cytogenet 28: 55–76

Fridovich I (1978) The biology of oxygen radicals. Science 201: 875–880

Frischer H, Chu LK, Ahmad T, Justice P, Smith GF (1981) Superoxide dismutase and glutathione peroxidase abnormalities in erthyrocytes and lymphoid cells in Down syndrome. In: Brewer GJ (ed) The Red Cell: Fifth Ann Arbor Conference. AL Liss, New York, pp 269–283

Fuentes JJ, Genesca L, Kingsbury TJ, Cunningham KW, Perez-Riba M, Estivill X, de la Luna S (2000) DCSR1, overexpressed in Down syndrome, is an inhibitor of calcineurin-mediated signaling pathways. Hum Mol Genet 9: 1681–1690

Gilles L, Ferradini C, Foos J, Pucheault J, Allard D, Sinet PM, Jerome H (1976) The estimation of red cell superoxide dismutase activity by pulse radiolysis in normal and trisomic cells. Hum Genet 31: 197–202

Greber-Platzer S, Scatzmann-Turhani D, Wollenek G, Lubec G (1999a) Evidence against the current hypothesis of "gene dosage effects" of trisomy 21: ets-2, encoded on chromosome 21 is not overexpressed in hearts of patients with Down syndrome. Biochem Biophys Res Commun 254: 395–399

Greber-Platzer S, Schatzmann-Turhani D, Cairns N, Balcz B, Lubec G (1999b) Expression of the transcription factor ETS2 in brains of patients with Down Syndrome-evidence against the overexpression-gene dosage hypothesis. J Neural Transm 57: 270–281

Gulesserian T, Engidawork E, Fountoulakis M, Lubec G (2001a) Antioxidant proteins in fetal brain: superoxide dismutase-1 (SOD1) protein is not overexpressed in fetal Down syndrome. J Neural Transm 61: 71–84

Gulesserian T, Seidl R, Hardmeier R, Cairns N, Lubec G (2001b) Superoxide dismutase SOD1, encoded by chromosome 21, but not SOD2 is overexpressed in brains of patients with Down syndrome. J Invest Med 49: 41–46

Hall B (1965) Delayed ontogenesis in human trisomy syndromes. Hereditas (Lund) 52: 334–344

Holtzman DM, Bayney RM, Li Y, Khosrovi H, Berger CN, Epstein CJ, Mobley WC (1992) Dysregulation of gene expression in mouse trisomy 16, an animal model of Down syndrome. EMBO J 11: 619–627

Imlay JA, Chin SM, Linn S (1988) Toxic DNA damage by hydrogen peroxide through the Fenton reaction *in vivo* and *in vitro*. Science 240: 640–642

Kelner MJ, Bagnell R (1990) Alteration of growth rate and fibronectin by imbalances in superoxide dismutase and glutathione peroxidase activity. Biol Reactive Intermediates IV: 305–309

Kelner MJ, Bagnell R, Montoya M, Estes L, Uglik SF, Cerutti P (1995) Transfection with human copper-zinc superoxide dismutase induces bidirectional alterations in other antioxidant enzymes, proteins, growth factor response, and paraquat resistance. Free Rad Biol Med 18: 497–506

Kola I, Cristiano F, de Haan JB, Sumarsono S, Thomas R, Corrick C, Tymms M (1993) Genes, embryogenesis and Down syndrome. In: Moeloek F, Affandi B, Trounson AO (eds) Advances in human reproduction, vol 38. Parthenon Publishing Group, pp 309–320

Lemieux N, Malfoy B, Forrest GL (1993) Human carbonyl reductase (CBR) localized to band 21q22.1 by high-resolution fluorescence in situ hybridization displays gene dosage effects in trisomy 21 cells. Genomics 15: 169–172

Mann DMA, Esiri MM (1989) The pattern of acquisition of plaques and tangles in the brains of patients under 50 years of age with Down's Syndrome. J Neurol Sci 89: 169–179

Meyer M, Schreck R, Baeuerle PA (1993) H_2O_2 and antioxidants have opposite effects on activation of NF-$_{\kappa}$B and AP-1 in intact cells: AP-1 as secondary antioxidant-responsive factor. EMBO J 12: 2005–2015

Minc-Golomb D, Knobler H, Groner Y (1991) Gene dosage of CuZnSOD and Down's syndrome: diminished prostaglandin synthesis in human trisomy 21, transfected cells and transgenic mice. EMBO J 10: 2119–2124

Mirochnitchenko O, Inouye M (1996) Effect of overexpression of human Cu,Zn superoxide dismutase in transgenic mice on macrophage functions. J Immunol 156: 1578–1586

Nabarra B, Casanova M, Paris D, Nicole A, Toyama K, Sinet PM, Ceballos I, London J (1996) Transgenic mice overexpressing the human Cu/Zn-SOD gene: ultrastructural studies of a premature thymic involution model of Down's syndrome (Trisomy 21). Lab Invest 74: 67–626

Neve J, Sinet PM, Molle L, Nicole A (1983) Selenium, zinc and copper levels in Down's syndrome (trisomy 21): blood levels and relations with glutathione peroxidase and superoxide dismutase. Clin Chim Acta 133: 209–214

Neve RL, Finch EA, Dawes LR (1988) Expression of the Alzheimer amyloid precursor gene transcript in the human brain. Neuron 1: 669–677

Odetti P, Angelini G, Dapino D, Zaccheo D, Garibaldi S, Dagna-Bricarelli F, Piombo G, Perry G, Smith M, Traverso N, Tabaton M (1998) Early glycoxidation damage in brains from Down's syndrome. Biochem Biophys Res Commun 243: 849–851

Pallister C, Jung SS, Shaw I, Nalbantoglu J, Gauthier S, Cashman NR (1997) Lymphocyte content of amyloid precursor protein is increased in Down's syndrome and aging. Neurobiol Aging 18: 97–103

Pastor M-C, Sierra C, Dolade M, Navarro E, Brandi N, Cabre E, Mira A, Seres A (1998) Antioxidant enzymes and fatty acid status in erythrocytes of Down's syndrome patients. Clin Chem 44: 924–929

Patterson DH (1987) The causes of Down Syndrome. Sci Am 257: 42–49

Pritchard MA, Kola I (1999) The "gene dosage effect" hypothesis versus the "amplified developmental instability" hypothesis in Down syndrome. J Neural Transm 57: 293–303

Rehder H (1981) Pathology of trisomy 21, with particular reference to persistent common atrioventricular canal of the heart. In: Burgio GR, Fraccaro M, Tiepolo L, Wolf U (eds) Trisomy 21. An International Symposium. Springer, Berlin Heidelberg New York Tokyo, pp 57–73

Rudolf AM (1984) Oxygenation in the fetus and neonate — a perspective. Semin Perinatol 8: 158–167

Schwab M, Niemeyer C, Schwarzer U (1998) Down syndrome, transient myeloproliferative disorder, and infantile liver fibrosis. Med Pediatr Oncol 31: 159–165

Shapiro BL (1994) The environmental basis of the Down syndrome phenotype. Dev Med Child Neurol 36: 84–90

Sherman L, Levanon D, Lieman-Hurwitz J, Dafni N, Groner Y (1984) Human Cu/Zn superoxide dismutase gene: molecular characterization of its two mRNA species. Nucl Acids Res 12: 9349–9365

Siegel S (1956) In: Non-parametric statistics for the behavioural sciences. International Student edition. McGraw-Hill Kogakusha LTD, Tokyo, Japan

Sies H, de Groot H (1992) Role of reactive oxygen species in cell toxicology. Toxicol Lett 64–65: 547–551

Sinet PM, Michelson AM, Bazin A, Lejeune J, Jerome H (1975a) Superoxide dismutases activities of blood platelets in trisomy 21. Biochem Biophys Res Commun 67: 904–909

Sinet PM, Michelson AM, Bazin A, Lejeune J, Jerome H (1975b) Increase in glutathione peroxidase activity in erythrocytes from trisomy 21 subjects. Biochem Biophys Res Commun 67: 910–915

Stefani I, Galt J, Palmer A, Affara N, Ferguson-Smith M, Nevin NC (1988) Expression of chromosome 21 specific sequences in normal and Down's syndrome tissues. Nucl Acids Res 16: 2885–2896

Sumarsono SH, Wilson TJ, Tymms MJ, Venter DJ, Corrick CM, Kola R, Lahoud MH, Papas TS, Seth A, Kola I (1996) Down's syndrome-like skeletal abnormalities in Ets2 transgenic mice. Nature 379: 534–537

Tam CF, Walford RL (1980) Alteration in cyclic nucleotides and cyclase-specific activities in T lymphocytes of aging normal humans and patients with Down's syndrome. J Immunol 125: 1665–1670

Tan YH, Tischfield J, Ruddle FH (1973) The linkage of genes for the human interferon induced antiviral protein and indophenol oxidase-B traits to chromosome G-21. J Exp Med 137: 317–330

Wadhera S, Millar WT (1994) Second trimester abortions: trends and medical complications. Health Reports 6: 441–454

White BA, Bancoft FC (1982) Cytoplasmic dot hybridization. Simple analysis of relative mRNA levels in multiple small cell or tissue samples. J Biol Chem 257: 8569–8572

Wisniewski KE, Wisniewski HM, Wen GY (1985) Occurrence of neuropathological changes and dementia of Alzheimer's disease in Down's Syndrome. Ann Neurol 17: 278–282

Authors' address: J. B. de Haan, Baker Heart Research Institute, Diabetic Complications Group, Commercial Rd., Prahran, 3181 Australia, e-mail: judy.dehaan@baker.edu.au

Overexpression of C1-tetrahydrofolate synthase in fetal Down Syndrome brain

M. Fountoulakis[1], **T. Gulesserian**[2], and **G. Lubec**[2]

[1] Hoffmann-La Roche, Genomics and Proteomics Technologies, Basel, Switzerland
[2] Division Basic Research, Department of Pediatrics, University of Vienna, Austria

Summary. Trisomy 21, Down Syndrome, is the most common genetic cause of human mental retardation and results from non-disjunction of chromosome 21. Several reports have been linking folate metabolism to DS and indeed, chromosome 21 even encodes for a specific folate carrier. The availability of brain tissue along with the advent of proteomics enabled us to identify and quantify C1-tetrahydrofolate synthase (THF-S), a key element in folate metabolism in brain along with other enzymes involved in C1-metabolism.

Brains of controls and DS subjects at the 18th–19th week of gestation were homogenised and separated on 2 dimensional gel electrophoresis with subsequent in-gel digestion and mass spectrometrical identification and quantification with specific software.

THF-S was represented by three spots, possibly representing isoforms or posttranslational modifications. Two spots were significantly, about twofold, increased in fetal DS brain:

Controls [means $\pm$ SD: (spot 1) 2.55 $\pm$ 0.69; (spot 3) 1.39 $\pm$ 0.86] vs. Down syndrome [means $\pm$ SD: (spot 1) 4.25 $\pm$ 1.63; (spot 3) 4.43 $\pm$ 2.13].

These results were reproducible when THF-S levels were normalised versus the housekeeping protein actin and neuron specific enolase to compensate cell or neuronal loss.

C1-metabolism related enzymes ribose-phosphate pyrophosphokinase I, inositol monophosphate dehydrogenase, guanidine monophosphate synthease and S-adenosylmethionine synthase, gamma form, were comparable between groups.

Overexpression of this key enzyme in fetal DS brain at the early second trimester may indicate abnormal folate metabolism and may reflect folate deficiency. This may be of pathomechanistic relevance and thus extends and confirms the involvement of folate metabolism in trisomy 21.

Introduction

Down syndrome (DS, Trisomy 21), is the most common genetic cause of human mental retardation, with an incidence of 1 among 600–1,000 live births

(Smith and Berg, 1976) and is a major cause of premature pregnancy failure. It is estimated that 1 out of 500 conceptions have trisomy 21 and that 80% of these are lost during early pregnancy. The nondisjunction event resulting in two copies of chromosome 21 takes place in anaphase of meiosis I, during oocyte maturation before ovulation, and/or in anaphase of meiosis II, around the time of fertilization in the adult female (Picton et al., 1998; Hunt and Lemaire-Adkins, 2000) but the biochemical and molecular basis for meiotic nondisjunction is not fully understood yet.

Folate serves as substrate in a series of interconnected metabolic cycles involving thymidilate and purine biosynthesis (providing nucleotide precursors for DNA synthesis), methionine synthesis via homocysteine (Hcy) remethylation, serine and glycine interconversion, and metabolism of histidine and formate. Folate is involved in several methylation reactions via S-adenosylmethionine (SAM), directly or indirectly essential for cell function, division, and differentiation (van der Put et al., 2001).

Chronic folate/methylation deficiency in vivo and in vitro has been associated with abnormal DNA methylation (Fowler et al., 1998; Jacob et al., 1998), DNA strand breaks (Blount et al., 1997; Duthie, 1999), altered chromosome recombination (MacGregor et al., 1997) and aberrant chromosome segregation (Titenko-Holland et al., 1998; Xu et al., 1999). On the basis of this evidence, James et al. (1999) proposed that gene-nutrient interactions associated with abnormal folate metabolism and DNA hypomethylation might increase the risk of chromosome nondisjunction.

Vital cellular processes depend on folate-mediated one-carbon metabolism, i.e. the transfer of a carbon group. Folate acts as donor and acceptor of one-carbon units in a variety of critical enzymatic reactions involved in one-carbon metabolism. One-carbon units are principally derived from the β-carbon of serine, but also from glycine, methyl- and dimethylglycine, formate, and histidine (Pasternack et al., 1996). One carbon units are obtained from THF-mediated reactions required for several major cellular processes like nucleic acid biosynthesis, protein biosynthesis, amino acid metabolism, methyl-group biogenesis, and vitamin metabolism (Shane, 1989; Appling, 1991).

Since alteration of folate metabolism in mothers of DS children has been described as a maternal risk factor, we used a proteomics approach to compare protein expression of enzymes of and related to folate-dependent one-carbon metabolism in fetal brain tissue.

Materials and methods

Fetal brain tissue

Brain tissue (cerebral cortex) of aborted control fetuses with no obvious abnormalities (n = 7; 1 female and 6 males with mean gestational age of 18.8 ± 2.2 weeks) and DS (n = 9; 2 females and 7 males with mean gestational age of 19.6 ± 2.0 weeks) was used for the experiment. The brain samples were obtained from Drs. M. Diessen and J. C. Ferreres, Medical and Molecular Genetics Center-IRO, Hospital Duran i Reynals,

Barcelona, Spain. All procedures were performed with the consent of the hospital's ethics committee. All samples were stored at −70°C and the freezing chain was never interrupted.

Two dimensional gel electrophoresis (2-DE)

Brain tissue was suspended in 0.5 ml sample buffer composed of 40 mM Tris, 5 M urea (Merck, Darmstadt, Germany), 2 M thiourea (Sigma, St. Louis, MO, USA), 4% CHAPS (Sigma), 10 mM 1,4-dithioerythritol (Merck), 1 mM EDTA (Merck) and a mixture of protease inhibitors (1 mM PMSF and 1 μg each of pepstatin A, chymostatin, leupeptin and antipain). The suspension was sonicated for approximately 30 sec and centrifuged at 150,000 g for 45 min. The protein content in the supernatant was determined by the Coomassie blue method (Bradford, 1976).

2-DE was performed essentially as reported (Langen et al., 1997). Samples of 2 mg protein were applied on immobilized pH 3–10 nonlinear gradient strips in sample cups at their basic and acidic ends. Focusing started at 200 V and the voltage was gradually increased to 5,000 V at 3 V/min and kept constant for a further 24 h (approximately 180,000 Vh totally). The second-dimensional separation was performed on 9–16% sodium dodecylsulfate (SDS) gradient polyacrylamide gels. The gels (180 × 200 × 1.5 mm) were run at 40 mA per gel. The gels were stained with colloidal Coomassie blue (Novex, San Diego, CA, USA) for 48 h and destained with water following protein fixation for 12 h in 40% methanol containing 5% phosphoric acid. Molecular masses were determined by running standard protein markers (Gibco, Basel, Switzerland), covering the range 10–200 kDa. pI values were used as given by the supplier of the IPG strips. Excess of dye was washed out from the gels with water and the gels were scanned in an Agfa DUOSCAN densitometer (resolution 200). Electronic images of the gels were recorded using Photoshop (Adobe) and PowerPoint (Microsoft) software.

Matrix-associated laser desorption ionization mass spectroscopy (MALDI-MS)

MALDI-MS analysis was performed as described elsewhere (Fountoulakis and Langen, 1997) with some modifications. The spots were excised with a spot picker and placed into 96-well microtiter plates. Each spot was destained with 100 μl of 30% acetonitrile in 50 mM ammonium bicarbonate and dried in a speedvac evaporator. Each dried gel piece was re-hydrated with 4 μl of 3 mM Tris-HCl, pH 9.0, containing 50 ng trypsin (Promega, Madison, WI, USA). After 16 h at room temperature, 7 μl of distilled water was added to each gel piece and the samples were shaken in for 10 min. Four μl of 50% acetonitrile, containing 0.3% trifluoroacetic acid and the standard peptides des-Arg-bradykinin (Sigma, 904.4681 Da) and adrenocorticotropic hormone fragment 18–39 (Sigma, 2465.1989 Da), in water were added to each gel piece and shaken for 10 min. The application of the samples was performed with a SymBiot I sample processor (PE Biosystems, Framingham, MA, USA). 1.5 μl of the peptide mixture was simultaneously applied with 1 μl of matrix, consisting of a saturated solution of α-cyano-4-hydroxycinnamic acid (Sigma) in 50% acetonitrile, containing 0.1% trifluoroacetic acid. Samples were analyzed in a time-of-flight mass spectrometer (Reflex 3, Bruker Analytics, Bremen, Germany). An accelerating voltage of 20 kV was used. Peptide matching and protein searches were performed automatically. The peptide masses were compared to the theoretical peptide masses of all available proteins from all species. Monoisotopic masses were used and a mass tolerance of 0.0025% was allowed. The algorithm used for determining the probability of a false positive match with a given MS-spectrum is described elsewhere (Berndt et al., 1999).

Quantification of protein spots

Protein spots were outlined (first automatically and then manually) and quantified using the ImageMaster 2D Elite software (Amersham Pharmacia Biotechnology). For purpose of quantification, an area containing the protein of interest was chosen and percentage of volume of the spots representing the protein was determined in comparison with the total proteins present in the area.

Statistical analyses

Values are expressed as mean ± standard deviation of percentage of the spot volume in each particular gel after subtraction of the background values. Inter-group differences were analyzed using non-parametric Mann-Whitney U-test and level of significance was set at $P < 0.05$.

Results

Using two-dimensional gel electrophoresis with subsequent mass spectrometrical identification and quantification of proteins with specific software, we were able to quantify marker enzymes involved in the folate cycle (C1-tetrahydrofolate synthase, THF-S), de novo purine ribonucleotide synthesis (Ribose-phosphate-pyrophosphokinase I, RPPI, inositol-monophosphate dehydrogenase, IMP-DH; Guanidine monophosphate synthase, GMP-S) and methionine metabolism (S-adenosylmethionine synthetase gamma form, SAM-S).

On 2-D gel THF-S was detected as 3 spots with comparable molecular weights of 102 kDa, representing isoforms or posttranslational modifications. Spots 1 and 3 were significantly increased in DS. The central (spot number 2) spot was increased as well, although not reaching significance and the sum of the spots for THF-S revealed a significant increase in DS (Fig. 1 and Table 1).

C1-tetrahydrofolate synthase (1,2,3)

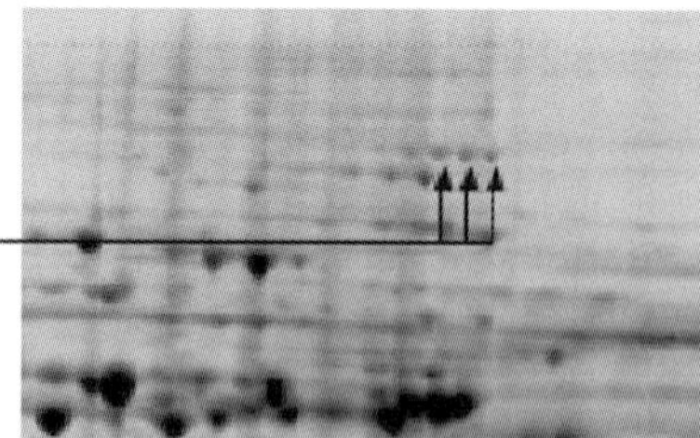

Fig. 1. Typical electrophoresis pattern of C1-tetrahydrofolate synthase (P11586) consisting of three spots in controls and DS patients. C1-tetrahydrofolate synthase on 2-DE gels from age-matched controls and patients with DS were quantified using ImageMaster 2D Elite software and compared to each other employing the non parametric Mann-Whitney U test in order to detect different expressional levels. Protein levels were determined as the percentage volume of the total proteins present in the gel part considered. Quantitative analysis was performed in brains of patients with DS compared with controls; C1-tetrahydrofolate synthase spot 1 and spot 3 in controls and DS patients showed significantly increased expression

Table 1. The levels of quantified one-carbon metabolism enzymes in control and DS fetal brain. Results (means ± SD, n = number of quantified samples) represent the percentage of the volume of the spot corresponding to the particular enzyme as compared to all proteins present in the particular area of interest. Quantification was performed using ImageMaster 2D Elite version 3.1 software. Between-group differences were calculated by Mann-Whitney U-test and the level of significance was set at $P < 0.05$*

Acc. Number	Protein	Spot	Control (n)	Down syndrome (n)
P09329	Ribose-phosphate pyrophosphokinase I		4.73 ± 0.84 (6)	3.53 ± 0.87 (4)
P11586	C1-tetrahydrofolate synthase	1	2.55 ± 0.69 (7)	4.25 ± 1.63 (4)*
		2	2.74 ± 1.06 (7)	4.26 ± 1.58 (4)
		3	1.39 ± 0.86 (5)	4.43 ± 2.13 (4)*
		summary	6.89 ± 1.86 (5)	12.95 ± 5.25 (4)*
P12268	IMP dehydrogenase 2		1.10 ± 0.80 (7)	1.84 ± 0.52 (7)
P31153	S-adenosylmethionine synthetase γ form		0.17 ± 0.13 (6)	0.26 ± 0.07 (4)
P49915	GMP synthase	1	0.07 ± 0.06 (7)	0.04 ± 0.02 (6)
		2	0.10 ± 0.06 (7)	0.08 ± 0.05 (6)
		summary	0.18 ± 0.08 (7)	0.14 ± 0.05 (6)

When normalised versus the housekeeping protein actin to rule out cell loss or neuron specific enolase (NSE) to rule out neuronal loss as confounding factor, THF-S levels remained about two-fold overexpressed (housekeeping proteins evaluated by western blotting, not shown).

SAM–S was observed as a single spot and was comparable between groups. From enzymes involved in de novo purine ribonucleotide synthesis we identified and quantified RPPI and IMP-DH and GMP-S. Quantification of the single spot identified as RPPI was not different either in DS fetal brain and IMP-DH was comparable between groups as well. Two spots were identified as GMP synthase, probably representing isoforms or different posttranslational modification, and quantification showed comparable expression. Data on numerical results are shown in Table 1 and on protein identification are listed in Table 2.

Discussion

The major outcome of the study is the remarkable and significant increase of THF-S in fetal DS brain, which has never been reported before in brain at the protein level, at the early second trimester of gestation. This finding may indicate deteriorated folate handling/metabolism induced by maternal or fetal defects that may well contribute to the host of impaired systems observed in DS (Epstein, 1992), in particular to abnormal morphology and wiring of the brain, which is well-known and unequivocal for neural tube defects and folate deficiency. This work supports our previous information on increased reduced

Table 2. Levels of qualified one-carbon metabolic enzymes normalised versus the housekeeping genes actin and neuron specific enolase, representing cell/neuronal density

Protein	Spot	Vs.	Controls	Down syndrome
Ribose-phosphate pyrophosphokinase I		Actin	1.48 ± 0.49	1.29 ± 0.72
		NSE	0.79 ± 0.41	0.96 ± 0.47
C1-tetrahydrofolate synthase	1	Actin	2.94 ± 1.47	4.76 ± 1.29*
		NSE	0.83 ± 0.42	1.70 ± 0.71*
	2	Actin	1.72 ± 0.75	3.17 ± 1.41*
		NSE	0.46 ± 0.40	1.33 ± 0.47*
	3	Actin	1.94 ± 0.94	4.17 ± 1.29*
		NSE	0.66 ± 0.57	1.77 ± 0.93*
IMP dehydrogenase 2		Actin	0.92 ± 0.49	1.07 ± 0.72
		NSE	0.83 ± 0.72	0.93 ± 0.67
S-adenosylmethionine synthetase γ form		Actin	0.17 ± 0.12	0.21 ± 0.16
		NSE	0.29 ± 0.14	0.30 ± 0.19
GMP synthase	1	Actin	0.02 ± 0.01	0.01 ± 0.01
		NSE	0.01 ± 0.01	0.02 ± 0.02
	2	Actin	0.06 ± 0.03	0.07 ± 0.04
		NSE	0.05 ± 0.04	0.06 ± 0.02

folate carrier protein, encoded on chromosome 21, in a comparable panel of fetal DS brains (Lubec et al., 2003).

Cytosolic C1-THF synthase is a trifunctional enzyme that catalyses the interconversion of reduced forms of folate to supply activated one-carbon units required for a variety of metabolic pathways. The enzyme activities include 10-formyl-tetrahydrofolate synthetase, 5,10-methylene tetrahydrofolate cyclohydrolase and 5,10-methylene tetrahydrofolate dehydrogenase (Wahls et al., 1993). Increased protein expression of C1-THF synthase detected in fetal DS brain tissue may be induced by functional folate deficiency in the fetus (and or mother) characterized by decreased concentration of THF. We have determined protein levels of the enzyme but it was reported that changes in yeast C1-THF synthase activity are due to changes in the steady-state protein concentration of C1-THF synthase, rather than to modulation of the activation of a pre-existing enzyme pool (Appling and Rabinowitz, 1985). Our finding of increased THF-S along with a series of alterations of THF-handling enzymes (Al-Gazali et al., 2001; Hassold et al., 2001; Hine and James, 2000; Botto and Yang, 2000; Rosenblatt, 1999), including methylene tetrahydrofolate reductase (Chadefaux-Vekemans et al., 2002; Hobbs et al., 2000), and MTRR polymorphism (O'Leary et al., 2002), are complementing and extending knowledge on a possible role for folate in pathomechanisms and pathogenesis of the DS phenotype.

The other enzymes involved in one carbon metabolism determined in our study were not deranged at the protein level, probably indicating limitation of the C1-defect to the folate cycle in DS per se. GMP synthase and SAM synthetase gamma form have never been reported at the protein level in brain before and we herewith demonstrate an analytical possibility to check these key enzymes of one carbon-handling enzymes. Ribose-phosphate pyrophosphokinase 1 was already reported in fetal brain of controls and DS individuals along with a large series of carbohydrate handling enzymes and re-analysis with normalisation versus actin and NSE was here included (Kitzmueller et al., 2001) to complete relevant marker enzymes of interconnected metabolic cycles. In vitro studies on lymphocytes of patients with trisomy 21 by Peeters and coworkers (1989) have shown that the IMP dehydrogenase system is not affected in DS and we add information that this enzyme is now identified and presenting with unchanged expression in the human brain with DS.

Taken together, we detected increased C1-tetrahydrofolate synthase in fetal DS brain at the early second trimester of gestation. Elevated levels of this enzyme may well be reflecting aberrant folate metabolism in DS and represent folate deficiency early in life.

Acknowledgements

G.L. is highly indebted to the Red Bull Company for generous financial support of the study and to M. Dierssen for providing fetal brain samples. We appreciate the secretarial hand of C. Avramovic for the preparation of the manuscript.

References

Al-Gazali LI, Padmanabhan R, Melnyk S, Yi P, Pogribny IP, Pogribna M, Bakir M, Hamid ZA, Abdulrazzaq Y, Dawodu A, James SJ (2001) Abnormal folate metabolism and genetic polymorphism of the folate pathway in a child with Down syndrome and neural tube defect. Am J Med Genet 103: 128–132

Alvarez L, Corrales F, Martin-Duce A, Mato JM (1993) Characterisation of a full-length cDNA encoding human liver S-adenosylmethionine synthetase: tissue-specific gene expression and mRNA levels in hepatopathies. Biochem J 293: 481–486

Appling DR (1991) Compartmentation of folate-mediated one-carbon metabolism in eukaryotes. FASEB J 5: 2645–2651

Appling DR, Rabinowitz (1985) Regulation of expression of the ADE3 gene for yeast C1-tetrahydrofolate synthase, a trifunctional enzyme involved in one-carbon metabolism. J Biol Chem 260: 1248–1256

Berndt P, Hobohm U, Langen H (1999) Reliable automatic protein identification from matrix assisted laser desorption/ionization mass spectrometric peptide fingerprints. Electrophoresis 20: 3521–3526

Blount BC, Mack MM, Wehr CM, Macgregor JT, Hiatt RA, Wickremasinghe RG, Everson RB, Ames BN (1997) Folate deficiency causes uracil misincorporation into human DNA and chromosome breakage: implications for cancer and neuronal damage. Proc Natl Sci USA 94: 3290–3295

Botto LD, Yang Q (2000) 5,10-Methylenetetrahydrofolate reductase gene variants and congenital anomalies: a HuGE review. Am J Epidemiol 151: 862–877

Bradford M (1976) A rapid and sensitive method for the quantification of microgram quantities of protein utilizing the principle of protein-dye binding. Anal Biochem 72: 248–254

Chadefaux-Vekemans B, Coude M, Muller F, Oury JF, Chabli A, Jais J, Kamoun P (2002) Methylenetetrahydrofolate reductase polymorphism in the etiology of Down syndrome. Pediatr Res 51: 766–767

Duthie SJ (1999) Folic acid deficiency and cancer: mechanisms of DNA instability. Br Med Bull 55: 578–592

Epstein CJ (1995) Down Syndrome. In: Scriver CR, Beaudet AL, Sly WS, Valle D (eds) The metabolic and molecular bases of inherited disease 7th edn. McGraw Hill, New York, pp 749–794

Fountoulakis M, Langen H (1997) Identification of proteins by matrix assisted laser desorption ionisation mass spectrometry following in-gel digestion in low salt, non-votalie buffer and simplified peptide recovery. Anal Biochem 250: 153–156

Fowler BM, Giuliano AR, Piyathilake C, Nour M, Hatch K (1998) Hypomethylation in cervical tissue: is there a correlation with folate status. Cancer Epidemiol Biomarkers Prev 7: 901–906

Hassold TJ, Burrage LC, Chan ER, Judis LM, Schwartz S, James SJ, Jacobs PA, Thomas NS (2001) Maternal folate polymorphisms and the etiology of human nondisjunction. Am J Hum Genet 69: 434–439

Hine RJ, James SJ (2000) Down syndrome and folic acid update. J Am Diet Assoc 100: 1004

Hobbs CA, Sherman SL, Yi P, Hopkins SE, Torfs CP, Hine RJ, Pogribna M, Rozen R, James SJ (2000) Polymorphisms in genes involved in folate metabolism as maternal risk factors for Down syndrome. Am J Hum Genet 67: 623–630

Hunt PA, Lemaire-Adkins R (1998) Genetic control of mammalian female meiosis. Curr Top Dev Biol 37: 359–381

Jacob RA, Gretz DM, Taylor PC, James SJ, Pogribny IP, Miller BJ, Henning SM, Swendseid ME (1998) Moderate folate depletion increases plasma homocysteine and decreases lymphocyte DNA methylation in postmenopausal women. J Nutr 128: 1204–1212

James SJ, Pogribna M, Pogribny IP, Melnyk S, Hine RJ, Gibson JB, Yi P, Tafoya DL, Swenson DH, Wilson VL, Gaylor DW (1999) Abnormal folate metabolism and mutation in the methylenetetragydrofolate reductase gene may be maternal risk factors for Down syndrome. Am J Clin Nutr 70: 495–501

Kitzmueller E, Greber S, Fountoulakis M, Lubec G (2001) Carbohydrate handling enzymes in fetal Down syndrome brain. J Neural Transm [Suppl] 61: 203–210

Langen H, Roeder D, Juranville J, Fountoulakis M (1997) Effect of protein application mode and acrylamide concentration on the resolution of protein spots separated by two-dimensional gel electrophoresis. Electrophoresis 18: 2085–2090

MacGregor JT, Wehr C, Hiatt RA, Peters B, Tucker JD, Langlois RG, Jacob RA, Jensen RH, Yager JW, Shigenaga MK, Frei B, Eynon BP, Ames BN (1997) "Spontaneous" genetic damage in man: evaluation of interindividual variability, relationship among markers of damage, and influence of nutritional status. Mutat Res 377: 125–135

O'Leary VB, Parle-McDermott A, Molloy AM, Kirke PN, Johnson Z, Conley M, Scott JM, Mills JL (2002) MTRR and MTHFR polymorphism: link to Down syndrome? Am J Med Genet 107: 151–155

Pasternack LB, Littlepage LE, Laude DA Jr, Appling DR (1996) 13C NMR analysis of the use of alternative donors to the tetrahydrofolate-dependent one-carbon pools in Saccharomyces cerevisae. Arch Biochem Biophys 326: 158–165

Peeters M, Rethore MO, de Kermadec S, Lejeune J (1989) Correlation between the effects of rT3 and IMP dehydrogenase inhibitors on normal and trisomic 21 lymphocyte cultures. Ann Genet 32: 211–213

Picton H, Briggs D, Gosden R (1998) The molecular basis of oocyte growth and development. Mol Cell Endocrinol 145: 27–37

Rosenblatt DS (1999) Folate and homocysteine metabolism and gene polymorphisms in the etiology of Down syndrome. Am J Clin Nutr 70: 429–430

Shane B (1989) Polyglutamate synthesis and role in the regulation of one-carbon metabolism. Vitam Horm 45: 263–335

Smith G, Berg J (1976) Down's Anomaly, 2nd edn. Churchill Livingstone, Edinburgh New York

Titenko-Holland N, Jacob RA, Shang N, Balaraman A, Smith MT (1998) Micronuclei in lymphocytes and exfoliated buccal cells of postmenopausal women with dietary change in folate. Mutat Res 417: 101–114
van der Put NM, van Straaten HW, Trijbels FJ, Blom HJ (2001) Folate, homocysteine and neural tube defects: an overview. Exp Biol Med 226: 243–270
Wahls WP, Song JM, Smith GR (1993) Single-stranded DNA binding activity of C1-tetrahydrofolate synthase enzymes. J Biol Chem 268: 23792–23798
Xu GL, Bestor TH, Bourchis D, Hsieh CL, Tommerup N, Bugge M, Hulten M, Qu XY, Russo TT, Veigas-Pequignot E (1999) Chromosome instability and immunodeficiency syndrome caused by mutations in a DNA methyltransferase gene. Nature 402: 187–191

Authors' address: Prof. Dr. G. Lubec, CChem, FRSC (UK), University of Vienna, Department of Pediatrics, Währinger Gürtel 18, A-1090 Vienna, Austria, e-mail: gert.lubec@akh-wien.ac.at

Increased expression of human reduced folate carrier in fetal Down syndrome brain

G. Lubec[1], **M. Bajo**[1,2], **M. S. Cheon**[1], **H. Bajova**[1], and **L. H. Matherly**[3]

[1]Department of Pediatrics, University of Vienna, Vienna, Austria
[2]Institute of Neuroimmunology of SAS, Bratislava, Slovak Republic
[3]Department of Pharmacology, Karmanos Cancer Institute, Wayne State University School of Medicine, Detroit, Mi, U.S.A.

Summary. Down syndrome (trisomy of chromosome 21) (DS) is the most common genetic cause of mental retardation. In our study we employed immunoblotting to evaluate protein expression of reduced folate carrier (hRFC), encoded by a gene localised on chromosome 21, in fetal DS brain. We observed increased expression of hRFC-immunoreactive band with an apparent MW of approximately 150 kDa, whereas the other bands (MWs ~60 and 50 kDa), were comparable to control. In conclusion, we suggest that aberrant hRFC expression may well have a role in the already observed deterioration of folate metabolism in DS. Moreover, no alterations of expression level of p53 and Sp1, supposed to play a role in the regulation of hRFC, suggest that regulation of hRFC expression in fetal life by these proteins is highly unlikely, at least by changes in their protein level.

Folate is an essential vitamin and reactions that depend on folate are collectively referred to as one-carbon metabolism, are required for DNA, RNA and protein synthesis. One-carbon metabolism is involved in several key cellular metabolic processes comprising purine and pyrimidine nucleotide synthesis, glycine, histidine and serine metabolism, and formation of the primary methylating agent — S-adenosylmethionine (SAM) (Bailey and Gregory, 1999).

Folate plays a very important role during the development of the central nervous system (CNS) as well as in adulthood. In children of women who are folate deficient during pregnancy increased incidence of neuronal tube defects is observed (Scholl and Johnson, 2000). Moreover, in children suffering from inborn errors of folate metabolism, degeneration of CNS, atrophy of cerebral cortex and mental retardation were described (for reviews see Erbe, 1975; Martin, 1988). In adults, low serum folate concentration has been related to cerebral atrophy, dementia and to poor cognitive functions, particularly in older adults (Snowdon et al., 2000; Wang HX et al., 2001).

Abnormal folate metabolism has been proposed to be a maternal risk factor for Down syndrome (DS) (Hobbs et al., 2000), the most common genetic cause of human mental retardation (Epstein, 2001). DS (trisomy of chromosome 21) presents with a series of impaired metabolic functions including folate and one-carbon metabolism related metabolic pathways

(Hobbs et al., 2000; Epstein, 2001; Pogribna et al., 2001; Al-Gazali et al., 2001). The role of triplicated chromosome 21 for the pathogenesis of DS, proposed as gene dosage effect, has not been fully elucidated yet.

Enzymes participating in one-carbon metabolism, the trifunctional enzyme with glycinamide ribonucleotide synthetase (GARS), aminoimidazole ribonucleotide synthetase (AIRS) and glycinamide ribonucleotide formyltransferase (GART) activities (GARS-AIRS-GART) involved in *de novo* purine synthesis, cystathionine β-synthase (CBS) catalyzing the conversion of homocysteine to cystathionine, and reduced folate carrier (hRFC) representing a major transport system of folates in mammalian cells, are encoded by genes localised on chromosome 21. While in fetal DS brain GARS-AIRS-GART showed comparable protein expression, a different expression pattern of the enzyme was described postnatally in DS (Cheon et al., 2003; Brodsky et al., 1997). Similary, no change in protein expression of CBS was found in DS brain prenatally, whereas increased CBS levels as well as activity were determined in DS patients after birth (Al-Gazali et al., 2001; Cheon et al., 2003; Chadefaux et al., 1985).

In our study, we evaluated protein expression of hRFC in fetal DS brain. hRFC, a typical transporter protein with 12 membrane-spanning domains, is a bidirectional anion transporter with high affinity for reduced folates and antifolates but low affinity for folic acid (Sirotnak and Tolner, 1999). In the CNS, hRFC was detected by immunohistochemistry in choroid plexus as well as in axons and dendrites (Wang Y et al., 2001). Protein epression of hRFC is regulated developmentally (Sirotnak and Tolner, 2001) and recently *in vitro* studies have shown downregulation of hRFC protein expression caused by increased expression of p53 in cells and by increased glucose concentration in a cell culture medium (Ding et al., 2001; Naggar et al., 2002). Therefore, we tested whether expression of hRFC is regulated by p53 expression also in brain of DS fetuses. One of the mechanism of transcriptional repression by p53 likely includes a protein-protein interaction with transcriptional factor Sp1 (Borellini and Glazer, 1993). Since Sp1 binding sites are present in the promoter of gene encoding hRFC and Sp1 plays a role in the regulation of transcription of gene encoding hRFC (Sirotnak and Tolner, 1999), we evaluated also protein expression of Sp1 in our samples.

Conceptus abortus fetal brain samples (cerebral cortex) of 4 controls (4 females with gestational age of 18.25 ± 0.5 weeks) and 4 DS (4 females with gestational age of 19 weeks) were obtained from Drs. Mara Dierssen and J. C. Ferreres from the Medical and Molecular Genetics Center-IRO, Hospital Duran i Reynals, Barcelona, Spain. Brain samples were taken in accordance with rules of the local Ethical Committee. All samples were stored at −70°C and the freezing chain was never interrupted.

Brain samples were homogenised in homogenisation buffer (10 mM Tris-HCl (pH 7.5), 150 mM NaCl, 1 mM PMSF and Protease Inhibitor Coctail (Roche, Basel, Switzerland) for 30 seconds (six strokes) at 440 rev/min in a Potter-Elvehjem homogeniser. The homogenate was centrifuged (10 minutes/4000 g/4°C) and the supernatant was used for measurement of protein concentration (BCA Protein Assay Kit (Pierce, USA)) and for Western blotting.

After mixing the samples with sample buffer (125.5 mM Tris-HCl base, 70 mM sodium dodecyl sulfate, 0.001% Bromphenol blue, 20% glycerol, and 2% 2-mercaptoethanol, pH 6.8) and incubation at 95°C for 5 minutes, the samples (containing 10 μg) were loaded onto a homogenous ExcelGel SDS gels (Amersham Pharmacia Biotech, Sweden). In the case of hRFC, instead of incubation at 95°C for 5 minutes, the samples were freeze and thawed 10 times. Electrophoresis and transfer of proteins onto PVDF membrane (MilliPore, USA) were performed with Multiphor II electrophoresis system and NovaBlot system (Amersham Pharmacia Biotech, Sweden), respectively, according to recommendations of the producer. Membranes were incubated in a blocking solution (10 mM Tris-HCl, pH 7.5, 150 mM NaCl, 0.1% Tween 20 and 1.5% non-fat dry milk) followed by incubation with diluted primary antibodies — rabbit polyclonal anti-hRFC — 1:5,000 (supplied by Prof. Matherly), mouse monoclonal anti-p53 — 1:2,000 (sc-126, Santa Cruz Biotechnology, Inc., USA), rabbit polyclonal anti-Sp1 — 1:200 (H-225, Santa Cruz Biotechnology, Inc., USA), mouse monoclonal anti-NSE (neuron specific enolase) 1:3,000 (MAB324, Chemicon, USA), mouse monoclonal anti-GFAP (glial fibrillary acidic protein) 1:5,000 (MAB360, Chemicon, USA), and mouse monoclonal anti-actin — 1:1,000 (A 4,700, Sigma, USA). After washing (3 × 15 minutes with blocking solution), membranes were probed with secondary antibodies coupled to horseradish peroxidase (Southern Biotechnology Associates, Inc., USA) for 1 hour. For visualisation, Western Lightning™ (PerkinElmer Life Sciences, Inc., USA) was used. The density of immunoreactive bands was measured by software program RFLP scan version 2.1 (Scanalytics, USA). Statistical calculations were carried out using the non-parametric Mann-Whitney *U* test and regression analysis. Statistical significance was set at $P < 0.05$.

In fetal cerebral cortex of control and DS fetuses we detected several RFC-immunoreactive bands. Besides three bands with MWs of approximately 50, 60 and 150 kDa, we identified a cluster of bands with MWs ranging from 50 − 60 kDa (Fig. 1). The pattern of the cluster did not allow us to measure optical density (OD) of the individual bands. Therefore, we measured only OD of the well separated bands. While bands with lower MW (50 and 60 kDa) did not show any differences between DS and control (control 2.10 ± 0.26 vs. DS 1.85 ± 0.17, $P = 0.34$, and 0.19 ± 0.05 vs. 0.27 ± 0.08, $P = 0.11$, respectively), increased levels of the 150 kDa band were observed in DS (1.82 ± 0.46 vs. 3.40 ± 0.58, $P = 0.03$).

Using anti-p53 antibody we detected a single band in all fetal brain samples. In the control group p53-immunoreactivity was comparable between the individuals samples, whereas in DS intragroup variability was higher, ranging from almost undetectable band (sample ds3) to an intensive band (sample ds4) (Fig. 1). Taking into account this variability, we did not find a significant change in protein level of p53 in fetal brains of control and DS individuals (0.93 ± 0.26 vs. 0.72 ± 0.64, $P < 0.49$).

In fetal brain, the protein expression pattern of transcription factor Sp1 was characterised by several immunoreactive bands (Fig. 1). None of the detected Sp1-immunoreactive band showed significant changes in protein

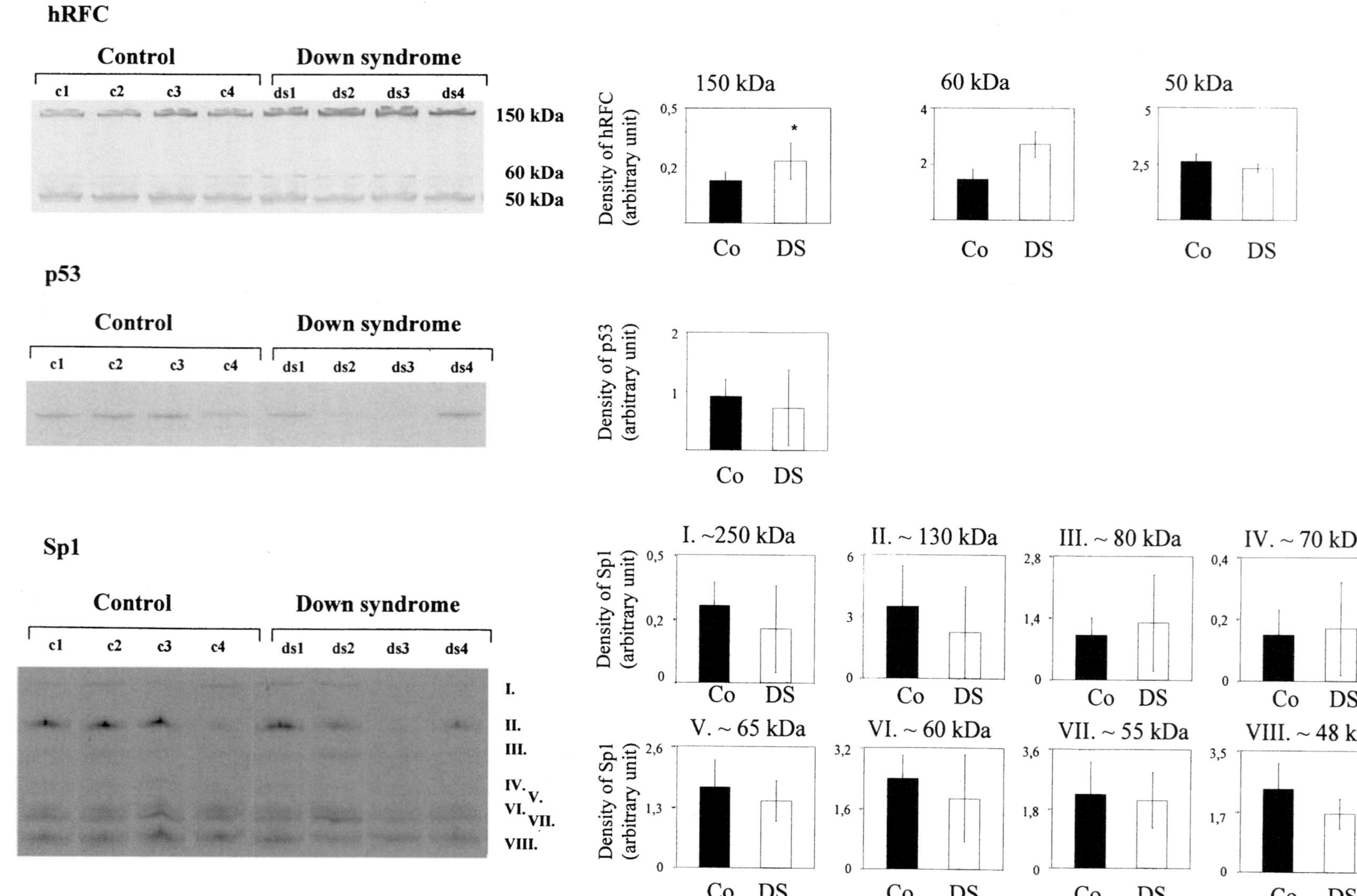
hRFC
Control
Down syndrome
c1 c2 c3 c4 ds1 ds2 ds3 ds4
150 kDa
60 kDa
50 kDa
Density of hRFC (arbitrary unit)
150 kDa
60 kDa
50 kDa
Co DS
p53
Control
Down syndrome
c1 c2 c3 c4 ds1 ds2 ds3 ds4
Density of p53 (arbitrary unit)
Co DS
Sp1
Control
Down syndrome
c1 c2 c3 c4 ds1 ds2 ds3 ds4
I.
II.
III.
IV.
V.
VI.
VII.
VIII.
Density of Sp1 (arbitrary unit)
I. ~250 kDa
II. ~ 130 kDa
III. ~ 80 kDa
IV. ~ 70 kDa
V. ~ 65 kDa
VI. ~ 60 kDa
VII. ~ 55 kDa
VIII. ~ 48 kDa
Co DS

expression between DS and control (~250 kDa band −0.21 ± 0.17 vs. 0.30 ± 0.085, P = 0.69; ~130 kDa −2.25 ± 2.20 vs. 3.48 ± 1.95, P = 0.69; ~80 kDa band −0.37 ± 0.31 vs. 0.29 ± 0.11, P = 0.89; ~70 kDa band −0.17 ± 0.15 vs. 0.15 ± 0.08, P < 0.99; ~65 kDa −0.10 ± 0.03 vs. 0.12 ± 0.04, P = 0.34; ~60 kDa −0.94 ± 0.57 vs. 1.21 ± 0.30, P = 0.34; ~55 kDa −1.04 ± 0.42 vs. 1.13 ± 0.48, P = 0.88; ~48 kDa −1.7 ± 0.44 vs. 2.41 ± 0.73, P = 0.11, in DS and control, respectively).

To rule out the influence of differences in cell composition and/or cell loss on our results, we evaluated protein expression of protein markers — neuron specific enolase (NSE) (marker for neuronal cell density), glial fibrillary acidic protein (GFAP) (marker for non-neuronal cell density) and actin (marker for cell density) (Fig. 2). Expression of none of these proteins was changed in fetal DS brain compared to control (NSE: 11.75 ± 1.44 vs. 10.40 ± 0.93, P = 0.34; GFAP: 4.51 ± 2.28 vs. 7.29 ± 1.98, P = 0.20; actin: 1.16 ± 0.30 vs. 0.89 ± 0.22, P = 0.20, in DS and control, respectively).

Linear regression analysis displayed no correlation between the expression level of observed proteins and gestational age (data not shown).

This is the first study showing data on protein expression pattern of human reduced folate carrier in brain during prenatal development. Our results revealed several RFC-immunoreactive bands in cerebral cortex of fetal brain. A similar expression pattern of human RFC, represented by broadly migrating bands, was reported in cultured cells (Wong et al., 1998). In brain, immunohistochemical studies performed on mice showed predominant localization of RFC in choroid plexus, but RFC was also detected in dendrites and axons as well (Wang Y et al., 2001). Likely explanations for the hRFC expression pattern in human fetal brain include different mRNA splice variants and/or post-translational modifications that have been reported in human RFC (Sirotnak and Tolner, 1999). Five mRNA splice variants of hRFC described in humans (variant I, II, IIa, IIb, and III) have major differences in their 5′ untranslated region that results in very similar MWs. Since MW of hRFC variant I and III predicted from ORF are ~65 kDa, and MW of the variants II is ~ 58 kDa (Sirotnak and Tolner, 1999), the RFC-immunoreactive bands with MW 50 and 60 kDa, and 50–60 kda cluster detected in our brain samples might correspond rather to variants II than variants I or III of human RFC in fetal brain.

In human K562 and CCRF-CEM tumor cells, human RFC was found to be heavily glycosylated, with the majority representing N-linked glycosides (Sirotnak and Tolner, 1999). Although there are a number of potential N-

Fig. 1. Results from immunoblotting of hRFC, p53 and Sp1. Denatured proteins were (10 μg) loaded, separated on a 7.5% (hRFC) or 12.5% (p53, Sp1) homogeneous gels and transferred onto PVDF membrane. The membranes were incubated with rabbit polyclonal anti-hRFC (human reduced-folate carrier), mouse monoclonal anti-p53 and rabbit polyclonal anti-Sp1, respectively. Visualization was performed by usage of chemiluminescence reagents. Statistical significance * (analysed by Mann-Whitney *U* test) was set at P < 0.05

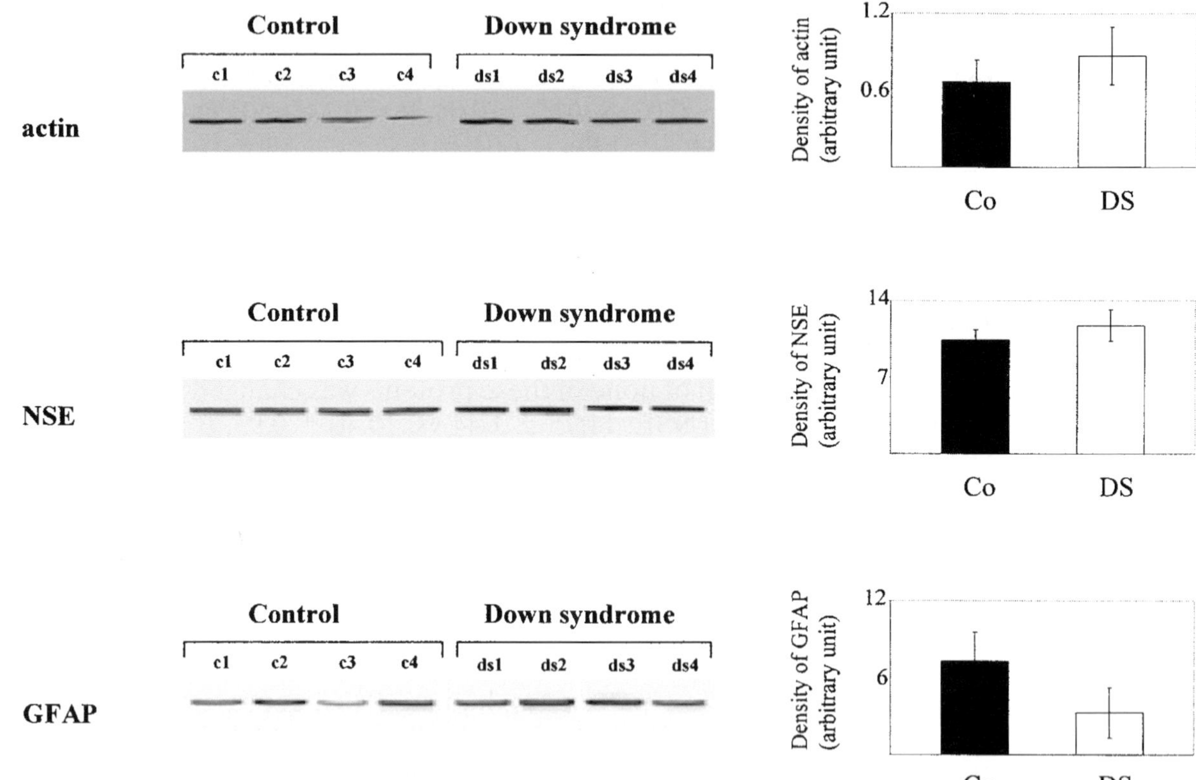

Fig. 2. Results from immunoblotting of actin, NSE and GFAP. Brain homogenate containing 10 μg protein/lane were separated on a 12.5% homogeneous gels and transferred onto PVDF membrane. The membranes were incubated with mouse monoclonal anti-actin, mouse monoclonal anti-NSE (neuronal specific enolase) and mouse monoclonal anti-GFAP (glial fibrillary acidic protein). Visualization was performed by usage of chemiluminescence reagents. Statistical significance * (analysed by Mann-Whitney *U* test) was set at $P < 0.05$

glycosylation sites on the molecule of hRFC, only one of them (N^{58}) is supposed to be able to contribute to N-glycosylation (Sirotnak and Tolner, 1999). In the study of Wong et al. (1998) carried out on cell cultures, the MW of N-glycosylated human RFC estimated by SDS-PAGE was ~85 kDa, whereas MW of de-glycosylated human RFC form was only ~65 kDa corresponding to MW predicted from ORF. From this point of view, we suggest that the RFC-immunoreactive bands with MW 50–60 kDa detected in our study likely represent de-glycosylated forms of hRFC, probably variants II that was reported to be the most abundant variant (Tolner et al., 1998). In our study, we were unable to identify any band with MW ~85 kDa corresponding to glycosylated hRFC forms, but instead of that band we detected an RFC-immunoreactive band with MW ~150 kDa. Since all of the former studies on human RFC were carried out in cell culture, we propose that the ~150 kDa RFC-immunoreactive band represents another, highly glycosylated, form of

RFC and that discrepancy might be caused by a tissue specific regulation of hRFC transcription (Tolner et al., 1998).

Unchanged levels of RFC-immunoreactive bands with low MW and increased protein expression of only ~ 150 kDa RFC-immunoreactive band in fetal DS brain are suprising. Since hRFC is encoded by a single gene localised on triplicated chromosome 21 (21q22.3) in DS individuals (Sirotnak and Tolner, 1999), all forms of hRFC were supposed to be increased. A possible reason of the expression pattern might be regulation of hRFC transcription by dual promoters. As mentioned above, particular hRFC splice variants have major differences in 5′UTR where a presence of two promoters were reported (Sirotnak and Tolner, 1999). Promoter A (P2) contains a consensus CRE/AP1 element that is recognized by diverse complexes of members of the b-ZIP superfamily and its tissue specific regulation has been suggested (Whetstine and Matherly, 2001). The second promoter, promoter B (P1) contains a highly conserved GC box recognized by Sp1 and Sp3 transcription factors that are considered to regulate constitutive transcription of the promoter (Matherly, 2001). Promoter A was reported to be more efficient in driving transcription of hRFC than promoter B (Tolner et al., 1998). But the mechanism that determine differential promoter activities and alternates promoter usage between cell or tissue types is an area that needs to be further explored (Matherly, 2001).

In our study, we focused on a role of p53 in regulation of hRFC protein expression in fetal brain. *In* vitro, Matherly and coworkers showed repression of hRFC gene expression by p53, a nuclear phosphoprotein involved in control of cell growth and apoptosis (Ding et al., 2001). In adult DS patients, increased protein expression of p53 was detected in brain (Seidl et al., 1999). The exact mechanism of p53-mediated transcriptional repression of hRFC is not established, but the involvement of direct or indirect effects on the transactivation of the hRFC promoter B is supposed (Ding et al., 2001). We here show that modulation of hRFC expression by p53 is highly unlikely.

For regulation of basal hRFC-B transcription (via GC-box), the role of Sp1 family of transcription factors has been established (Whetstine and Matherly, 2001). p53 binding to Sp1 resulting in repression of Sp1-regulated promoters has been reported (Wang Q et al., 1998). In fetal brain, we determined the presence of several Sp1-immunoreactive bands and we propose that the different Sp1-immunoreactive bands may result from heterogeneous Sp1 mRNAs, phosphorylation and O-glycosylation (Black et al., 2001). Unchanged levels of any of the several Sp1-immunoreactive bands in fetal DS brain indicates that Sp1, similarly to p53, may not be involved in the regulation of hRFC by changes in its protein level.

Although a functional significance of the increased expression of the major ~ 150 kDa hRFC immunoreactive protein in folate metabolism in DS is unclear, we suggest that altered hRFC expression in DS fetuses may indicate deranged folate metabolism in DS during prenatal development. Main biological mechanism responsible for development of the neuropathology caused by low folate levels or deranged folate-dependent metabolism seems to be increased production of homocysteine due to the decreased methylation reac-

tion (Bottiglieri, 1996). Recently, in vitro studies showed that homocysteine can be directly toxic to neurons via the activation of NMDA receptors or apoptosis triggered by DNA damage (Lipton et al., 1997; Kruman et al., 2000). In DS individuals, increased levels and activity of cystathione-beta-synthase, encoded by a gene localised on chromosome 21, are considered to be a major mechanism leading to increased homocysteine production detected in DS individuals (Pogribna et al., 2001).

Conclusion

In our study, we observed an increased level of ~150 kDa immunoreactive hRFC in brain of DS fetuses suggesting alteration of folate metabolism in DS during prenatal development. Aberrant hRFC expression may well have a role and our observations contribute to the already published deterioration of folate metabolism in DS. We furthermore show that regulation of hRFC expression in fetal life by p53 and Sp1 is highly unlikely at least by changes in protein level.

References

Al-Gazali L, Padmanabhan R, Melnyk S, Yi P, Pogribny IP, Bakir M, Hamid ZA, Abdulrazzaq Y, Dawodu A, James SJ (2001) Abnormal folate metabolism and genetic polymorphism of the folate pathway in a child with Down syndrome and neuroal tube defect. Am J Med Genet 103: 128–132

Bailey LB, Gregory JF 3rd (1999) Folate metabolism and requirements. J Nutr 129: 779–782

Black AR, Black JD, Azizkhan-Clifford J (2001) Sp1 and Krüpepel-like factor family of transcription factors in cell growth regulation and cancer. J Cell Physiol 188: 143–160

Borellini F, Glazer RI (1993) Induction of Sp1-p53 DNA-binding heterocomplexes during granulocyte/macrophage colony-stimulating factor-dependent proliferation in human erythroleukemia cell line TF-1. J Biol Chem 268: 7923–7928

Bottiglieri T (1996) Folate, vitamin B12, and neuropsychiatric disorders. Nutr Rev 54: 382–390

Brodsky G, Barnes T, Bleskan J, Becker L, Cox M, Patterson D (1997) The human GARS-AIRS-GART gene encodes two proteins which are differentially expressed during human brain development and temporally overexpressed in cerebellum of individuals with Down syndrome. Hum Mol Genet 6: 2043–2050

Chadefaux B, Rethore MO, Raoul O, Ceballos I, Poissonnier M, Gilgenkranz S, Allard D (1985) Cystathione beta synthase: gene dosage effect in trisomy 21. Biochem Biophys Res Commun 128: 40–44

Cheon MS, Bajo M, Kim SH, Claudio JO, Patterson D, Kruger WD, Kondoh H, Lubec G (2003) Protein levels of genes encoded on chromosome 21 in fetal Down syndrome brain: challenging the gene dosage effect hypothesis, part II. Amino Acids 24: 119–125

Ding BC, Whetstine JR, Witt TL, Schuetz JD, Matherly LH (2001) Repression of human reduced folate carrier gene expression by wild type p53. J Biol Chem 276: 8713–8719

Epstein CJ (2001) Down syndrome (trisomy 21). In: Scriver SR, Beaudet AL, Sly WS, Valle D (eds) The metabolic and molecular bases of inherited disease, 8th edn. McGraw-Hill, New York, pp 1223–1256

Erbe RW (1975) Inborn errors of folate metabolism, part II. N Engl J Med 293: 807–812
Hobbs CA, Sherman SL, Yi P, Hopkins SE, Torfs CP, Hine RJ, Pogribna M, Rozen R, James SJ (2000) Polymorphism in genes involved in folate metabolism as maternal risk factors for Down syndrome. Am J Hum Genet 67: 623–630
Kruman II, Culmsee C, Chan SL, Kruman Y, Guo Z, Penix L, Mattson MP (2000) Homocysteine elicits a DNA damage response in neurons that promotes apoptosis and hypersensitivity to excitotoxicity. J Neurosci 20: 6920–6926
Lipton SA, Kim WK, Choi YB, Kumar S, D'Emilia DM, Raydu PV, Arnelle DR, Stamler JS (1997) Neurotoxicity associated with dual actions of homocysteine at the N-methyl-D-aspartate receptor. Proc Natl Acad Sci USA 94: 5923–5928
Martin DC (1988) B12 and folate deficiency demencia. Clin Geriatr Med 4: 841–852
Matherly LH (2001) Molecular and cellular biology of the human reduced folate carrier. Prog Nucl Acid Res Mol Biol 67: 131–162
Naggar H, Ola MS, Moore P, Huang W, Bridges CC, Ganapathy V, Smith SB (2002) Downregulation of reduced-folate transporter by glucose in cultured RPE cells and in RPE of diabetic mice. Invest Ophtalmol Vis Sci 43: 556–563
Pogribna M, Melnyk S, Pogribny IP, Chango A, Yi P, James SJ (2001) Homocysteine metabolism in children with Down syndrome: in vitro modulation. Am J Hum Genet 69: 88–95
Scholl TO, Johnson WG (2000) Folic acid: influence on the outcome of pregnancy. Am J Clin Nutr 71: 1295S–1303S
Seidl R, Fang-Kircher S, Bidmon B, Cairns N, Lubec G (1999) Apoptosis-associated proteins p53 and APO-1/Fas (CD95) in brains of adult patients with Down syndrome. Neurosci Lett 260: 9–12
Sirotnak FM, Tolner B (1999) Carrier-mediated membrane transport of folates in mammalian cells. Annu Rev Nutr 19: 91–122
Snowdon DA, Tully CL, Smith CD, Riley KP, Markesbery WR (2000) Serum folate and the severity of atrophy of the neocortex in Alzheimer disease: findings from the Nun Study. Am J Clin Nutr 71: 993–998
Tolner B, Roy K, Sirotnak FM (1998) Structural analysis of the human RFC-1 gene encoding a folate transporter reveals multiple promoters and alternatively spliced transcripts with 5′end heterogeneity. Gene 211: 331–341
Wang HX, Wahlin A, Basun H, Fastbom J, Winblad B, Fratiglioni L (2001) Vitamin B(12) and folate in relation to the development of Alzheimer's disease. Neurology 56: 1188–1194
Wang Q, Beck WT (1998) Transcriptional suppression of multidrug resistence-associated protein (MRP) gene expression by wild-type p53. Cancer Res 58: 5762–5769
Wang Y, Zhao R, Russell RG, Goldman ID (2001) Localization of the murine reduced folate carrier as assed by immunohistochemical analysis. Biochim Biophys Acta 1513: 49–54
Whetstine JR, Matherly LH (2001) The basal promoters for the human reduced folate carrier gene are regulated by a GC-box and a cAMP-response element/AP-1-like element. Basis for tissue-specific gene expression. J Biol Chem 276: 6350–6358
Wong SC, Zhang L, Proefke SA, Matherly LH (1998) Effects of the loss of capacity for N-glycosylation on the transport activity and cellular localization of the human reduced folate carrier. Biochim Biophys Acta 1375: 6–12

Authors' address: Prof. Dr. G. Lubec, CChem, FRSC (UK), Department of Pediatrics, University of Vienna, Währinger Gürtel 18, A-1090 Vienna, Austria, e-mail: gert.lubec@akh-wien.ac.at

Chromosome 21 KIR channels in brain development

E. Thiery[1], S. Thomas[1], S. Vacher[2], A.-L. Delezoide[3], J. M. Delabar[1], and N. Créau[1]

[1] EA3508, Université Denis Diderot, Paris, France
[2] Unité de Différentiation cellulaire, ISREC, Epalinges, Switzerland
[3] Biologie du Développement, Hôpital Robert Debré, Paris, France

Summary. Two KIR (K+ Inwardly Rectifying) channel genes have been identified on chromosome 21, in a region associated with important phenotypic features of trisomy 21, including mental retardation: KIR3.2 (GIRK2) and KIR4.2. We analysed the expression of these channel genes in developing human and mouse brains to determine the possible role of the corresponding channels in brain development and function. KIR3.2, which has been extensively studied in the mouse, was found to be expressed in the human cerebellum during development. The KIR4.2 channel is expressed later in development in both mice and humans. We compared the expression of these channels in terms of RNA and protein levels and discussed the potential synergy and consequences of the overexpression of these channels in Down's syndrome brain development.

Introduction

KIR (K+ Inwardly Rectifying) channels form functional tetramers (homo- or heterotetramers) in the membrane that allow the selective influx of K+ into cells. They are involved in many functions depending on the specificity of the cell: maintenance of resting membrane potential and control of excitability, hormonal secretion, regulations of K+ homeostasis and of cellular volume. Two KIR channel genes have been identified on chromosome 21, in a region associated with important phenotypic features of trisomy 21, including mental retardation (Gosset et al., 1997; Dahmane et al., 1998). These genes are located 300kb apart in a head to head orientation. KIR3.2 (GIRK2, *KCNJ6*) is the more proximal of the two and was the first to be cloned (Lesage et al., 1994). The more distal gene is called KIR4.2 (*KCNJ15*) (Gosset et al., 1997). The discovery that the *weaver* mutation corresponds to a point mutation affecting the pore domain of the KIR3.2 channel (Patil et al., 1995) has led to extensive studies of its function (for review: Mark and Herlitze, 2000). The mouse KIR3.2 mutant displays abnormal cerebellar development, severe ataxia, tremor, and homozygote male sterility due to a loss of K+ specificity of the channel (Slesinger et al., 1997). KIR3 channels are activated by G proteins and, in the brain, KIR3.1 is the major partner of KIR3.2 in the functionnal channel. The KIR3.2 gene is large and gives rise to multiple transcripts, which are translated in the brain to give proteins with different C-terminal ends (Wei et al., 1998; Inanobe et al., 1999). The properties of KIR4.2 have been studied *in vitro*: this channel is pH-sensitive and regulated by protein kinase C (Pearson et al., 1999; Pessia et al., 2001). It is expressed mainly in the kidney (Gosset et al., 1997; Shuck et al., 1997) though northern blot analysis showed that mRNA for this channel is also produced in the human brain (Gosset et al., 1999). The KIR4.2 gene has a highly complex structure with constant and alternative exons in the 5′ region giving rise to numerous alternative ends (Gosset et al., 1997, 1999 and unpublished data).

We investigated the possible involvement of these KIR channels in brain development and function, to determine whether they are potential candidates for involvement in generating the abnormalities observed in trisomy 21 (Down's syndrome).

Material and methods

In situ hybridization (ISH) on mouse brain sections

Mice were killed using CO_2 at the P7, P18, P30 and adult stages. Brains were removed and directly frozen in cryomoulds containing Tissuetek slowly on liquid nitrogen. Cryosections (8 μm) were cut and fixed in 4% paraformaldehyde for 10 min. The murine cDNA used as a probe for ISH has been described elsewhere (Thiery et al., 2000). This probe detected only KIR4.2 transcripts on northern blots.

In situ hybridization (ISH) on foetal tissue sections

Human foetal tissues (17 and 33 weeks of gestation) were collected from legally terminated pregnancies, in agreement with French law and Ethics Committee recommendations. Tissues and paraffin sections were prepared and hybridization carried out as previously described (Rachidi et al., 2000). We used a 450 bp riboprobe amplified from KIR 3.2 exon 4 (with oligonucleotides ggtacttttgtacatgctggg and ggtgtaaaatatctacctgcc), subcloned into the PCR2.1 vector (Invitrogen). This fragment is unique in the human genome and does not cross-hybridize with other members of the KIR3 family. It is homologous to a fragment of murine GIRK2 present in all brain transcripts (Wei et al., 1998).

Hybridization with ^{35}S-labelled riboprobes was performed according to standard protocols (Peuchmaur et al., 1990). Slides were counterstained with haematoxylin and photographed under a Nikon stereomicroscope or a Zeiss PhomiIII microscope, using bright- and dark-field methods and Kodak 320T or 64T film.

Immunohistochemistry

Brains were fixed in 4% paraformaldehyde for 24 h, immersed in 30% sucrose in phosphate-buffered saline (PBS), and then frozen and stored at −80°C. Brains were embedded in Tissuetek and then 40 μm sections were cut on a cryostat at −20°C. The sections were rinsed in PBS and immunohistochemical procedure was then carried out on floating sections. Sections were washed and blocked and antibodies diluted in 0.1% Triton X-100 −1% skimmed milk powder in PBS. We used the following purified rabbit antibodies: anti-GIRK2 antibodies (Alomone: directed against a 41-amino acid sequence in the C-terminal part of the protein) which detect all GIRK2 proteins, and anti-KIR4.2 antibodies (directed against a 15-amino acid sequence in the N-terminal part of the protein). We used western blotting to check that the anti-KIR4.2 recognized a protein at the expected size. Antibody binding was detected by incubation with biotinylated donkey anti-rabbit Ig (Amersham), streptavidin-horse radish peroxidase and DAB+ (Dako) according to standard protocols.

Results

Expression of murine kir4.2 and human KIR4.2 during brain development

In previous *in situ* hybridization studies of *kir4.2* on sections of mouse embryos from E10.5 to E17.5, no signal was detected in the brain (Thiery et al., 2000). Postnatal expression was then analysed at various stages: P7, P18, P30 and adult. A signal was first observed at P7 in the olfactory bulb: it was stronger in the accessory olfactory bulb (AOB) than in the main bulb, but was restricted to the mitral cells in both cases (Fig. 1A–D). The signal in the AOB was still strong at P18 and P30 (Fig. 1E–H). Although a weak signal began to be detected in various parts of the brain particularly the hippocampus by P30, signals were only clearly visualised at the adult stage. In the adult brain, *kir4.2* expression was detected in the Purkinje cells of the cerebellum, the hippocampus, the pontine nuclei (not shown) and other large neurons throughout the brain (Fig. 1I–N).

We previously showed by northern blotting that human KIR4.2 is expressed in various regions of the brain (Gosset et al., 1999). We also isolated two cDNAs from foetal brain (11 to 14 week-olds foetuses) by the cDNA selection method (Gosset et al., 1997). One of these cDNAs consisted of part of the coding region, whereas the other identified a new 5′ non-coding exon (AC: Y13895), which was not found in cDNAs from other tissues or in EST databases (Gosset et al., 1999). We therefore carried out PCR amplification of foetal brain cDNAs using two primers that bound to constant 5′ non-coding exons on either side of this particular exon (MTC, Clontech; 20 to 25 week-olds foetuses). A specific band was obtained after two rounds of 30 cycles of PCR (data not shown). This transcript may therefore be considered to be produced in small quantities in human foetal brain at these stages.

In situ hybridization of KIR3.2 on human foetal cerebellum and brainstem

We compared the pattern of expression of KIR3.2 in human foetuses with that in rat and mice embryos, by hybridizing a cDNA probe, specific for exon 4, with sections of cerebellum and brainstem (Fig. 2). At 17 weeks of develop-

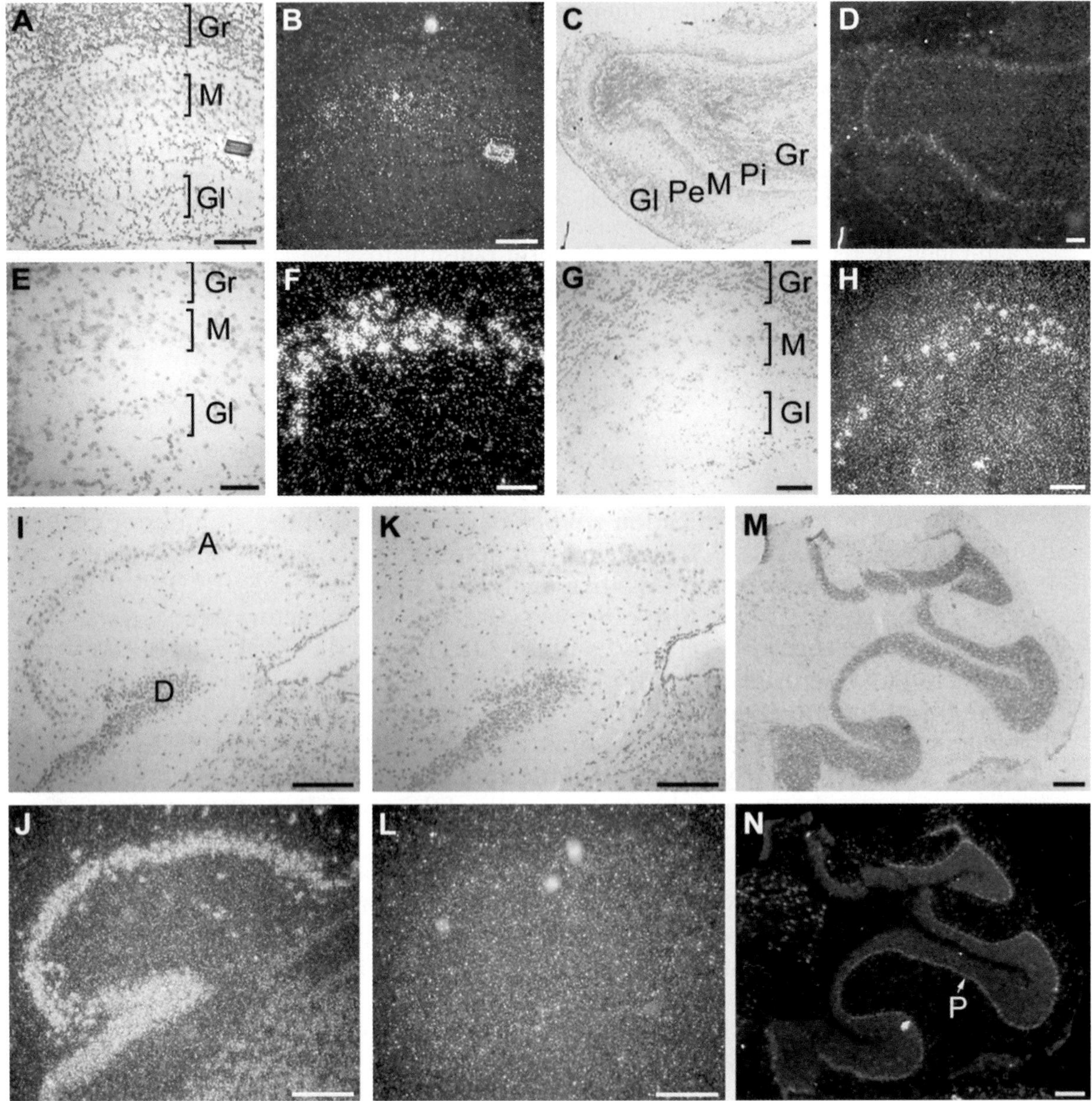

Fig. 1. Distribution of KIR4.2 mRNA in the postnatal mouse brain. P7 (**A–D**); P18 (**E**, **F**); P30 (**G**, **H**); Adult (**I–N**). Accessory olfactory bulb: **A**, **B**; **E–H**. Main olfactory bulb: **C**, **D**. Hippocampus: **I**, **J**; with sense probe **K**, **L**. Cerebellum: **M**, **N**, *A* Ammon's horn: *D* dentate gyrus; *Gr* granular cell layer; *Gl* Glomerular layer; *M* Mitral cell layer; *P* Purkinje cells; *Pe* external plexiform layer; *PI* internal plexiform layer. Bright field: **A**, **C**, **E**, **G**, **I**, **K**, **M**; Dark field: **B**, **D**, **F**, **H**, **J**, **L**, **N**. Scale bars **A–H**: 100 μm; **I–N**: 200 μm

ment (Fig. 2A–F), a signal was detected in the reticular formation of the pons and in the cerebellar hemispheres. At this stage, the signal was strongest in the external granular layer, and was stronger on the internal than on the external side of this layer. Granule cells migrating towards the internal granular layer

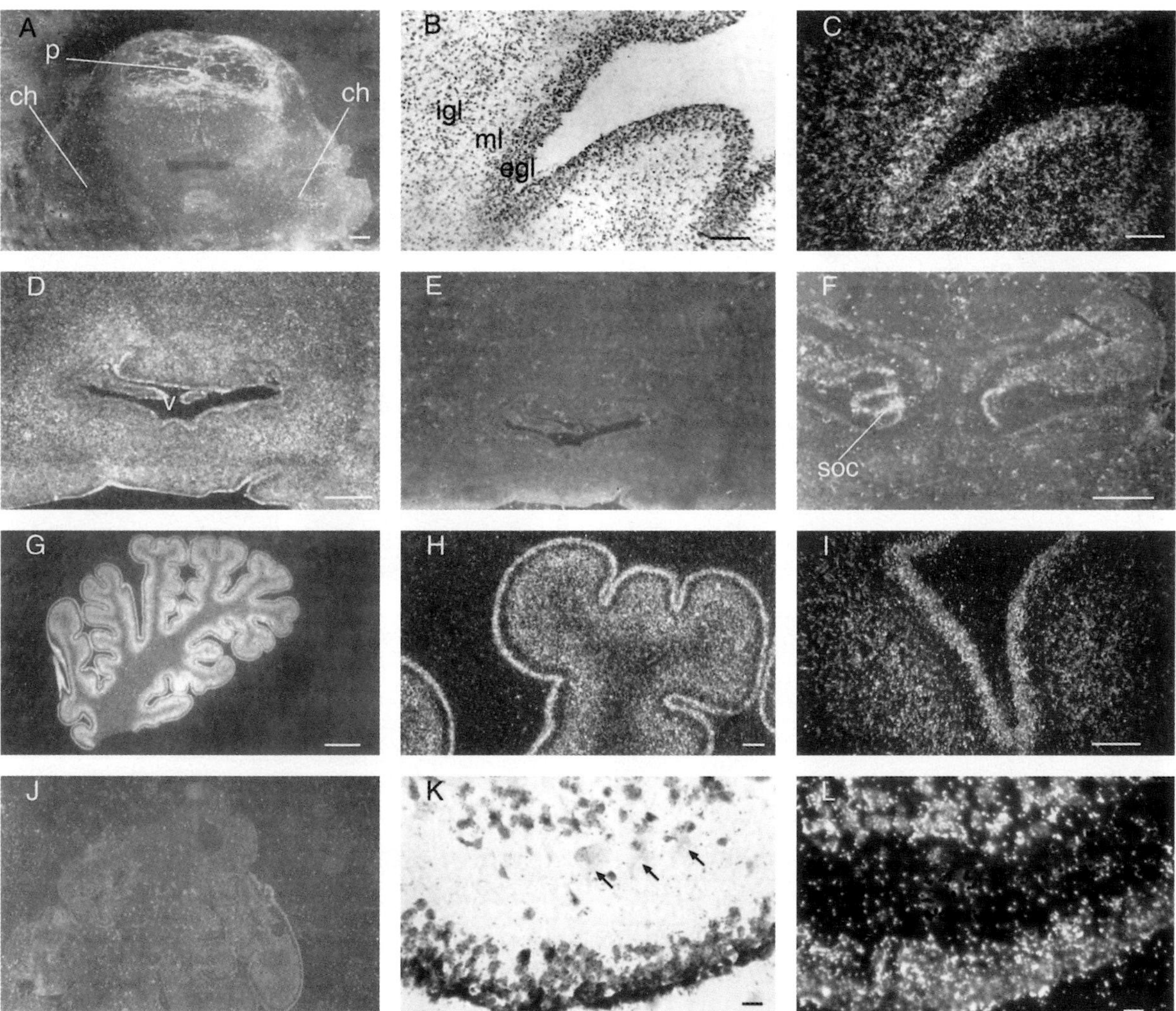

Fig. 2. Distribution of Kir3.2 mRNA in the cerebellum and brainstem at 17 (**A–F**) and 33 (**G–L**) weeks of development. Bright-field photomicrographs (**B**, **K**) and dark-field photomicrographs (**A**, **C–J**, **L**) **A** Pons (p) and cerebellar hemispheres (ch); **B**, **C** enlargement of a section of cerebellar cortex (*egl* external granular layer, *ml* molecular layer, *igl* internal granular layer); **D**, **E** transverse section of the vermis plus 4th ventricule (v); **F** transverse section of the medulla oblongata with the superior olivary complex (soc); **G–L** transverse section of the cerebellum and enlargement of lobulas; in **L**, arrows indicate cell bodies of Purkinje cells. **E**, **J**, Sense probe

were also labelled and the density of labelled cells was similar in the molecular layer and in the internal granular layer. A horizontal section of the myelencephalon, cut at a lower level than the previous sections (Fig. 2F), showed that KIR3.2 was also strongly expressed in the superior olivary complex.

At 33 weeks of development (Fig. 2G–L), a transverse section of the cerebellum showed strong labelling of the cerebellar cortex with clear labelling of the external and internal granular layers. At this stage, there was a clear difference between this area and the molecular layer which showed only

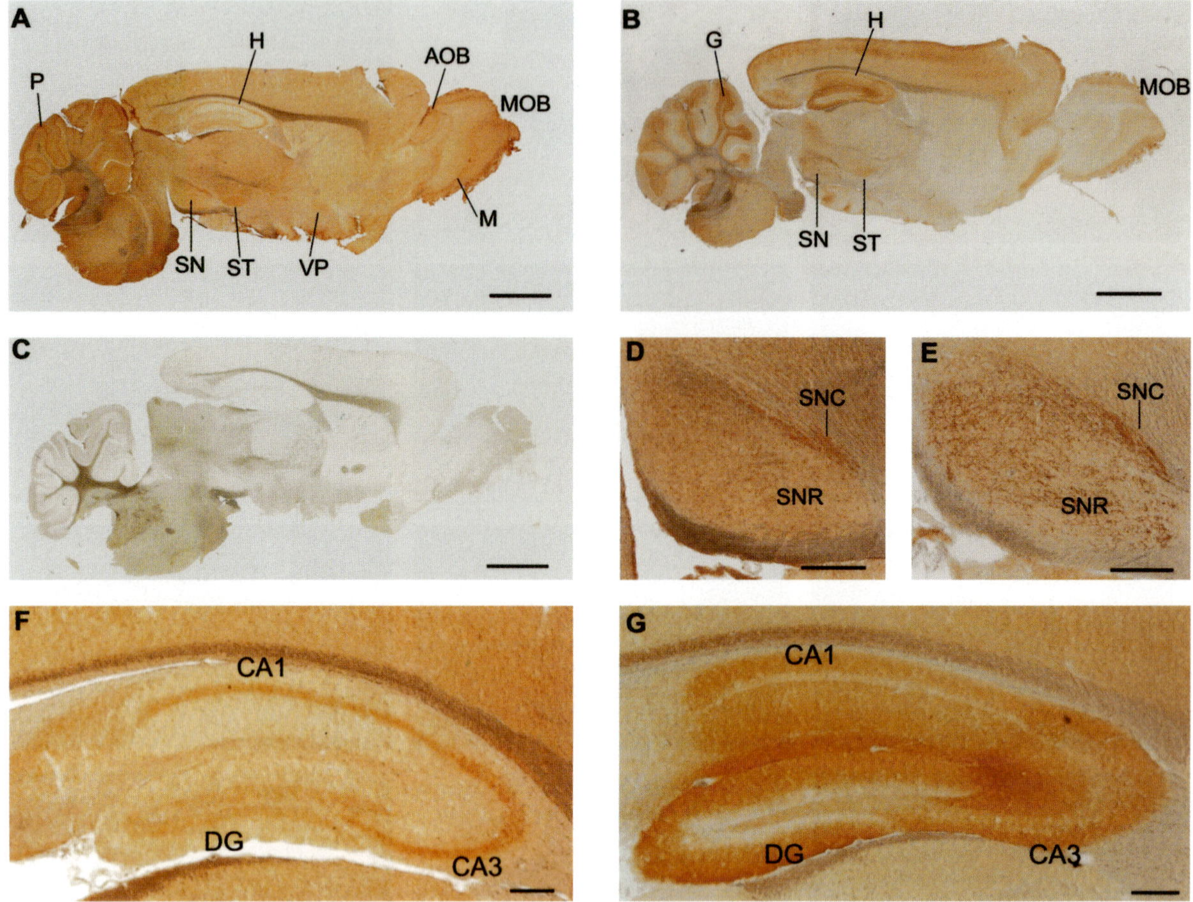

Fig. 3. Immunohistochemistry for KIR4.2 and GIRK2 on mouse brain sections Anti-KIR4.2 (**A**, **D**, **F**) anti-GIRK2 (**B**, **E**, **G**) antibodies. **C** Control, with the biotinylated anti-rabbit Ig. *AOB* Accessory olfactory bulb; *DG* dentate gyrus; *CA1* and *CA3* regions of Ammon's horn; *G* granular cells; *H* hippocampal formation; *P* Purkinje cells; *M* mitral cells; *MOB* main olfactory bulb; *SNC* substantia nigra pars compacta; *SNR* substantia nigra pars reticula; *ST* sub-thalamic nucleus; *VP* ventral pallidum. Scale bars **A–C**: 175 μm; **D–G**: 25 μm

faint labelling. The internal side of the external layer was again more strongly labelled than the external side. In the molecular layer the Purkinje cells appeared to be unlabelled or very weakly labelled (Fig. 2K,L).

Immunohistochemistry on adult mouse brain

Figure 3 shows the signals obtained with the anti-KIR4.2 antibody and with the commercial anti-GIRK2 (KIR3.2) antibody on close sections.

The cellular distribution of the KIR4.2 protein was identical to that of the corresponding transcripts. In addition, the protein was mainly present in the cell bodies (Fig. 3A,D,F). The protein was present in the Purkinje cells of the cerebellum, and in the substantia nigra (Fig. 3D), the subthalamic nucleus and the ventral pallidum which form part of the basal ganglia. In the hippocampus, Ammon's horn was more strongly labelled than the dentate gyrus (Fig. 3F). The protein was present in the mitral cells of both the main and the accessory olfactory bulbs.

The distribution of KIR3.2 was as previously described (Liao et al., 1996; Inanobe et al., 1999), with this protein present in larger amounts in the dendrites than in the cell bodies (Fig. 3B,E,G). This pattern is particularly clear in the hippocampus (CA1 and dentate gyrus: Fig. 3G) and in the substantia nigra (Fig. 3E).

Discussion

Comparison of KIR expression in mice and humans

KIR4.2 appears to be weakly expressed during early development in mice and humans. The expression of KIR4.2 in the olfactory bulbs (accessory and main), in mouse postnatal stages, in combination with the detection of transcripts in the vomeronasal and the olfactory epithelium during mouse embryogenesis suggests for this channel an important role in the olfactory system (Thiery et al., 2000). Changes in olfactory event-related potentials and deficits in olfactory function were found in Down's syndrome and these deficits increase with age (Wetter and Murphy, 1999; Nijjar and Murphy, 2002). In rodents, the olfactory system is very well developped and has important behaviour involvements (for review: Firestein, 2001; Del Punta et al., 2002). Little is known about the role of this system in human development though Winberg and Porter (1998) have suggested a role in neonatal behaviour, and it is unclear whether humans have a vomeronasal organ though its structure may be different from that in rodents (Johnson, 1998; Smith et al., 2002). KIR4.2 is expressed throughout the brain to various extents, but exclusively in neurons. Neurons located in regions associated with motor functions (basal ganglia, cerebellum) and cognitive functions (hippocampus) were strongly labelled, suggesting that this channel may interfere in these functions, which present abnormalities in Down's syndrome. Thus, the production of a specific antibody for the detection, by immunohistochemistry, of KIR4.2 on mouse brain should make it possible to determine the precise location of this channel in the nervous systems of mice and humans and identify its partners by co-immunolocalization or co-immunoprecipitation.

KIR3.2 is produced early in development in the mouse brain (at E15.5), in the migrating granuloblasts. At P0, this channel is detected in various nuclei of the brainstem, including the pontine nucleus (Chen et al., 1997). At 17 weeks of development, human foetuses sections display strong signals in the reticular formation, in the cerebellar hemispheres and in the vermis. At this stage, and in the brainstem, strong labelling seems to be restricted to specific nuclei such as the superior olivary complex. This complex is involved in auditory function, alterations in which are among the phenotypic features of Down's syndrome.

The pattern in humans is similar to that observed in mice, with a strong signal in the granule cells migrating from the border of the 4th ventricle and from the external granular layer (EGL). At 17 weeks of development, both the mitotically active matrix layer and the posmitotic premigrating layer are labelled. At 33 weeks of development, high levels of expression are still detected in the EGL, with a weak signal observed in the Purkinje cell layer that seems to be due to migrating granule cells rather than to Purkinje cells. This observation is consistent with the results obtained in mice by Wei et al. (1997). The labelling of EGL seemed to be stronger on the internal side than on the external side in the postmitotic premigrating granule cells. As seen at P10 in mouse cerebellum, the 33-weeks sections of human cerebellum showed weak labelling of the molecular layer and strong labelling of the internal granular layer.

Comparison of KIR4.2 and KIR3.2 expression patterns

The pattern of expression of the two channels differs during development. KIR3.2 is produced earlier in the developing brain, in the cerebellum both in mice and humans suggesting that it may play a role in the development of this structure. In adult mouse brain, KIR4.2 and KIR3.2 mRNA and protein were detected in various cell layers of the cerebellum. In the cerebral hemispheres, in addition to the cortex, several regions expressed both channels: the pontine nuclei, the substantia nigra and the subthalamic nucleus. We therefore cannot rule out the possibility of synergy between the two channels although their distributions differ slightly: KIR3.2 was more strongly expressed in the dendrites than in the cell bodies, whereas this was not the case of KIR4.2.

Potential role of KIR overexpression in trisomy 21 brain

Overproduction of a particular KIR subunit may have various consequences. First, in cells with large numbers of heterotetramers, an increase of one of the subunits, modifying the ratio of the various subunits may lead to physiological changes in the functional channel (activation, pH sensitivity, phosphorylation, sensitivity to PIP2, etc.). Second, it has been shown in mice deficient for KIR3.2 that the amount of protein for the partner of this subunit, KIR3.1, also decreases in the brain (Signorini et al., 1997). Thus, it is expected that the overproduction of a KIR subunit may lead to changes in the levels of mRNA or protein for its partners. This is the case in the Shaker overexpression model in the mouse (Sutherland et al., 1999): other shaker subunits which form multimers were unaffected, overproduced or underproduced, as shown by mRNA levels in the hippocampus. These modifications also led to changes in neuron excitability. Third, the production of larger numbers of transcripts may result in the production of a larger number of channels at the membrane. However, this appears unlikely. The formation of a functional channel is also regulated by other proteins, such as the PDZ domain proteins, which are known to mediate complex formation at the membrane, or G proteins subunits, which are known to regulate the gating of KIR3 channels. However, as has been shown in *Xenopus* oocytes (Peleg et al., 2002), we cannot exclude the possibility of an increase in the number of channels in conditions in which there is no deficit or limiting quantity of interacting proteins. Various types of PDZ domain proteins interact with the KIR-chr21 channels. CIPP (Kurschner et al., 1998) was shown by the yeast two-hybrid method to interact with only KIR4.1 and KIR4.2 subunits, no interaction with KIR3 subunits being detected. CIPP is present in large amounts in the kidney and in the Purkinje cells of the cerebellum. However, PSD95 has been shown to interact with members of the KIR3 (GIRK) subfamily members. Thus, even if KIR3.2 and KIR4.2 display similar distributions, as in Ammon's horn, they may not form heterotetramers, because the formation of functional channels requires different interacting proteins. It is also known that most functional channels in the brain consist in KIR3.1–KIR3.2 or of KIR3.2 alone. The partners of KIR4.2 are not known: one candidate is the KIR7.1 channel expressed in Purkinje cells and the pyramidal cell layer of the hippocampus (Krapivinski et al., 1998), as KIR4.1 is expressed only in the glial cells (Higashi et al., 2001). Another potential partner is KIR5.1, which has been shown to form functional channels in vitro (Pessia et al., 2001), but the distribution of this protein is poorly defined. Fourth, Burrone et al. (2002), demonstrated *in vitro* that the overexpression of a KIR channel in a neurone may lead to changes in the number and quality of synapses made with other neurones. Finally, Baxter et al. (2000) showed that in trisomy 21 and its mouse model, Ts65Dn, there is a 10–15% reduction of cerebellar volume and a significant decrease in granular density. These results suggest that genes expressed in the developping cerebellum and in three copies in the Ts65Dn model may lead to this phenotype. Thus, KIR3.2 is a good candidate for involvement in these phenotypes of Down's syndrome. The consequences of KIR4.2 and KIR3.2 overexpression need first to be analyzed in details, KIR channels being potential drug targets (Curran, 1998).

Acknowledgements

We would like to thank I. Berthuy and Z. Chettouh for technical assistance. This work was supported by grants from EC (QLRT-2001-00816) and CNRS. E.T. held a fellowship from the Fondation de la Recherche Médicale.

References

Baxter LL, Moran TH, Richtsmeier JT, Troncoso J, Reeves RH (2000) Discovery and genetic localization of Down syndrome cerebellar phenotypes using the Ts65Dn mouse. Hum Mol Genet 9(2): 195–202

Burrone J, O'Byrne M, Murthy VN (2002) Multiple forms of synaptic plasticity triggered by selective suppression of activity in individual neurons. Nature 420(6914): 414–418

Chen SC, Ehrhard P, Goldowitz D, Smeyne RJ (1997) Developmental expression of the GIRK family of inward rectifying potassium channels: implications for abnormalities in the weaver mutant mouse. Brain Res 778(2): 251–264

Curran ME (1998) Potassium ion channels and human disease: phenotypes to drug targets? Curr Opin Biotech 9: 565–572

Dahmane N, Ghezala GA, Gosset P, Chamoun Z, Dufresne-Zacharia MC, Lopes C, Rabatel N, Gassanova-Maugenre S, Chettouh Abramowski V, Fayet E, Yaspo ML, Korn B, Blouin JL, Lehrach H, Poutska A, Antonarakis SE, Sinet PM, Sinet PM, Creau N, Delabar JM (1998) Transcriptional map of the 2.5-Mb CBR-ERG region of chromosome 21 involved in Down syndrome. Genomics 48(1): 12–23

Firestein S (2001) How the olfactory system makes sense of scents. Nature 413(6852): 211–218

Gosset P, Ghezala GA, Korn B, Yaspo ML, Poutska A, Lehrach H, Sinet PM, Creau N (1997) A new inward rectifier potassium channel gene (KCNJ15) localized on chromosome 21 in the Down syndrome chromosome region 1 (DCR1). Genomics 44(2): 237–241

Gosset P, Ait-Ghezala G, Sinet PM, Creau N (1999) Isolation and analysis of chromosome 21 genes potentially involved in Down syndrome. J Neural Transm [Suppl] 57: 197–209

Higashi K, Fujita A, Inanobe A, Tanemoto M, Doi K, Kubo T, Kurachi Y (2001) An inwardly rectifying K(+) channel, Kir4. 1, expressed in astrocytes surrounds synapses and blood vessels in brain. Am J Physiol Cell Physiol 281(3): C922–C931

Hill CE, Briggs MM, Liu J, Magtanong L (2002) Cloning, expression, and localization of a rat hepatocyte inwardly rectifying potassium channel. Am J Physiol Gastrointest Liver Physiol 282(2): G233–G240

Inanobe A, Yoshimoto Y, Horio Y, Morishige KI, Hibino H, Matsumoto S, Tokunaga Y, Maeda T, Hata Y, Takai Y, Kurachi Y (1999) Characterization of G-protein-gated K+ channels composed of Kir3.2 subunits in dopaminergic neurons of the substantia nigra. J Neurosci 19(3): 1006–1017

Johnson EW (1998) CaBPs and other immunohistochemical markers of the human vomeronasal system: a comparison with other mammals. Microsc Res Tech 41(6): 530–541

Krapivinsky G, Medina I, Eng L, Krapivinsky L, Yang Y, Clapham DE (1998) A novel inward rectifier K+ channel with unique pore properties. Neuron 20(5): 995–1005

Kurschner C, Mermelstein PG, Holden WT, Surmeier DJ (1998) CIPP, a novel multivalent PDZ domain protein, selectively interacts with Kir4.0 family members, NMDA receptor subunits, neurexins, and neuroligins. Mol Cell Neurosci 11(3): 161–172

Lesage F, Duprat F, Fink M, Guillemare E, Coppola T, Lazdunski M, Hugnot JP (1994) Cloning provides evidence for a family of inward rectifier and G-protein coupled K+ channels in the brain. FEBS Lett 353(1): 37–42

Liao YJ, Jan YN, Jan LY (1996) Heteromultimerization of G-protein-gated inwardly rectifying K+ channel proteins GIRK1 and GIRK2 and their altered expression in weaver brain. J Neurosci 16(22): 7137–7150

Mrk MD, Herlitze S (2000) G-protein mediated gating of inward-rectifier K+ channels. Eur J Biochem 267(19): 5830–5836

Nijjar RK, Murphy C (2002) Olfactory impairment increases as a function of age in persons with Down syndrome. Neurobiol Aging 23(1): 65–73

Patil N, Cox DR, Bhat D, Faham M, Myers RM, Peterson AS (1995) A potassium channel mutation in weaver mice implicates membrane excitability in granule cell differentiation. Nat Genet 11(2): 126–129

Pearson WL, Dourado M, Schreiber M, Salkoff L, Nichols CG (1999) Expression of a functional Kir4 family inward rectifier K+ channel from a gene cloned from mouse liver. J Physiol 514: 639–653

Peleg S, Varon D, Ivanina T, Dessauer CW, Dascal N (2002) G(alpha)(i) controls the gating of the G protein-activated K(+) channel, GIRK. Neuron 33(1): 87–99

Pessia M, Imbrici P, D'Adamo MC, Salvatore L, Tucker SJ (2001) Differential pH sensitivity of Kir4.1 and Kir4.2 potassium channels and their modulation by heteropolymerisation with Kir5.1. J Physiol 532(Pt 2): 359–367
Peuchmaur M, Emilie D, Crevon MC, Solal-Celigny P, Maillot MC, Lemaigre G, Galanaud P (1990) IL-2 mRNA expression in Tac-positive malignant lymphomas. Am J Pathol 136(2): 383–390
Rachidi M, Lopes C, Gassanova S, Sinet PM, Vekemans M, Attie T, Delezoide AL, Delabar JM (2000) Regional and cellular specificity of the expression of TPRD, the tetratricopeptide Down syndrome gene, during human embryonic development. Mech Dev 93(1–2): 189–193
Shuck ME, Piser TM, Bock JH, Slightom JL, Lee KS, Bienkowski MJ (1997) Cloning and characterization of two K+ inward rectifier (Kir) 1.1 potassium channel homologs from human kidney (Kir1. 2 and Kir1.3). J Biol Chem 272(1): 586–593
Signorini S, Liao YJ, Duncan SA, Jan LY, Stoffel M (1997) Normal cerebellar development but susceptibility to seizures in mice lacking G protein-coupled, inwardly rectifying K+ channel GIRK2. Proc Natl Acad Sci USA 94(3): 923–927
Slesinger PA, Stoffel M, Jan YN, Jan LY (1997) Defective gamma-aminobutyric acid type B receptor-activated inwardly rectifying K+ currents in cerebellar granule cells isolated from weaver and Girk2 null mutant mice. Proc Natl Acad Sci USA 94(22): 12210–12217
Smith TD, Bhatnagar KP, Shimp KL, Kinzinger JH, Bonar CJ, Burrows AM, Mooney MP, Siegel MI (2002) Histological definition of the vomeronasal organ in humans and chimpanzees, with a comparison to other primates. Anat Rec 267(2): 166–176
Sutherland ML, Williams SH, Abedi R, Overbeek PA, Pfaffinger PJ, Noebels JL (1999) Overexpression of a Shaker-type potassium channel in mammalian central nervous system dysregulates native potassium channel gene expression. Proc Natl Acad Sci USA 96(5): 2451–2455
Thiery E, Gosset P, Damotte D, Delezoide AL, de Saint-Sauveur N, Vayssettes C, Creau N (2000) Developmentally regulated expression of the murine ortholog of the potassium channel KIR4.2 (KCNJ15). Mech Dav 95(1–2): 313–316
Wei J, Dlouhy SR, Bayer S, Piva R, Verina T, Wang Y, Feng Y, Dupree B, Hodes ME, Ghetti B (1997) In situ hybridization analysis of Grik2 expression in the developing central nervous system in normal and weaver mice. J Neuropathol Exp Neurol 56(7): 762–771
Wei J, Hodes ME, Piva R, Feng Y, Wang Y, Ghetti B, Dlouhy SR (1998) Characterization of murine Girk2 transcript isoforms: structure and differential expression. Genomics 51(3): 379–390
Wetter S, Murphy C (1999) Individuals with Down's syndrome demonstrate abnormal olfactory event-related potentials. Clin Neurophysiol 110(9): 1563–1569

Authors' address: N. Créau, PhD, EA3508, Université Denis Diderot, Case 7104, 2, place Jussieu, F-75251 Paris Cedex 05, France, e-mail: creau@paris7.jussieu.fr

Reduction of chromatin assembly factor 1 p60 and C21orf2 protein, encoded on chromosome 21, in Down Syndrome brain

K. S. Shim[1], **J. M. Bergelson**[2], **M. Furuse**[3], **V. Ovod**[4], **T. Krude**[5], and **G. Lubec**[1]

[1] Department of Pediatrics, University of Vienna, Vienna, Austria
[2] Division of Immunologic and Infectious Diseases, Children's Hospital of Philadelphia, Philadelphia, USA
[3] Department of Cell Biology, Faculty of Medicine, Kyoto University, Kyoto, Japan
[4] MAbs Development and Production, FIT Biotech Oyj Plc, Tampere, Finland
[5] Department of Zoology, University of Cambridge, Cambridge CB2 3EJ, United Kingdom

Summary. Trisomy 21 (Down syndrome, DS) is the most common genetic cause of mental retardation, resulting from triplication of the whole or distal part of human chromosome 21. Overexpression of genes located on chromosome 21, as a result of extra gene load, has been considered a central hypothesis for the explanation of the DS phenotype. This gene dosage hypothesis has been challenged, however. We have therefore decided to study proteins whose genes are encoded on chromosome 21 in brain of patients with DS and Alzheimer's disease (AD), as all patients with DS from the fourth decade show Alzheimer-related neuropathology. Using immunoblotting we determined Coxsackievirus and adenovirus receptor (CAR), Claudin-8, C21orf2, Chromatin assembly factor 1 p60 subunit (CAF-1 p60) in frontal cortex from DS, AD and control patients. Significant reduction of C21orf2 and CAF-1 p60, but comparable expression of CAR and claudin-8 was observed in DS but all proteins were comparable to controls in AD, even when related to NSE levels to rule out neuronal cell loss or actin to normalise versus a housekeeping protein. Reduced CAF-1 p60 may reflect impaired DNA repair most probably due to oxidative stress found as early as in fetal life continuing into adulthood. The decrease of C21orf2 may represent mitochondrial dysfunction that has been reported repeatedly and also data on CAR and claudin-8 are not supporting the gene-dosage hypothesis at the protein level. As aberrant expression of the four proteins was not found in brains of patients with AD, decreased CAF and C21orf2 can be considered specific for DS.

Introduction

Trisomy 21 (Down syndrome, DS) is the most frequent chromosomal disorder that occurs in approximately 1 in 800 to 1,000 live births. DS invariably presents with mental retardation and a typical phenotype (Epstein et al., 1995). From the fourth decade all DS patients develop Alzheimer-like neuropathology and a significant number of DS patients, however not all develop dementia (Cairns, 1999; Epstein et al., 1995). According to the gene dosage hypothesis overexpression of genes encoded on chromosome 21 are responsible for the DS phenotype (Epstein et al., 2001). This hypothesis has been already challenged, however and a series of genes encoded on chromosome 21 and proteins whose genes are encoded on chromosome 21 ("chromosome 21 proteins") have shown brain expressional levels comparable to controls including T-oligomycin sensitivity conferring protein, alpha-A-crystallin, peptide 19 (Cheon et al., 2003a,b,c), SOD1 (Gulesserian et al., 2001) or even decreased as in the case of collagen type VI (Engidawork et al., 2001) or the SAM and SH3 containing protein HACS-1 (syn.: SMSN). Some "chromosome 21 proteins" were shown to be increased (Cheon et al., 2003c; Yasuhiro et al., 2002) although the increase cannot be clearly assigned to overexpression of the corresponding genes but may well be due to different mechanisms. It is also not clear whether deranged "chromosome 21 proteins" per se are leading to brain damage in DS or are innocent bystanders or are leading to aberrant expression of other protein classes not encoded on chromosome 21 and there is a host of dysregulated proteins in DS brain that may be responsible for neuropathological changes in DS (Lubec et al., 2002).

In this study we tested four proteins that were considered candidates for a role in the complex pathomechanisms involved in adult DS brain: Coxsackievirus and adenovirus receptor (CAR; 21q21.1) involved in brain development, Claudin-8 (22q22.3) as a tight junction-related protein, C21orf2 (21q22.3) as a mitochondrial protein and chromatin assembly factor 1 p60 subunit (CAF-1 p60; 21q22.2) as a marker for recombination and DNA repair and indeed, significant and remarkable reduction of CAF and C21orf2 were observed in frontal cortex of DS discriminating adult DS with AD-neuropathology from AD, thus indicating a specific dysregulation of these proteins in DS brain.

Materials and methods

Brain samples

Postmortem human adult frontal cortex samples (superior frontal gyrus) were obtained from Dr. N. J. Cairns (MRC London Brain Bank for Neurodegenerative Disease, Institute of Psychiatry, King's College, UK) (Table 1). AD patients fulfilled the National Institute of Neurological Disorders and Stroke and Alzheimer's disease and Related Disorders Association criteria for probable AD (Tierney et al., 1998). The neuropathological diagnosis of "definite AD" was confirmed using the CERAD criteria (Mirra et al., 1991). All DS patients were karyotyped and possessed trisomy 21. A formal cognitive assessment of dementia in DS was not performed. In all DS brains there were abundant and extensive beta-amyloid deposits, neurofibrillary tangles and neuritic plaques. Normal brains obtained from individuals with no history of neurological or psychiatric illness were used as controls. The major cause of death was bronchopneumonia in AD and DS

Table 1. Autopsy data used in this study

	Control	AD	DS
N*	6	6	6
Age (years)	60.17 ± 9.28	59.33 ± 6.44	57.83 ± 8.18
Postmortem interval (hrs)	34.00 ± 12.87	36.50 ± 27.46	31.17 ± 23.13

* All samples were male

and heart disease in controls. After dissection, coronal slices were snap frozen and stored at −70°C until use.

Antibodies

The polyclonal rabbit anti-CAR antiserum recognized the extracellular region of human CAR (Cohen et al., 2001); the polyclonal rabbit antiserum was specific for the COOH-terminal cytoplasmic domains of Claudin-8 (Morita et al., 1999); the polyclonal mouse antiserum react with 20–23 amino acid (QRLSELYLRRNRI) of C21orf2 (Krohn et al., 1997); the polyclonal rabbit antibody against CAF-1 p60 (Marheineke et al., 1998) were used for this study. Two antibodies for neuron specific enolase (NSE, Chemicon, UK) and actin (Sigma, USA) were purchased. Commercially available anti-rabbit and anti-mouse secondary antibodies coupled to horseradish peroxidase (Southern Biotechnology Associates, Inc., Alabama, USA) were also used.

Western blotting

Frozen brain samples were ground with a mortar under liquid nitrogen, directly suspended and homogenised with homogenisation buffer containing 10 mM Tris-HCI (pH 7.5), 150 mM NaCl, 0.05% (v/v) Tween 20, 1 mM PMSF (phenylmethylsulfonyl fluoride; Sigma, Austria) and 1 tablet of protease inhibitor cocktail (Roche, Austria). The homogenised samples were centrifuged at 8,000 × g for 10 minutes at 4°C. The BCA protein assay kit (Pierce, USA) was used to determine the concentration of protein in the supernatant. Samples (10 ug) were mixed with the same volume of sample buffer (60 mM Tris-HCl, pH 6.8, 2% SDS, 0.1% bromophenol blue, 25% glycerol, 14.4 mM 2-mercaptoethanol), incubated at 95°C for 15 minutes and loaded onto a 12.5% ExcelGel SDS homogeneous gel (Amersham Pharmacia Biotech, Sweden). Electrophoresis was performed on Multiphor II Electrophoresis System (Amersham Pharmacia Biotech, Sweden). Proteins separated on the gel were transferred onto PVDF membrane (Millipore, USA) and membranes were blocked in blocking buffer (10 mM Tris-HCl, pH 7.5, 150 mM NaCl, 2% non-fat dried milk and 0.1% Tween 20). Membranes were incubated for 2 hours at room temperature with diluted primary antibodies (1:2,000 for Claudin-8 and CAR, 1:800 for C21orf2, 1:1,500 for CAF-1p60, 1:3,000 for actin and NSE). After 4 times washing for 10 min with blocking buffer, membranes were probed with 1:2,000 diluted secondary antibodies for 1 hour. Membranes were washed 4 times for 10 minutes and developed with the Western blot Chemiluminescene reagent (PerkinElmer Life Science, Inc., USA).

Analysis and statistics

The densities of detected bands were measured using RFLP scan version 2.1 software program (Scanalytics, USA) and expressed as arbitrary units. Between group differences were calculated by non-parametric Mann-Whitney U test. Correlation between protein expression and age and post-mortem intervals of brain samples was assessed by Pearson correlation. The statistical analysis was performed by GraphPad Instat 2 software version 2.05 and the level of significance was set at $P < 0.05$. All results were presented as mean $\pm$ standard deviation (SD).

Results

We investigated the expression level of four proteins encoded on chromosome 21 (CAR, Claudin-8, C21orf2, CAF-1 p60) in adult brains with DS and AD compared to controls by Western blot (Fig. 1). We detected one band at 46 kDa and double bands at 31–34 kDa with anti-CAR antiserum. The anti-claudin-8 antibody recognized three strong bands at 23–30 kDa, and the antibody against C21orf2 reacted with one major band at 25 kDa and a weak band at 30 kDa. The CAF-1 p60 antibody detected one band at 60 kDa. Two proteins, NSE and actin, were used as reference proteins for neuronal and cellular density, and showed comparable expression levels in DS, AD and controls (Fig. 1).

The density of immunoreactive bands of two proteins (CAR, Claudin-8) was comparable between DS, AD and controls. However, the protein level of

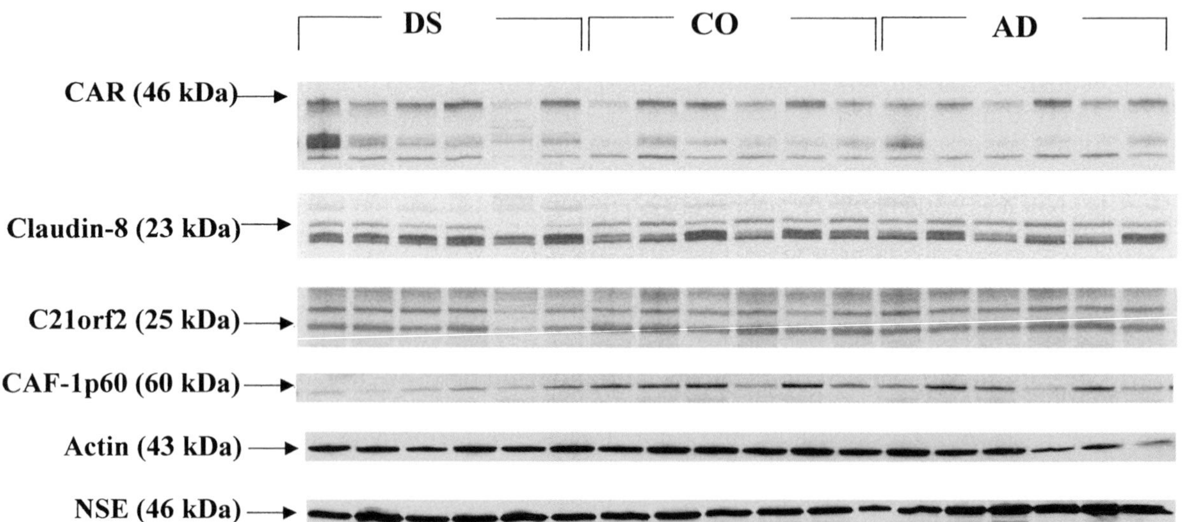

Fig. 1. Western blot of four proteins encoded on chromosome 21 and reference proteins in adult brain of DS compared with control (CO). Ten micrograms of protein were separated on 12.5% homogeneous gel and transferred onto PVDF membranes. The membranes were immunoreacted with primary and secondary antibody. Specific immunoreactive band of CAR, Claudin-8, C21orf2, CAF-1 p60, actin and NSE were detected at 46 kDa, 23 kDa, 25 kDa, 60 kDa, 43 kDa and 46 kDa were detected in all samples

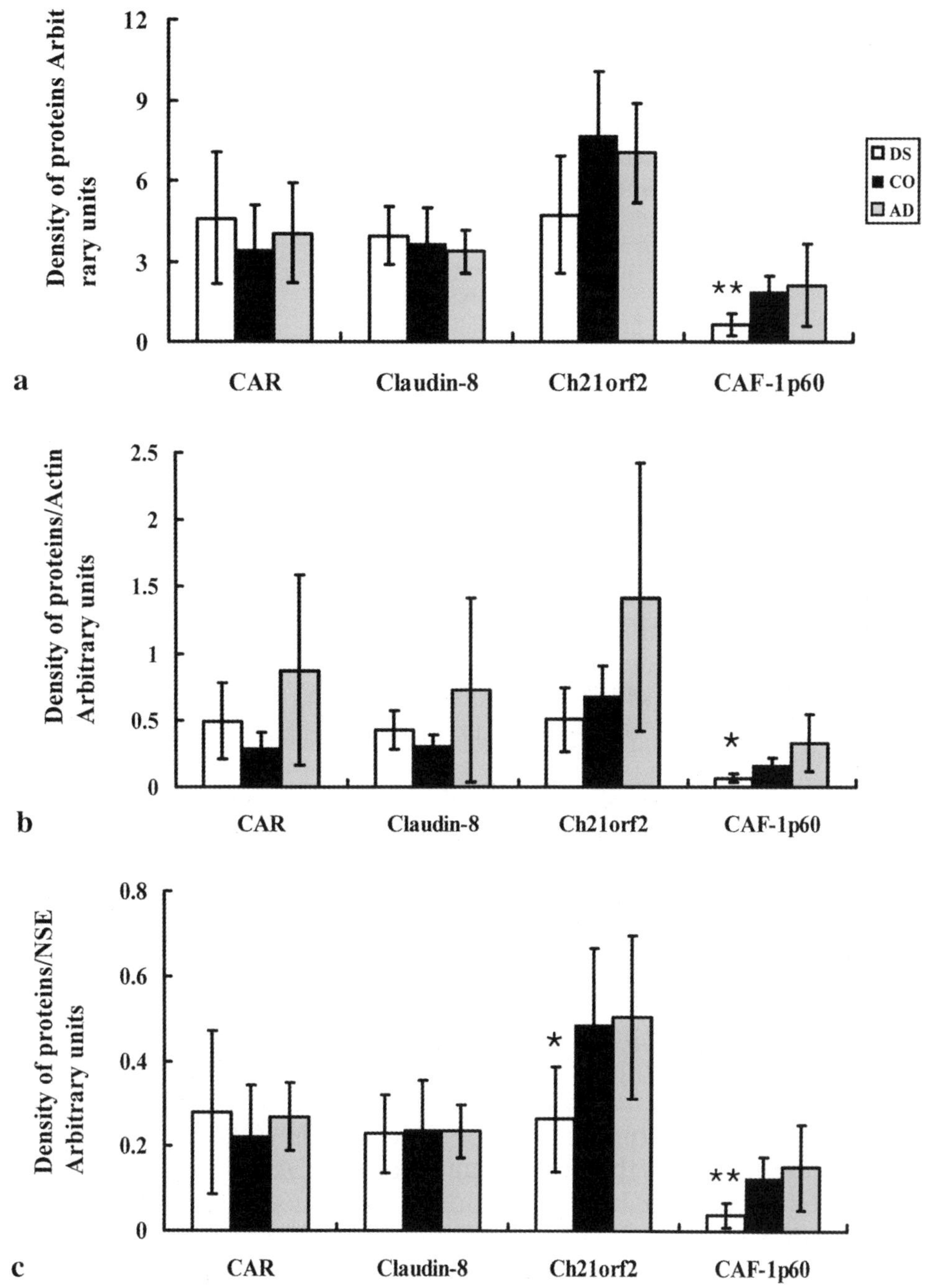

Fig. 2. **a** Expression levels of four proteins. The density of immunoreactive bands for each protein was normalized with that of either actin or NSE. * $P < 0.05$, ** $P < 0.01$ **b** Actin normalized protein levels. **c** NSE normalized protein levels

CAF-1 p60 was significantly decreased only in DS as shown in Fig. 2a. There was no statistically significant difference between DS ($P = 0.006$) and AD and protein levels of C21orf2 in DS showed reduced patterns compared to controls (Fig. 2a). When the levels of four proteins were normalized with those of actin, the specific reduction of CAF-1 p60 in DS was observed (Fig. 2b). We also determined NSE levels in order to normalize the four protein levels with neuronal density. The reduction of C21orf2 and CAF-1 p60 in DS was statistically significant and even more pronounced when related to NSE levels (Fig. 2c). There was no correlation between protein expression of four proteins and either age or postmortem intervals of brain samples (data not shown).

Discussion

Here, we show that the protein level of C21orf2, CAF-1 p60 was down-regulated, but CAR and Claudin-8 were comparable in brain of patients with DS compared to controls. Normalization versus beta actin and NSE rules out the potential influence of neuronal or simply cell loss in the brain, which has been demonstrated in patients with DS.

Coxsackievirus and adenovirus receptor (CAR) is a 46 kDa transmembrane glycoprotein responsible for coxsackie B virus (CVB) infection of human cells and functions in adenovirus attachment influencing adenovirus-mediated gene delivery (Tomko et al., 1997; Bergelson et al., 1997), but its physiologic functions have not been elucidated. Using CAR antiserum, we detected one band at 46 kDa which may be full-length of CAR and a doublet at 34–31 kDa which may be truncated or isoforms. Emerging reports suggest that this observation is credible, and alternative splice products encoded by the human CAR gene have been identified (Fechner et al., 1999; Thoelen et al., 2001). The band of 46 kDa was comparable between DS, AD and controls, and the density of doublet was highly variable among individual sample.

CAR is localized to homotypic intercellular contacts, mediating homotypic cell aggregation as an adhesion molecule, and recruiting tight junction protein ZO-1 to sites of cell-cell contact in epithelial cells (Cohen et al., 2001). Structurally CAR belongs to the immunoglobulin superfamily, similar to ICAM-1 and N-CAM (van Raaij et al., 2000). The extracellular domain of CAR facilitates intercellular adhesion and the transmembrane and intracellular domains of CAR lead to modulation of the cell cycle regulatory proteins p21 and Rb, which inhibits growth of cancer cells (Okegawa et al., 2001). These studies indicated that membrane bound CAR may elicit signals to modulate cell cycle regulators contributing to contact-dependent regulation of cell growth. Cell surface adhesion molecules are thought to play a crucial role in neuronal development function, including cell migration and sorting, axon guidance, formation of neuronal connection and synaptic plasticity (Engidawork et al., 2001). Recently, it was demonstrated that CAR is strongly expressed in heart and brain of mice until the newborn phase; its expression subsequently become decreased in adult mice (Ito et al., 2000). In primary neurons from hippocampi of mouse embryos, the expression of mouse CAR was observed throughout the cells including those in cell bodies, neurite and growth cones on immunohistochemistry (Honda et al., 2000). These observations imply that the expression of CAR exhibit developmental changes and CAR could be involved in neuro-network formation in the developing nervous system through cell contact. However, the role of CAR which is a member of the immunoglobulin superfamily involved in neural network formation, has not been studied yet in adult brain with cognitive defects. Unaltered CAR in adult DS and AD brain compared to controls, suggests that CAR is not responsible for deficient neuronal outgrowth and abnormal wiring in DS brain.

Tight junction (TJ) is a specialized membrane structure of polarized endo- and epithelial cells, directly involved in barrier and fence functions to regulate diffusion of solutes through the paracellular pathway (Schneeberger et al., 1992). The molecular composition of TJ has been recently investigated and only three tight junction proteins, occulin, the claudin family (claudin-3 to -15) and junctional adhesion molecule, are known to be transmembrane proteins (Furuse et al., 1998). Claudin was identified as a major component localized exclusively in TJ strands (Morita et al., 1999). When cDNA of claudin was introduced into fibroblasts lacking TJ of claudin expression, a huge network of TJ strands was induced between stable transfectants (Furuse et al., 1998). In vertebrate brain, two types of cells are known to bear TJ or TJ-like structures, vascular endothelial cells and oligodendrocytes. In endothelial cells, a typical TJ strand network is well developed, which is thought to be responsible for the blood-brain barrier (Rubin et al., 1991). In myelinated nerve fibers of central nervous system (CNS), each internodal segment between two consecutive nodes of Ranvier is composed of a cellular process of oligodendrocytes with its myelin lamellae. CNS myelin has a distinctive structural feature, the so-called "radial component" which is a series of radially arranged intralamellar strands that span the myelin sheath. These intralamellar strands look like TJ, and recently have been shown to contain claudin-11. Claudin-11 was originally detected as an oligodendrocyte-specific protein and, specifically expressed not only in oligodendrocytes but also in Sertoli cells forming the blood-testis barrier (Morita et al., 1999). Additionally, Claudin-1 and 2 were detected in the chorioid plexus epithelium which the blood-CSF barrier and is responsible for the secretion of CSF from blood (Wolburg et al., 2001). In brain, Claudin-5 together with occulin was described to be present in endothelial cells tight junctions forming the blood-brain barrier (Lippoldt et al., 2000). Recently, it was demonstrated that the expression of claudin-8 mRNA only is specifically and progressively decreased in hippocampi of kindled compared to control rats as neuronal excitability is enhanced (Lamas et al., 2002). These studies suggest that the modulation of expression of a TJ protein could affect permeability of blood-brain barrier (BBB). Interestingly, 3 members of the claudin gene family (claudin-8, -14 and 17) are encoded on chromosome 21, triplication of which is leading to abnormalities of Down syndrome (DS) and altered BBB permeability was proposed to be immatured in neuropathology of DS (Emerson et al., 1995). Therefore, claudins as constitutive proteins of TJ are likely candidates for dysfunction of blood-brain barrier under pathological states in DS. In this study, we found three bands at 23–30 kDa using anti-claudin-8 antiserum, representing post-translational modifications or isoforms in brain and indeed claudin-8, using the phosphorylated sequence prediction program (Nikolaj et al., 1999), contains several predicted phosphorylated sites (data not shown). The density of three bands was comparable in DS, AD and controls even when normalized with NSE and beta-actin, maybe reflecting integrity of TJ for BBB function in adult DS.

C21orf2 is encoded on chromosome 21q22.3 known to contain genes for non-syndromic autosomal recessive deafness (DFNB8/10) and autoimmune polyendocrinopathy-candidiasis-ectodermal dystrophy (APECED) (Ahonen et al., 1990; Veske et al., 1996). On western blotting, C21orf2 protein is detected at 25 kDa. It appears to be targeted to mitochondria as immunofluorescence studies and co-staining with a mitochondrial-specific dye suggested the subcellular localization of C21orf2 to mitochondria (Krohn et al., 1997). Although mutations in components of the mitochondrial genetic system have been implicated in hearing impairment, the TMPRSS3 gene which encodes a transmembrane serine protease was found to be responsible for two non-syndromic recessive deafness (DFNB8/10) loci located on Chromosome 21 (Watternhofer et al., 2002). C21orf21 mRNA shows ubiquitous expression in fetal and adult tissues studied, including heart, brain, skeletal muscle and pancreas, known to be a rich source for mitochondria (Scott et al., 1998), but the function of this protein remains completely unknown so far. In a previous study, we reported the considerable increase of C21orf2 protein expression in fetal DS brain (Cheon et al., 2002c). However, the C21orf2 level was statistically decreased in adult brain from DS, not in AD, compared to control. While increased C21orf2 in fetal DS may reflect increased mitochondrial proliferation, decreased C21orf2 may point to mitochondrial disfunction which in turn maybe leading to neuronal vulnerability (Busciglio et al., 2002; Kim et al., 2001; Krapfenbauer et al., 1999) and maybe oxidative stress in DS.

Chromatin assembly factor 1 (CAF-1) was identified as a histone chaperone, involved in nucleosome assembly of replicating DNA and repair of DNA damage (Smith et al., 1989). CAF-1 exists as a multiprotein complex, p150, p60 and p48, in vivo (Smith et al., 1991). Among them, p60 is essentially required for DNA replication-dependent assembly of nucleosomes (Kaufman et al., 1995) and for nucleosome assembly in DNA repair. Following UV irradiation in vivo, phosphorylated p60 is specifically recruited to chromatin undergoing DNA repair during interphase. The kinetics of appearance of the phosphorylated p60 was parallel to the dose of UV irradiation and the number of UV lesion/repair sites in DNA, consistent with the timing of early repair events (Martini et al., 1998). In addition, purified p60 can be directly phosphorylated by purified cyclin A/Cdk2, cyclin E/Cdk2, and cyclin B1/Cdk1, but not by cyclin D/Cdk4 complexes in vitro. CAF-1 containing hyperphosphorylated p60 prepared from mitotic cells is inactive in nucleosome assembly and becomes activated by dephosphorylation in vitro (Keller et al., 2000). These data propose that p60 could serve as a signal for the state of chromatin assembly during DNA repair process and depend on its phosphorylated state. Furthermore, human CAF-1 p60 is shown to function synergistically with histone chaperone anti-silencing function 1 (Asf1) to assemble nucleosomes during nucleotide excision repair (NER) (Mello et al., 2002). NER is a fundamental mechanism repairing many different kinds of

DNA damage by genotoxic agents and NER related genes are responsible for repairing DNA lesions resulting from the oxidative stress. Expression of excision repair-cross-complementing proteins, p80 and p89, was significantly higher in frontal and temporal cortex from DS, compatible with increased oxidative DNA damage in vivo (Hermon et al., 1998). Additionally, DNA-repair gene ERCC2 and ERCC3 (excision-repair-cross-complementing) for nucleotide excision repair and XRCC1 (X-ray-repair-cross-complementing) for X-ray-repair, were increased in DS brain at the transcriptional level (Fang-Kircher et al., 1999; Sanford et al., 1993), indicating that cells of patients with DS may be more susceptible to ionizing irradiation as well as DNA-damaging agents by deficient DNA repair process. However, the function of CAF-1 p60 related to DNA repair has not been elucidated so far. In this study, we present that CAF1 p60 was statistically decreased in adult brain of patients with DS, but not in AD, compared to control when related to NSE and actin levels. DS cells showed lower DNA-repair efficiency and also an accelerated decline in DNA-repair capacity with age (Raji et al., 1998). These results propose that deteriorated DNA-repair potential could be one of the probable pathomechanisms of this chromosomal disorder.

In conclusion, our data showing comparable protein levels of CAR and claudin-8 in frontal lobe of patients with DS, provide evidence against the gene dosage hypothesis for the development of the DS phenotype, suggesting that the DS phenotype can not be simply explained by overexpression of genes residing on chromosome 21. Additionally, downregulated protein levels of C21orf2 and CAF-1 p60 in adult brain of patients with DS in contrast to AD, indicate that both proteins can be considered specific for the neuropathological/developmental changes in DS per se rather than AD-neuropathological alterations found regularly in aged DS patients.

Acknowledgements

We thank Dr. N. J. Cairns (MRC London Brain Bank for Neurodegenerative Disease, Institute of Psychiatry, King's College, UK), Dr. M. Dierssen (Medical and Molecular Genetics Center-IRO, Hospital Durani Reynals, Barcelona, Spain) and Dr. J. C. Ferreres (Department of Pathology UDIAT-CD, Corporacis Sani'ria Parc Tauli, Sabadell, Barcelona, Spain) for postmortem human brain tissues. We are highly indebted to the Red Bull Company, Salzburg, for generous financial support.

References

Ahonen P, Myllarniemi S, Sipila I, Perheentupa J (1990) Clinical variation of autoimmune polyendocrinopathy-candidiasis-ectodermal dystrophy (APECED) in a series of 68 patients. N Engl J Med 322: 1829–1836

Bergelson JM, Cunningham JA, Droguett G, Kurt-Jones EA, Krithivas A, Hong JS, Horwitz MS, Crowell RL, Finberg RW (1997) Isolation of a common receptor for Coxsackie B viruses and adenoviruses 2 and 5. Science 275: 1320–1323

Cairns NJ (1999) Neuropathology of Down syndrome. J Neural Transm [Suppl] 57: 61–74

Cheon MS, Kim SH, Yaspo ML, Blasi F, Aoki Y, Melen K, Lubec G (2003a) Protein levels of genes encoded on chromosome 21 in fetal Down syndrome brain: challenging the gene dosage effect hypothesis, part I. Amino Acids 24: 111–117

Cheon MS, Bajo M, Kim SH, Claudio JO, Stewart AK, Patterson D, Kruger WD, Kondoh H, Lubec G (2003b) Protein levels of genes encoded on chromosome 21 in fetal Down syndrome brain: challenging the gene dosage effect hypothesis, part II. Amino Acids 24: 119–125

Cheon MS, Kim SH, Ovod V, Kopitar-Jerala N, Morgan I, Hatefi Y, Ijuin T, Takenawa T, Lubec G (2003c) Protein levels of genes encoded on chromosome 21 in fetal Down syndrome brain: chanllenging the gene dosage effects hypothesis, part III. Amino Acids 24: 127–134

Cohen CJ, Shieh JT, Pickles RJ, Okegawa T, Hsieh JT, Bergelson JM (2001) The coxsackievirus and adenovirus receptor is a transmembrane component of the tight junction. Proc Natl Acad Sci USA 26: 15191–15196

Emerson JF, Kesslak JP, Chen PC, Lott IT (1995) Magnetic resonance imaging of the aging brain in Down syndrome. Prog Clin Biol Res 393: 123–138

Engidawork E, Bajic N, Fountoulakis M, Dierssen M, Greber-Platzer S, Lubec G (2001) Beta-amyloid precursor protein, ETS-2 and collagen alpha 1 (VI) chain precursor, encoded on chromosome 21, are not overexpressed in fetal Down syndrome: further evidence against gene dosage effect. J Neural Transm [Suppl] 61: 335–346

Epstein CJ (1995) Down Syndrome (trisomy 21). In: Scriver SR, Beaudet AL, Sly WS, Valle D (eds) The metabolic and molecular bases of inherited disease, 7th edn. McGraw Hill, New York, pp 749–794

Epstein CJ (2001) Down Syndrome (trisomy 21). In: Scriver SR, Beaudet AL, Sly WS, Valle D (eds) The metabolic and molecular bases of inherited disease, 8th edn. McGraw Hill, New York, pp 1223–1256

Fang-Kircher SG, Labudova O, Kitzmueller E, Rink H, Cairns N, Lubec G (1999) Increased steady state mRNA levels of DNA-repair genes XRCC1, ERCC2 and ERCC3 in brain of patients with Down syndrome. Life Sci 64: 1689–1699

Fechner H, Haack A, Wang H, Wang X, Eizema K, Pauschinger M, Schoemaker R, Veghel R, Houtsmuller A, Schultheiss HP, Lamers J, Poller W (1999) Expression of coxsackie adenovirus receptor and alphav-integrin does not correlate with adenovector targeting in vivo indicating anatomical vector barriers. Gene Ther 9: 1520–1535

Furuse M, Fujita K, Hiiragi T, Fujimoto K, Tsukita S (1998) Claudin-1 and -2: novel integral membrane proteins localizing at tight junctions with no sequence similarity to occluding. J Cell Biol 141: 1539–1550

Gulesserian T, Engidawork E, Fountoulakis M, Lubec G (2001) Antioxidant proteins in fetal brain: superoxide dismutase-1 (SOD-1) protein is not overexpressed in fetal Down syndrome. J Neural Transm [Suppl] 61: 71–84

Hermon M, Cairns N, Egly JM, Fery A, Labudova O, Lubec G (1998) Expression of DNA excision-repair-cross-complementing proteins p80 and p89 in brain of patients with Down syndrome and Alzheimer's disease. Neurosci Lett 251: 45–48

Honda T, Saitoh H, Masuko M, Katagiri-Abe T, Tominaga K, Kozakai I, Kobayashi K, Kumanishi T, Watanabe YG, Odani S, Kuwano R (2000) The coxsackievirus-adenovirus receptor protein as a cell adhesion molecule in the developing mouse brain. Brain Res Mol Brain Res 77: 19–28

Ito M, Kodama M, Masuko M, Yamaura M, Fuse K, Uesugi Y, Hirono S, Okura Y, Kato K, Hotta Y, Honda T, Kuwano R, Aizawa Y (2000) Expression of coxsackievirus and adenovirus receptor in hearts of rats with experimental autoimmune myocarditis. Circ Res 86: 275–280

Kaufman PD, Kobayashi R, Kessler N, Stillman B (1995) The p150 and p60 subunits of chromatin assembly factor I: a molecular link between newly synthesized histones and DNA replication. Cell 81: 1105–1114

Keller C, Krude T (2000) Requirement of Cyclin/Cdk2 and protein phosphatase 1 activity for chromatin assembly factor 1-dependent chromatin assembly during DNA synthesis. J Biol Chem 275: 35512–35521

Kim SH, Vlkolinsky R, Cairns N, Fountoulakis M, Lubec G (2001) The reduction of NADH ubiquinone oxidoreductase 24- and 75-kDa subunit in brains of patients with Down syndrome and Alzheimer's disease. Life Sci 68: 2741–2750

Krapfenbauer K, Yoo BC, Cairns N, Lubec G (1999) Differential display reveals deteriorated mRNA levels of NADH3 (complemented) in cerebellum of patients with Down syndrome. J Neural Transm [Suppl] 57: 211–220

Krohn K, Ovod V, Vilja P, Heino M, Scott H, Kyriakou DS, Antonarakis S, Jacobs HT, Isola J, Peterson P (1997) Immunochemical characterization of a novel mitochondrially located protein encoded by a nuclear gene within the DFNB8/10 critical region on 21q22.3. Biochem Biophys Res Commun 238: 806–810

Lamas M, Gonzalez-Mariscal L, Gutierrez R (2002) Presence of claudins mRNA in the brain. Selective modulation of expression by kindling epilepsy. Brain Res Mol Brain Res 104: 250–254

Lippoldt A, Liebner S, Andbjer B, Kalbacher H, Wolburg H, Haller H, Fuxe K (2000) Organization of choroids plexus epithelial and endothelial cell tight junctions and regulation of claudin-1, -2 and -5 expression by protein kinase C. Neuro Report 11: 1427–1431

Lubec G, Engidawork E (2002) The brain in Down syndrome (TRISOMY 21). J Neurol 249: 1347–1356

Marheineke K, Krude T (1998) Nucleosome assembly activity and intracellular localization of human CAF-1 changes during the cell division cycle. J Biol Chem 273: 15279–15286

Martini E, Roche DM, Marheineke K, Verreault A, Almouzni G (1998) Recruitment of phosphorylated chromatin assembly factor 1 to chromatin after UV irradiation of human cells. J Cell Biol 143: 563–575

Mello JA, Sillje HH, Roche DM, Kirschner DB, Nigg EA, Almouzni G (2002) Human Asf1 and CAF-1 interact and synergize in a repair-coupled nucleosome assembly pathway. EMBO Rep 3: 329–334

Mirra SS, Heyman A, McKeel D, Sumi SM, Crain BJ, Brownlee LM, Vogel FS, Hughes JP, van Belle G, Berg L (1991) The Consortium to Establish a Registry for Alzheimer's Disease (CERAD), part II. Standardization of the neuropathologic assessment of Alzheimer's disease. Neurology 41: 479–486

Morita K, Furuse M, Fujimoto K, Tsukita S (1999) Claudin multigene family encoding four-transmembrane domain protein components of tight junction strands. Proc Natl Acad Sci USA 96: 511–516

Morita K, Sasaki H, Fujimoto K, Furuse M, Tsukita S (1999) Claudin-11/OSP-based tight junctions of myelin sheaths in brain and Sertoli cells in testis. J Cell Biol 145: 579–588

Nicolaj B, Steen G, Soren B (1999) Sequence and structure-based prediction of eukaryotic protein phosphorylation sites. J Mol Biol 294: 1351–1362

Okegawa T, Pong RC, Li Y, Bergelson JM, Sagalowsky AI, Hsieh JT (2001) The mechanism of the growth-inhibitory effect of coxsackie and adenovirus receptor (CAR) on human bladder cancer: a functional analysis of car protein structure. Cancer Res 17: 6592–6600

Raji NS, Rao KS (1998) Trisomy 21 and accelerated aging: DNA-repair parameters in peripheral lymphocytes of Down's syndrome patients. Mech Ageing Dev 100: 85–101

Rubin LL (1991) The blood-brain barrier in and out of cell culture. Curr Opin Neurobiol 1: 360–363

Sanford KK, Parshad R, Price FM, Tarone RE, Schapiro MB (1993) X-ray induced chromatid damage in cells from Down syndrome and Alzheimer disease patients in relation to DNA repair and cancer proneness. Cancer Genet Cytogenet 70: 25–30

Schneeberger EE, Lynch RD (1992) Structure, function, and regulation of cellular tight junctions. Am J Physiol 262: L647–L661

Scott HS, Kyriakou DS, Peterson P, Heino M, Tahtinen M, Krohn K, Chen H, Rossier C, Lalioti MD, Antonarakis SE (1998) Characterization of a novel gene, C21orf2, on human chromosome 21q22.3 and its exclusion as the APECED gene by mutation analysis. Genomics 47: 64–70

Smith S, Stillman B (1989) Purification and characterization of CAF-I, a human cell factor required for chromatin assembly during DNA replication in vitro. Cell 58: 15–25

Smith S, Stillman B (1991) Immunological characterization of chromatin assembly factor I, a human cell factor required for chromatin assembly during DNA replication in vitro. J Biol Chem 266: 12041–12047

Thoelen I, Magnusson C, Tagerud S, Polacek C, Lindberg M, van Ranst M (2001) Identification of alternative splice products encoded by the human coxsackie-adenovirus receptor gene. Biochem Biophys Res Commun 287: 216–222

Tierney MC, Fisher RH, Lewis AJ, Zorzitto ML, Snow WG, Reid DW, Nieuwstraten P (1998) The NINCDSADRDA Work Group criteria for the clinical diagnosis of probable Alzheimer's disease: a clinicopathologic study of 57 cases. Neurology 38: 359–364

Tomko RP, Xu R, Philipson L (1997) HCAR and MCAR: the human and mouse cellular receptors for subgroup C adenoviruses and group B coxsackieviruses. Proc Natl Acad Sci USA 94: 3352–3356

van Raaij MJ, Chouin E, van der Zandt H, Bergelson JM, Cusack S (2000) Dimeric structure of the coxsackievirus and adenovirus receptor D1 domain at 1.7 A resolution. Struct Fold Des 11: 1147–1155

Veske A, Oehlmann R, Younus F, Mohyuddin A, Muller-Myhsok B, Mehdi SQ, Gal A (1996) Autosomal recessive non-syndromic deafness locus (DFNB8) maps on chromosome 21q22 in a large consanguineous kindred from Pakistan. Hum Mol Genet 5: 165–168

Wattenhofer M, Di Iorio MV, Rabionet R, Dougherty L, Pampanos A, Schwede T, Montserrat-Sentis B, Arbones ML, Iliades T, Pasquadibisceglie A, D'Amelio M, Alwan S, Rossier C, Dahl HH, Petersen MB, Estivill X, Gasparini P, Scott HS, Antonarakis SE (2002) Mutations in the TMPRSS3 gene are a rare cause of childhood nonsyndromic deafness in Caucasian patients. J Mol Med 80: 124–131

Wolburg H, Wolburg-Buchholz K, Liebner S, Engelhardt B (2001) Claudin-1, claudin-2 and claudin-11 are present in tight junctions of choroid plexus epithelium of the mouse. Neurosci Lett 307: 77–80

Yasuhiro A, Takeshi I, Tadanomi T, Laurence EB, Sachio T (2002) Excessive expression of synaptojanin in brains with Down syndrome. Brain Dev 24: 67–72

Authors' address: Prof. Dr. G. Lubec, CChem, FRSC (UK), Department of Pediatrics, University of Vienna, Währinger Gürtel 18, A-1090 Vienna, Austria, e-mail: gert.lubec@akh-wien.ac.at

The MNB/DYRK1A protein kinase: neurobiological functions and Down syndrome implications*

B. Hämmerle[1], **C. Elizalde**[1], **J. Galceran**[1], **W. Becker**[2], and **F. J. Tejedor**[1]

[1] Instituto de Neurociencias, Unidad de Neurobiologia del Desarrollo, CSIC y Universidad Miguel Hernandez, Campus de San Juan, San Juan (Alicante), Spain
[2] Institut für Pharmakologie und Toxikologie, Medizinische Fakultät, RWTH Aachen, Aachen, Germany

Summary. Major attention is being paid in recent years to the genes harbored within the so called Down syndrome Critical Region of human chromosome 21. Among them, those genes with a possible brain function are becoming the focus of intense research due to the numerous neurobiological alterations and cognitive deficits that Down syndrome individuals have. MNB/DYRK1A is one of these genes. It encodes a protein kinase with unique genetic and biochemical properties, which have been evolutionarily conserved from insects to humans. MNB/DYRK1A is expressed in the developing brain where it seems to play a role in proliferation of neural progenitor cells, neurogenesis, and neuronal differentiation. Although at a lower level, MNB/DYRK1A is also expressed in the adult brain where, as judged by the phenotype of mutant and transgenic animals, it may be involved in learning and memory. Nevertheless, most of the molecular mechanisms underlying these functions remain to be unraveled.

In this review we compile and discuss experimental evidences, which support the involvement of MNB/DYRK1A in several neuropatologies and cognitive deficits of Down syndrome.

Introduction

The Down syndrome (DS) is the major genetic cause of mental retardation. It is well known that the brain of DS patients presents diverse macroscopic and microscopic alterations such as reduction in the size of several brain regions (cortex, cerebellum, brainstem, etc), decreased neuronal number/density, abnormal neuronal differentiation processes and precocious Alzheimer-like

* DYRK1A is the official gene symbol approved by the HUGO Gene Nomenclature Committee. Nevertheless, MNB has been also frequently used by several authors. Thus, we use MNB/DYRK1A in this review to refer to these orthologous genes in contrast to other members of the DYRK family of protein kinases.

neurodegeneration (reviewed by Becker et al., 1991; Coyle et al., 1986; Korenberg et al., 1994). Nevertheless, the molecular bases of all these alterations remain unknown. DS is normally caused by the trisomy of chromosome 21 (HC21). Nevertheless, genetic studies in cases with partial trisomy suggest that a limited region, called the Down Syndrome Critical Region (DSCR), may be mainly responsible for the etiology of DS (Rahmani et al., 1989). The sequencing of HC21 has allowed the identification of the genes harbored within this region (Hattori et al., 2000). Among these genes, those with a possible brain function can be of great interest to understand the neurobiological alterations of DS. One of these genes is MNB/DYRK1A (Guimerá et al., 1996; Song et al., 1996), the orthologue of the minibrain (mnb) gene of Drosophila (Tejedor et al., 1995). Several homologues have been identified in diverse organisms, from yeast to humans (Becker et al., 1998; Becker and Joost, 1999). They encode related protein kinases with unique structural and biochemical properties (see Galceran et al., 2003, the next article in this issue, for a detailed review of the molecular properties of MNB/DYRK1A kinases). In this review we will address the relation of MNB/DYRK1A to brain function and development, and its possible implications in DS.

Expression of Mnb/Dyrk1A in the brain

The close structural and functional similarity between Drosophila mnb and vertebrate Mnb/Dyrk1A makes the fly an attractive model for the genetic and functional analysis. mnb is distinctly expressed in the CNS of Drosophila throughout development. In the embryo mnb is clearly present in the CNS but not in the PNS. During postembryonic development, mnb is preferentially expressed in the proliferative centers of the CNS, the tissue where adult neurons are generated. In the adult brain, mnb is expressed in several neuropile areas. Particularly interestingly is the high expression in the mushroom bodies, a central brain structure involved in sensory integration and learning/memory (Tejedor et al., 1995). In contrast to the preferential expression in the Drosophila brain, Northern and PCR analysis indicate that Mnb/Dyrk1A mRNA is ubiquitously expressed in different tissues and in most brain regions of rodents and humans (Guimerá et al., 1996, 1999; Song et al., 1996; Okui et al., 1999). In situ hybridization of rodents show preferential expression in brain. This is prominent in olfactory bulb, cerebellum, cortex, hippocampus and hypothalamus (Guimerá et al., 1996; Song et al., 1996). MNB/DYRK1A immunostaining partially contrasts with the hybridization data since it is strong in olfactory bulb, cerebellum, spinal cord and most motor nuclei of midbrain and brain stem (Marti et al., 2003). Western-blot and Northern-blot analysis also show contrasting results for the developmental pattern of expression. Thus, the levels of MNB/DYRK1A protein are high during development but gradually decreases to very low level during postnatal stages whereas mRNA levels do not seem to change so much (Okui et al., 1999). This implies that either the translation of mRNA or the turn over of the protein are differentially regulated.

The pattern of expression of Mnb/Dyrk1A during vertebrate brain development is rather complex and interesting. Two waves of expression have been observed. In early embryos, it is expressed before the onset of neurogenesis in the three general locations where neuronal precursors originate: neuroepithelia of the neural tube, neural crest, and cranial placodes (Hämmerle et al., 2002). At the cellular level, this consist of a transient expression during a single cell cycle of progenitor cells at the transition from proliferating to neurogenic divisions. Very interestingly, Mnb/Dyrk1A mRNA is asymmetrically localized during the mitosis of these cells and inherited by one of the sibling cells after division (Hämmerle et al., 2002). A second wave of Mnb/Dyrk1A expression has been found in chick embryonic brain (Hämmerle et al., 2003). This takes place in the CNS of intermediate and late vertebrate embryos. In these stages, Mnb/Dyrk1A expression appears to take place during late stages of differentiation of certain populations of neurons such as spinal and brainstem motor neurons, cochlear and vestibular neurons of the rhombencephalon, parahippocampic, hippocampic, dorsolateral cortex, Purkinje cells in the cerebellum, and mytral cells of the olfactory bulb. The expression appears to be restricted to neurons since no consistent expression was detected in astroglial or oligodendroglial cells. Interestingly, the beginning of Mnb/Dyrk1A expression in these neurons seems to precede the onset of dendritic tree differentiation and is initiated by a transient translocation from the cytoplasm to the nucleus (Hämmerle et al., 2003). Afterwards, MNB/DYRK1A is transported to the growing dendritic tree, where it co-localizes with Dynamin 1, a possible MNB/DYRK1A substrate (Chen-Hwang et al., 2002). Although this study has been carried out only in chicken, the evolutionary conservation of Mnb/Dyrk1A in vertebrates makes its conclusions most probably valid for rodent and human brain too. In any case, the availability of a detailed spatio-temporal pattern of Mnb/Dyrk1A expression during the development of the mouse brain will be important for the detailed analysis of DS mouse models. Finally, it is interesting to point that although Mnb/Dyrk1A genes of Drosophila and mammals produce several isoforms by alternative splicing (Tejedor et al., 1995; Guimerá et al., 1999), there is no data available about differential spatio-temporal expression of splice isoforms within the brain. However, one of the human 5′ splice transcripts appears to be expressed only in heart and skeletal muscle as judged by Northern analysis (Guimerá et al., 1999).

Functional roles of Mnb/Dyrk1A on brain development

A large part of what is known about the function of Mnb/Dyrk1A comes from the study carried out in Drosophila. mnb mutants exhibit a reduced brain size, specially in the optic lobes and central brain hemispheres. This appears to be due to a decrease in the number of cells generated during the proliferative processes of postembryonic development as compared to wild type flies. In accordance to this, mnb is expressed in the proliferative centers of the larval brain and its loss of function causes alterations in the stereotyped arrangement of neuroblasts in these centers (Tejedor et al., 1995). Altogether, these data show the involvement of mnb in neurogenesis. During pupal stages mnb mutants exhibit an increased number of degenerating neurons in the optic lobe. Although this could be subsequent to the decrease in the number of neurons, it can not be rule out that it can be a consequence of alterations in neuronal differentiation. In addition, mnb mutants also exhibit behavioral defects (Heisenberg et al., 1985; Tejedor et al., 1995). Several of these phenotypes like the poor visual pattern fixation can be well correlated to the loss of neurons in given brain regions (optic lobes). However, other phenotypes like the poor odor-discrimination learning are more likely to be due to lack of mnb function in adult brain, particularly in the mushroom bodies (Tejedor et al., 1995). These results contrast to the genetic analysis of mbk-1, the closest Mnb/Dyrk1A related gene of C.elegans. mbk-1 mutants do not show any neurodevelopmental alterations. This is coherent with the developmental pattern of expression, since mbk-1 expression seems to begin after the proliferation has finished (Raich et al., 2003). Thus, these diverging functional data together with the comparative structural analysis (Galceran et al., 2003) make mbk-1 a more divergent homologue of Mnb/Dyrk1A.

Very interestingly, Dyrk1A$^{-/-}$ mice show a large reduction in the embryo size, which appears to be due to a developmental structural delay (Fotaki et al., 2002). This reduction does not seem to be as brain specific as in Drosophila. This probably reflects the fact that, as mentioned before, the expression of Mnb/Dyrk1A is more ubiquitous in vertebrates than in Drosophila. The brain of Dyrk1A$^{-/-}$ mice shows a reduced number of postmitotic neurons but the early embryonic lethality precluded the analysis of possible phenotypes in proliferation and neurogenesis (Fotaki et al., 2002). Nevertheless, haploinsufficient Dyrk1A$^{+/-}$ mice are viable and present a significant reduction in brain size which seems to be region-specific since it is more pronounced in midbrain and hindbrain than in forebrain, and in ventral (hypothalamus, pons, medulla oblongata) than in dorsal (neocortex and cerebellum) structures (Fotaki et al., 2002). Strikingly, these alterations do not seem to correlate with the developmental pattern of Mnb/Dyrk1A expression found by other authors (Song et al., 1996; Hämmerle et al., 2002). Despite the brain size reduction, most brain regions of Dyrk1A$^{+/-}$ mice lack clear changes in neuronal components and cytoarchitecture. Thus, the reduced brain size could be explained by a decrease in the neuropile areas rather than by a reduction of neuronal number. The impairment of neurite outgrowth produced by transfection of a mutant MNB/DYRK1A kinase in hippocampal neurons (Yang et al., 2001) fits well with this possibility and supports a role in neuronal differentiation. Furthermore, the localization of MNB/DYRK1A in the growing dendritic tree of differentiating neurons specifically points to a role in dendrite differentiation (Hämmerle et al., 2003).

Transgenic mice overexpressing Mnb/Dyrk1A have motor abnormalities and cognitive deficits (Altafaj et al., 2001). They exhibit a delay in the acquisition of mature locomotor patterns, hyperactivity in stressful situations, learning deficit related to hippocampal activity, and alterations in reference memory. The fact that these transgenic mice do not show gross neurohistological modifications indicates that the phenotypes are more probably due to alterations in functions that Mnb/Dyrk1A plays during postnatal stages and adulthood.

In summary the available information demonstrates that Mnb/Dyrk1A plays important roles in neurogenesis and neuronal differentiation during brain development and in learning/memory in adulthood. However, most of the molecular mechanisms underlying these functions remain to be unraveled.

Possible implications of MNB/DYRK1A in Down syndrome

The most commonly accepted hypothesis for the molecular origins of DS is that the extra dose of several HC21 genes contribute to DS phenotypes (Antonarakis, 2001). It is also widely considered that neurobiological alterations caused by DS are in a great part due to modifications in brain development processes such as neurogenesis, neuronal differentiation, myelinization, and synaptogenesis (reviewed by Becker et al., 1991; Coyle et al., 1986). Therefore, those HC21 genes with a possible role in brain development have been the focus of DS related research in recent years. Among them, MNB/DYRK1A appears as a very interesting candidate since it is located within the DSCR (Guimerá et al., 1996; Song et al., 1996), it is expressed in DS-related brain regions such as cortex, hippocampus, and cerebellum (Guimerá et al., 1996; Song et al., 1996; Okui et al., 1999), and it is overexpressed in DS tissues (Guimera et al., 1999).

In Table 1, we summarize those DS neurobiological disorders that may be related to alterations in MNB/DYRK1A functions. The reduced brain size of DS individuals is apparently due to the decrease in neuron number (Becker et al., 1991; Coyle et al., 1986). The role of MNB/DYRK1A in neurogenesis can account for this deficit. For instance, one can predict that due to the possible role as a cell determinant of neurogenesis in vertebrates (Hämmerle et al., 2002), the overexpression of MNB/DYRK1A in progenitor cells at early embryonic stages would produce premature switch of progenitor cells to neurogenic divisions. As schematically represented in Fig. 1, this should be accompanied by a shortening of the proliferative period and, consequently,

Table 1. Possible relation of Down Syndrome neurobiological disorders to alterations in MNB/DYRK1A functions

Neurodevelopmental alterations	Possible MNB/DYRK1A function
Reduced number of neurons	Neurogenesis
Dendritic atrophy	Dendrite development
Neuromuscular dysfunction	Distinct expression in motor neurons
Audiovestibular dysfunction	Distinct expression in chloquear and vestibular neurons
Visual dysfunction	Distinct expression in developing retina

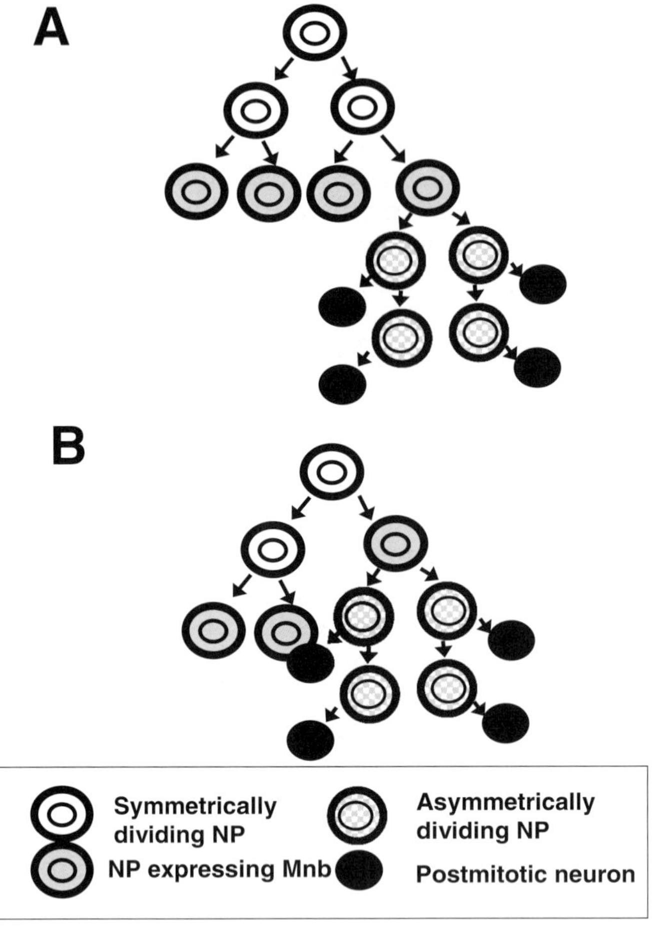

Fig. 1. A working model for the possible effect of the overexpression of MNB/DYRK1A on neurogenesis. **A** Schematic representation of the normal expression of MNB/DYRK1A during the cellular pattern of proliferation and neurogenesis. Founder progenitor cells divide symmetrically in the beginning, yielding more progenitor cells. At a given time in this sequence of division, MNB/DYRK1A is expressed in some progenitor cells, which in their next division switch to asymmetric (neurogenic) divisions to generate postmitotic neurons. In a simplified manner, the final number of cells (Tc) produced by a system with these two consecutive types of divisions is a function of the original number of founder cells (F) and the number of proliferative (P) and neurogenic (N) divisions following the equation: $Tc = F \times 2^{P} \times 2N$. **B** The overexpression of MNB/DYRK1A may cause a precocious switch to asymmetric divisions in some progenitor cells. This results in a reduced number of progenitor cells and, consequently, of postmitotic neurons. Based on this simplified model, the final number of cells (Tc) can be now estimated by introducing in the equation two new parameters: M, the number of progenitor cells that precociously switch to neurogenic divisions and R, the number of cycles that the precocious switch takes place in advance: $Tc = F \times (2^{P-M\times R} + M) \times 2N$

could result in a reduction of the neuronal progeny. The premature neurogenesis (Sweeney et al., 1989) and the reduced number of progenitor cells found in the ventricular neuroepithelium in trisomy 16 mice (Haydar et al., 2000), an animal model for DS, support this idea. It must be mentioned that mouse MNB/DYRK1A maps to chromosome 16 in the region of synteny with human chromosome 21.

A second function of MNB/DYRK1A that can be possibly related to DS neuropathologies is the involvement in dendrite development. It is known that cortical DS neurons exhibit dendritic shortening or atrophy (reviewed by Kaufman and Moser, 2000). This is due to alterations in dendrite growth during late development. It is well known that there is a close relationship between dendrite development and synaptogenesis. Not surprisingly, the dendritic field somehow determines the number and pattern of synaptic contacts received by each neuron (reviewed by Cline, 2001). Thus, a reduction in the dendritic field in DS brain will have an important impact in the neuronal connectivity and, consequently, in cognition.

Finally, the distinct expression of MNB/DYRK1A in neuronal groups such as motor neurons, chloquear/vestibular neurons, and in the developing retina may deserve the effort to study the possible relationship of MNB/DYRK1A to the increased incidence of neuromuscular, audiovestibular, and visual dysfunctions in DS individuals.

Altogether, the data already discussed help to build very attractive framework hypothesis for the developmental etiology of several DS neuropathologies based on alterations of MNB/DYRK1A kinase functions. Nevertheless, we should not forget, as previously mentioned, that there are relevant data supporting a function of MNB/DYRK1A on the adult brain and, therefore, it is likely that the alteration of this function may partially contribute to cognitive deficits. If so, this would open interesting possibilities for pharmacological intervention.

The challenge for the near future is to advance in the knowledge of the molecular mechanisms underlying MNB/DYRK1A kinase functions in order to provide further insights into the origins of DS neuropathologies and cognitive deficits.

References

Altafaj X, Dierssen M, Baamonde C, Marti E, Visa J, Guimera J, Oset M, Gonzalez JR, Florez J, Fillat C, Estivill X (2001) Neurodevelopmental delay, motor abnormalities and cognitive deficits in transgenic mice overexpressing Dyrk1A (minibrain), a murine model of Down's syndrome. Hum Mol Genet 10(18): 1915–1923

Antonarakis SE (2001) Chromosome 21: from sequence to applications. Curr Opin Genet Dev 11(3): 241–246

Becker W, Joost HG (1999) Structural and functional characteristics of Dyrk, a novel subfamily of protein kinases with dual specificity. Prog Nucl Acid Res Mol Biol 62: 1–17

Becker L, Mito T, Takashima S, Onodera K (1991) Growth and development of the brain in Down Syndrome. Prog Clin Biol Res 373: 133–152

Becker W, Weber Y, Wetzel K, Eirmbter K, Tejedor FJ, Joost HG (1998) Sequence characteristics, subcellular localization, and substrate specificity of DYRK-related kinases, a novel family of dual specificity protein kinases. J Biol Chem 273: 25893–25902

Chen-Hwang MC, Chen HR, Elzinga M, Hwang YW (2002) Dynamin is a minibrain kinase/dual specificity Yak1-related kinase 1A substrate. J Biol Chem 277(20): 17597–17604

Cline HT (2001) Dendritic arbor development and synaptogenesis. Curr Opin Neurobiol 11(1): 118–126

Coyle JT, Oster-Granite ML, Gearhart JD (1986) The neurobiologic consequences of Down syndrome. Brain Res Bull 16(6): 773–787

Fotaki V, Dierssen M, Alcantara S, Martinez S, Marti E, Casas C, Visa J, Soriano E, Estivill X, Arbones ML (2002) Dyrk1A haploinsufficiency affects viability and causes developmental delay and abnormal brain morphology in mice. Mol Cell Biol 22(18): 6636–6647

Galceran J, de Graaf K, Tejedor FJ, Becker W (2003) The MNB/DYRK1A protein kinase: genetic and biochemical properties. J Neural Transm [Suppl] (this volume)

Guimera J, Casas C, Pucharcos C, Solans A, Domenech A, Planas AM, Ashley J, Lovett M, Estivill X, Pritchard MA (1996) A human homologue of Drosophila minibrain (MNB) is expressed in the neuronal regions affected in Down syndrome and maps to the critical region. Hum Mol Genet 5(9): 1305–1310

Guimera J, Casas C, Estivill X, Pritchard M (1999) Human minibrain homologue (MNBH/DYRK1): characterization, alternative splicing, differential tissue expression, and overexpression in Down syndrome. Genomics 57: 407–418

Hämmerle B, Vera E, Spreicher S, Arencibia R, Martínez S, Tejedor FJ (2002) Mnb/Dyrk1A is transiently expressed and asymmetrically segregated in neural progenitor cells at the transition to neurogenic divisions. Dev Biol 246: 259–273

Hämmerle B, Carnicero A, Elizalde C, Cerón J, Martínez S, Tejedor FJ (2003) Expression patterns and subcellular localization of the Down Syndrome candidate protein MNB/DYRK1A suggest a role in late neuronal differentiation. Eur J Neurosci 17: 2277–2286

Hattori M, Fujiyama A, Taylor TD, Watanabe H, Yada T, Park HS, Toyoda A, Ishii K, Totoki Y, Choi DK, Groner Y, Soeda E, Ohki M, Takagi T, Sakaki Y, Taudien S, Blechschmidt K, Polley A, Menzel U, Delabar J, Kumpf K, Lehmann R, Patterson D, Reichwald K, Rump A, Schillhabel M, Schudy A, Zimmermann W, Rosenthal A, Kudoh J, Schibuya K, Kawasaki K, Asakawa S, Shintani A, Sasaki T, Nagamine K, Mitsuyama S, Antonarakis SE, Minoshima S, Shimizu N, Nordsiek G, Hornischer K, Brant P, Scharfe M, Schon O, Desario A, Reichelt J, Kauer G, Blocker H, Ramser J, Beck A, Klages S, Hennig S, Riesselmann L, Dagand E, Haaf T, Wehrmeyer S, Borzym K, Gardiner K, Nizetic D, Francis F, Lehrach H, Reinhardt R, Yaspo ML; The chromosome 21 mapping and sequencing consortium (2000) The DNA sequence of human chromosome 21. Nature 405: 311–309

Haydar TF, Nowakowski RS, Yarowsky PJ, Krueger BK (2000) Role of founder cell deficit and delayed neuronogenesis in microencephaly of the trisomy 16 mouse. J Neurosci 20(11): 4156–4164

Heisenberg M, Borst A, Wagner S, Byers D (1985) Drosophila mushroom body mutants are deficient in olfactory learning. J Neurogenet 2(1): 1–30

Kaufmann WE, Moser HW (2000) Dendritic anomalies in disorders associated with mental retardation. Cereb Cortex 10(10): 981–991

Korenberg JR, Chen XN, Schipper R, Sun Z, Gonsky R, Gerwehr S, Carpenter N, Daumer C, Dignan P, Disteche C (1994) Down syndrome phenotypes: the consequences of chromosomal imbalance. Proc Natl Acad Sci USA 91: 4997–5001

Marti E, Altafaj X, Dierssen M, de la Luna S, Fotaki V, Alvarez M, Perez-Riba M, Ferrer I, Estivill X (2003) Dyrk1A expression pattern supports specific roles of this kinase in the adult central nervous system.Brain Res 964(2): 250–263

Okui M, Ide T, Morita K, Funakoshi E, Ito F, Ogita K, Yoneda Y, Kudoh J, Shimizu N (1999) High-level expression of the Mnb/Dyrk1A gene in brain and heart during rat early development. Genomics 62: 165–171

Rahmani Z, Blouin JL, Creau-Goldberg N, Watkins PC, Mattei JF, Poissonnier M, Prieur M, Chettouh Z, Nicole A, Aurias A, Sinet PM, Delabar JM (1989) Critical role of the D21S55 region on chromosome 21 in the pathogenesis of Down syndrome. Proc Natl Acad Sci USA 86: 5958–5962
Raich WB, Moorman C, Lacefield CO, Lehrer J, Bartsch D, Plasterk RH, Kandel ER, Hobert O (2003) Characterization of caenorhabditis elegans homologs of the Down syndrome candidate gene DYRK1A. Genetics 163: 571–580
Song WJ, Sternberg LR, Kasten-Sportes C, Keuren ML, Chung SH, Slack AC, Miller DE, Glover TW, Chiang PW, Lou L, Kurnit DM (1996) Isolation of human and murine homologues of the Drosophila minibrain gene: human homologue maps to 21q22.2 in the Down syndrome "critical region". Genomics 38: 331–339
Sweeney JE, Hohmann CF, Oster-Granite ML, Coyle JT (1989) Neurogenesis of the basal forebrain in euploid and trisomy 16 mice: an animal model for developmental disorders in Down syndrome. Neuroscience 31(2): 413–425
Tejedor F, Zhu XR, Kaltenbach E, Ackermann A, Baumann A, Canal I, Heisenberg M, Fischbach KF, Pongs O (1995) Minibrain: a new protein kinase family involved in postembryonic neurogenesis in Drosophila. Neuron 14: 287–301
Yang EJ, Ahn YS, Chung KC (2001) Protein kinase Dyrk1 activates cAMP response element-binding protein during neuronal differentiation in hippocampal progenitor cells. J Biol Chem 276(43): 39819–39824

Authors' address: F. J. Tejedor, Instituto de Neurociencias, Unidad de Neurobiologia del Desarrollo, CSIC y Universidad Miguel Hernandez, Campus de San Juan, 03550 San Juan (Alicante), Spain, e-mail: f.tejedor@umh.es

The MNB/DYRK1A protein kinase: genetic and biochemical properties*

J. Galceran[1], **K. de Graaf**[2], **F. J. Tejedor**[1], and **W. Becker**[2]

[1] Instituto de Neurociencias, Unidad de Neurobiologia del Desarrollo, CSIC y Universidad Miguel Hernandez, Campus de San Juan, San Juan (Alicante), Spain
[2] Institut für Pharmakologie und Toxikologie, Medizinische Fakultät, RWTH Aachen, Aachen, Germany

Summary. The "Down syndrome critical region" of human chromosome 21 has been defined based on the analysis of rare cases of partial trisomy 21. Evidence is accumulating that DYRK1A, one of the 20 genes located in this region, is an important candidate gene involved in the neurobiological alterations of Down syndrome. Both the structure of the DYRK1A gene and the sequence of the encoded protein kinase are highly conserved in evolution. The protein contains a unique assembly of structural motifs outside the catalytic domain, including a nuclear localization signal, a PEST region, and a repeat of 13 consecutive histidines. MNB/DYRK1A* and related kinases are unique among serine/threonine-specific protein kinases in that their activity depends on tyrosine autophosphorylation in the catalytic domain. Also, evidence is accumulating that mRNA levels of MNB/DYRK1A are subject to tight regulation. A number of putative substrates of MNB/DYRK1A have emerged in the recent years, the majority of them being transcription factors. Although the function of MNB/DYRK1A in intracellular signalling and regulation of cell function is still poorly defined, current evidence suggests that the kinase may play a role in the regulation of gene expression.

Introduction

Human chromosome 21 trisomy (HC21) is the cause of Down syndrome (DS), the major cause of mental retardation with genetic origin in humans. Down syndrome patients present a wide range of alterations that include mental retardation, congenital heart disease, early onset of Alzheimer's disease and characteristic facial traits. Human chromosome 21 harbours more than 235

---

* DYRK1A is the official gene symbol approved by the HUGO Gene Nomenclature Committee. Nevertheless, MNB has been also frequently used by several authors. Thus, we use MNB/DYRK1A in this review to refer to these orthologous genes in contrast to other members of the DYRK family of protein kinases.

genes (http://www.rzpd.de/general/html/Chrom21/chr21_mar_2002.html). It is plausible that DS is the result of the overexpression of multiple individual genes, each one contributing to multiple or singular aspects of the syndrome. The genetic analysis of rare patients with partial trisomy 21 has allowed the definition of the so-called "Down syndrome critical region" (DSCR) as the minimal chromosomal region that, when present in three copies, would be necessary and sufficient to cause a major portion of the DS phenotype (Delabar et al., 1993). The DSCR encompasses a genetic interval of 2.5 Mb from CBR1 [carbonyl reductase (NADPH) 1] to ERG [transforming protein ERG] and comprises approximately 20 annotated or predicted genes and 5 pseudogenes. DYRK1A is one of the genes (SIM2 and ETS2 being the other ones) in the DSCR that have been analysed by means of transgenic mouse models for their potential relevance in DS. Transgenic mice overexpressing DYRK1A show alterations in brain function, e.g. learning behaviour (Smith et al., 1997; Altafaj et al., 2001), suggesting that its overexpression in trisomy of human chromosome 21 may contribute to mental retardation in DS. DYRK1A is the human ortholog of the Drosophila minibrain (mnb) gene. mnb mutants show reduction of up to 50% of the volume of specific regions of the brain (Tejedor et al., 1995). This phenotype indicates the relevance of MNB/DYRK1A in the formation of the central nervous system, as reviewed in the previous article of this issue (Hämmerle et al., 2003b). Here we summarize the present knowledge about the molecular biology of MNB/DYRK1A and its biochemical properties.

Genomic organization of the MNB/DYRK1A gene

DYRK1A is located between DSCR3 and KCNJ6 in the DSCR of human chromosome 21 and in the syntenic region of mouse chromosome 16. The gene comprises 11 coding and 3 non-coding exons and spans 150 kb (Wang et al., 1998; Guimerá et al., 1999). Remarkably, the exon/intron distribution of the coding regions is identical in the MNB/DYRK1A genes from mammals and fish, and is still largely conserved in Drosophila (Fig. 1A). Two alternative promoters in the human gene direct the expression of transcripts with different 5′-untranslated regions. Exon 1-containing transcripts are restricted to skeletal muscle and heart, whereas exon 2-containing transcripts were found to be ubiquitously expressed in all tissues tested, including 14 different parts of the brain (Guimerá et al., 1999). A second site of alternative splicing affects exon 6 and results in the presence of two variants of MNB/DYRK1A that either contain or lack a segment of 9 amino acids (Kentrup et al., 1996; Guimerá et al., 1999). So far, no functional difference between these splicing variants is known.

Structural characteristics of the MNB/DYRK1A gene product

MNB/DYRK1A encodes a protein kinase with an unusual domain structure (Fig. 1). The MNB/DYRK1A protein contains a catalytic domain with the typical features of serine/threonine-specific protein kinases (Kentrup et al., 1996) and is distantly related with the MAP kinase family. Within the protein kinase superfamily, DYRK1A and MNB belong to a small family of protein kinases (DYRK family) that are characterized by specific sequence motifs in the catalytic domain (Kentrup et al., 1996; Becker et al., 1998; Becker and Joost, 1999). MNB/DYRK1A harbours several unusual structural features outside the catalytic domain (Fig. 1). The amino-terminal domain contains a nuclear localization signal which has been shown to target DYRK1A to the nucleus (Song et al., 1997; Becker et al., 1998). Inside the nucleus, MNB/DYRK1A exhibits a punctate pattern of staining that resembles the subnuclear localization of splicing factors (Becker et al., 1998; Marti et al., 2003; Hämmerle et al., 2003a). Directly adjacent to the catalytic core domain, all kinases of the DYRK family contain a conserved sequence element which is particularly rich in acidic residues (DYRK homology box, DH box; Becker and Joost 1999). Molecular modelling suggests that the DH box and the catalytic domain form a common tertiary structure (Himpel et al., 2001). A PEST region (rich in proline, glutamic acid, serine and threonine) is located carboxy-terminal of the catalytic domain. This motif is believed to signal proteins for rapid degradation, however this hypothesis has so far not been tested for MNB/DYRK1A. Furthermore, a stretch of 13 consecutive histidine residues and a serine/threonine repeat are also conserved in the vertebrate orthologs. These repeats are not present in the closely homologue DYRK1B, which is encoded by a gene located on human chromosome 19 (19q13.2) and therefore is not affected by HC21 trisomy (Leder et al., 1999).

The sequence of the catalytic domain plus the DH-box and the nuclear localization signal is well conserved in MNB/DYRK1A orthologues in some invertebrates (Fig. 1B), suggesting that this kinase has an evolutionarily conserved function in cellular regulation. Interestingly, Ciona MNB/DYRK1A exhibits some similarity in the PEST region (Fig.1B), and Drosophila MNB contains a GAS domain which is supposed to play in invertebrates a similar role to the PEST domains in vertebrates (Tejedor et al., 1995). MNB/DYRK1A orthologs from invertebrates (Drosophila, Ciona) lack the histidine repeat and the Ser/Thr repeats but show significant sequence conservation of the C-terminus (Fig. 1B). Recently, a Caenorhabditis elegans homologue of MNB/DYRK1A has been characterized: mbk-1 (Raich et al., 2003). The sequence of MBK-1 is remarkably conserved in the catalytic domain and the DH-box (78% identity with human MNB/DYRK1A), but differs in the nuclear localization signal and the C-terminus. Taking also into account its different functional role (see Hämmerle et al., 2003b for a discussion of this issue), MBK-1 appears to be a more divergent homologue of MNB/DYRK1A.

Regulation of MNB/DYRK1A

The tight control of catalytic activity is crucial for the proper function of protein kinases. The knowledge of these regulatory mechanisms can provide

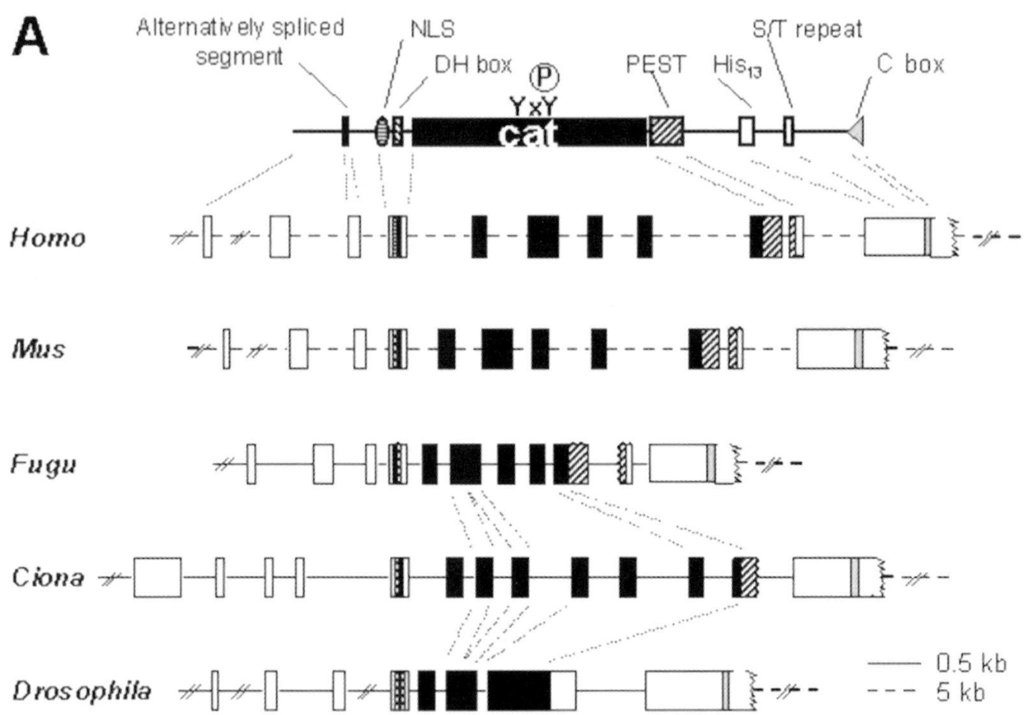

B

```
                          NLS                          DH box
Human       VYYAKKKRRHQQGQGDD.SSHKKERK.VY.NDGYDDDNYDYIVKNGEK
Chicken     VYYAKKKRRHQQGQGDD.SSHKKERK.VY.NDGYDDDNYDYIVKNGEK
Xenopus     VYYAKKKRRHQQGQGDD.SSHKKERK.VY.NDGYDDDNYDYIVKNGEK
Fugu        VYYAKKKRRHQQGQGED.SSHKKERK.VF.NDGYDDDNYDYIVKNGEK
Ciona       VYYAKKKRRAQQQSSEE.TRPQKERR.VF.NDGYDDVNYDYIIKNGEK
Drosophila  VYYAKKKRRAQQTQGDDDSSNKKERK.LY.NDGYDDDNHDYIIKNGEK
```

```
                   PEST-rich region
Human       TADEGTNTSNSVSTSPAMEQSQSSGTTSSTSSSSGGSSGTSNSG
Chicken     TADEGTNTSNSVSTSPAMEQSQSSGTTSSTSSSSGGSSGTSNSG
Xenopus     TADEGTNTSNSVSTSPAMEQSQSSGTTSSTSSSSGGSSGTSNSG
Fugu        TADEGTNTSSSVSTSPALEQSQSSGTTSSTSSSSGGSSGTSTSG
Ciona       TADGCTNTAAAPAQDLSSSSTGSSSSSGRARSDPGAHPVFTTTA
```

```
                Histidine repeat                       Serine/Threonine repeat
Human       NALHHHHGNSSHHHHHHHHHHHHHGQQAL<25>HSHHSMTSLSSSTTSSSTSSSSTGN
Chicken     NALHHHHGNSSHHHHHHHHHHHHHGQQAL<25>HSHHSMTSLASSTTSSSTSSSSTGN
Xenopus     NVPHHH.GNGS...HHHHHHHHHHGQHVL<25>HSQHSMTSLSSSNTSSSTSSSSTGN
Fugu        KALHPH....S......HTHHHHHGQ.MM<25>HGHLSMTSLSSSASSSSTSSSSTGN
```

```
               C-terminus
Human       EESPMTGVCVQQSPVASS
Chicken     EESPMTGVCVQQSPVASS
Xenopus     EDSPMTGVCVQHSPVASS
Fugu        EESPMTGVCVQQSSMASS
Ciona       QDSPMTGVQVQQSPVPSS
Drosophila  SSSPMVGVCVQQNPVVIH
```

important insights into the role of a kinase in cellular signalling. The full activation of MNB/DYRK1A (Himpel et al., 2001) and other DYRK-related kinases (DYRK1B, DYRK3, and Yak1p; Wiechmann et al., 2003 and references therein) depends on the autophosphorylation of the second tyrosine of the Tyr-X-Tyr motif in the so-called activation loop (Kentrup et al., 1996; Himpel et al., 2001). This activation mechanism by tyrosine autophosphorylation differs strikingly from that of the MAP kinases, which are activated by dual phosphorylation of a Thr-x-Tyr motif by specific upstream kinases. Recent data suggest that the tyrosine autophosphorylation of DYRK1A may be subject to regulation by extracellular signals. In the hippocampal progenitor cell line H19-7, tyrosine phosphorylation and kinase activity of DYRK1A was reported to be induced by fibroblast growth factor (Yang et al., 2001). A different mode of regulation has been described by Zhang et al. (2001) who reported that MNB/DYRK1A interacts with the adenovirus E1A oncogen and found a stimulatory effect of E1A on the kinase activity in vitro. Further work will be necessary to precisely define the signalling pathway(s) upstream of MNB/DYRK1A.

In addition to this short term control of enzymatic activity, MNB/DYRK1A appears also to be regulated at the level of gene expression. Approaches of global gene expression analysis have identified changes in the abundance of MNB/DYRK1A mRNA in several experimental systems of cellular differentiation and proliferation, e.g. in activated T-cells, in human neutrophils exposed to bacteria, and in the course of melanoma tumour progression. Also, it is interesting that MNB/DYRK1A is transiently expressed during a single cell cycle of neural progenitor cells of early developing brain at the transition from proliferative to neurogenic divisions (Hämmerle et al., 2002a). Although the relevant regulatory mechanisms remain unknown, these studies clearly indicate that mRNA levels of MNB/DYRK1A are subject to tight regulation.

Fig. 1. Comparative gene and protein structure of MNB/DYRK1A. A Schematic representation of the human DYRK1A protein and alignment of the gene structures from human (Homo), mouse (Mus), pufferfish (Fugu), sea squirt (Ciona), and fruitfly (Drosophila). Coding exons are depicted as boxes and introns by lines. Correspondence between protein sequence and coding exons is marked by dotted lines. Human and mouse introns are drawn at a 10 fold smaller scale than the rest of the sequences. **B** Evolutionary conservation of sequence motifs outside the catalytic domain. Alignments of the PEST-rich region, the sequence around the histidine repeat (His_{13}), and the serine/threonine repeat (S/T repeat) are based on published cDNA sequences of human and chicken DYRK1A (Guimerá et al., 1996; Hämmerle et al., 2002), genomic sequences from Drosophila (Tejedor et al., 1995), Fugu and Ciona intestinalis, and expressed sequence tags from Xenopus laevis. The activating phosphorylation of the Tyr-x-Tyr (YxY) motif in the catalytic domain (cat) is indicated. Shaded boxes indicate the defining residues of the bipartite nuclear localization signal (NLS) and the DH box

Downstream effectors of MNB/DYRK1A

The identification of the physiological substrates of a protein kinase is a necessary step towards an understanding of its biological function. Except for its ability to autophosphorylate on tyrosine, MNB/DYRK1A phosphorylates only serine and threonine residues in substrate peptides or proteins (Himpel et al., 2000). In assays with synthetic peptides, MNB/DYRK1A resembles the MAP kinases and cyclin-dependent kinases in its preference for phosphorylation sites with a proline in the position C-terminal of the target residue (Himpel et al., 2000). The presence of an arginine residue two or three positions N-terminal of the phosphorylation site appears to be a frequent feature of the target site (Himpel et al., 2000; Woods et al., 2001b) although it is absent in some proposed MNB/DYRK1A substrates (e.g., STAT3, CREB).

Several putative substrates of MNB/DYRK1A have been identified in recent years (Table 1). Four of these proteins are transcription factors (FKHR, CREB, STAT3, GLI1), which exhibit enhanced activity upon phosphorylation. All of these proteins play key roles in cell development. Different mechanisms have been described by which MNB/DYRK1A modulates the activity of these transcription factors. FKHR is an important regulator of cell survival and proliferation. MNB/DYRK1A phosphorylates FKHR specifically on Ser-329, an in vivo phosphorylation site that is conserved in related homologues (FKHR-L1 and AFX; Woods et al., 2001a). However, stimulation of FKHR-dependent promoter activity was found to be independent of MNB/DYRK1A's kinase activity (von Groote-Bidlingmaier et al., 2003). This surprising observation suggests that MNB/DYRK1A may act as a scaffolding protein and ascribes a potential function to its non-catalytic domain. CREB is a transcription factor that plays a role in neuronal development and differentiation, e.g. synaptic plasticity. Yang et al. (2001) reported that MNB/DYRK1A directly phosphorylates CREB and thereby promote neuronal differentiation of the hippocampal progenitor cell line, H19-7. STAT3 is a downstream effector of cytokines such as interleukin-6 (IL-6), leukemia inhibitory factor (LIF), or oncostatin M, and in many systems controls genes that determine the decision from cell growth to differentiation. STAT3 is phosphorylated by MNB/DYRK1A on Ser-727 in the transactivation domain (Matsuo et al., 2001; Wiechmann et al., 2003). Phosphorylation of Ser-727 is known to modulate the transcriptional activity of STAT3, but the effect of MNB/DYRK1A on STAT3-dependent promoter activity remains to be shown. GLI proteins are key downstream signaling components of the *hedgehog* signalling pathway, which controls cell proliferation and pattern formation. GLI1-dependent promoter activity was enhanced by MNB/DYRK1A in part through retaining GLI1 in the nucleus, but also by stimulating GLI1 transcriptional activity directly (Mao et al., 2002). In all these cases, further work will be necessary to define the role of MNB/DYRK1A in transcriptional regulation in more detail.

Three cytoplasmic proteins have also been shown to be in vitro-substrates of MNB/DYRK1A: the protein-synthesis initiation factor eIF2Bε, the microtubule-associated protein tau, and dynamin 1 (Table 1). In tau and eIF2Bε,

Table 1. Substrates and binding partners of DYRK1A

Substrate or Interacting Protein	Gene symbol	Substrates		Interaction	Effect	References
		Evidence	P-site[a]	Evidence[b]		
Forkhead in rhabdomyosarcoma (FKHR)	FOXO1A	in vitro	Ser-329	Co-IP	Stimulation of nuclear export[c] Stimulation of transcription[d]	Woods et al. (2001a) von Groote-Bidlingmaier et al. (2003)
Signal transducer and activator of transcription 3 (STAT3)	STAT3	in vivo[e]	Ser-727	—	—	Matsuo et al. (2001); Wiechmann et al. (2003)
cAMP responsive element binding protein (CREB)	CREB1	in vitro in vivo[e]	Ser-133	Y2H, Co-IP	Stimulation of transcription[d]	Yang et al. (2001)
Glioma-associated oncogene (GLI1)	GLI	in vitro	n.d.	—	Stimulation of transcription[d]	Mao et al. (2002)
Adenovirus E1A oncoprotein		—	—	Pulldown-assay	E1A stimulates activity of DYRK1A	Zhang et al. (2001)
eIF2Bε	EIF2B5	in vitro	Ser-539	—	Priming phosphorylation for GSK3	Woods et al. (2001b)
tau	MAPT	in vitro	Thr-212	—	Priming phosphorylation for GSK3	Woods et al. (2001b)
Dynamin	DNM1	in vitro	n.d.	Pulldown assay	Regulation of protein/protein interaction	Cheng-Hwang et al. (2002)

[a] phosphorylation site (n.d., not determined); [b] Co-IP, co-immunoprecipitation; Y2H, yeast-two-hybrid system; [c] enhanced nuclear localization of a point mutant of Ser-329; [d] co-transfection of DYRK1A causes enhanced activity in reporter gene assays; [e] co-transfection of DYRK1A causes enhanced phosphorylation

phosphorylation by MNB/DYRK1A creates recognition sites for subsequent phosphorylation by glycogen synthase kinase 3 (GSK3). Notably, tau and eIF2Bε are equally well phosphorylated by DYRK2 and may thus be substrates of this cytoplasmic kinase rather than of MNB/DYRK1A (Woods et al., 2001b). The phosphorylation of dynamin 1 by MNB/DYRK1A modulates its capacity to interact with components of the endocytotic apparatus. (Cheng-Hwang et al., 2002). Interestingly, MNB/DYRK1A and dynamin 1 colocalize in the growing dendritic tree of differentiating neurons. This finding suggests that phosphorylation of dynamin 1 by MNB/DYRK1A may be related to the molecular processes underlying neuronal differentiation (Hämmerle et al., 2003a). This fits with recent evidence indicating that a fraction of the MNB/DYRK1A molecules is located outside the nucleus (Hämmerle et al., 2003a; Marti et al., 2003). It is likely that new putative substrates of MNB/DYRK1A will be identified; and the task will be to determine which of them are the key mediators of its functions on cell differentiation and proliferation.

Conclusions and perspectives

Among the genes of the DSCR, the consideration of MNB/DYRK1A as a "candidate gene" for mental retardation is relatively well supported by the phenotype of transgenic mice that overexpress it, and by the analysis of its neurodevelopmental roles (Hämmerle et al., 2003b). In recent years, our understanding of the molecular function of MNB/DYRK1A has been enhanced by the identification of putative substrates and by the elucidation of its mode of activation. The present knowledge provides a starting point for future studies aimed to establish a link between the function of MNB/DYRK1A in cellular signalling and its role in brain development and brain function, and eventually to substantiate the presumed functional consequences of its overexpression in Down syndrome.

References

Altafaj X, Dierssen M, Baamonde C, Marti E, Visa J, Guimerá J, Oset M, Gonzalez JR, Florez J, Fillat C, Estivill X (2001) Neurodevelopmental delay, motor abnormalities and cognitive deficits in transgenic mice overexpressing Dyrk1A (minibrain), a murine model of Down's syndrome. Hum Mol Genet 10: 1915–1923

Becker W, Joost HG (1999) Structural and functional characteristics of Dyrk, a novel subfamily of protein kinases with dual specificity. Progr Nucl Acids Res Mol Biol 62: 1–17

Becker W, Weber Y, Wetzel K, Eirmbter K, Tejedor FJ, Joost HG (1998) Sequence characteristics, subcellular localization, and substrate specificity of DYRK-related kinases, a novel family of dual specificity protein kinases. J Biol Chem 273: 25893–25902

Chen-Hwang MC, Chen HR, Elzinga M, Hwang YW (2002) Dynamin is a minibrain kinase/dual specificity Yak1-related kinase 1A substrate. J Biol Chem 277: 17597–17604

Delabar JM, Theophile D, Rahmani Z, Chettouh Z, Blouin JL, Prieur M, Noel B, Sinet PM (1993) Molecular mapping of twenty-four features of Down syndrome on chromosome 21. Eur J Hum Genet 1: 114–124

Guimerá J, Casas C, Estivill X, Pritchard M (1999) Human minibrain homologue (MNBH/DYRK1): characterization, alternative splicing, differential tissue expression, and overexpression in Down syndrome. Genomics 57: 407–418

Hämmerle B, Vera-Samper E, Speicher S, Arencibia R, Martinez S, Tejedor FJ (2002) Mnb/Dyrk1A is transiently expressed and asymmetrically segregated in neural progenitor cells at the transition to neurogenic divisions. Dev Biol 246: 259–273

Hämmerle B, Carnicero A, Elizalde C, Cerón J, Martínez S, Tejedor FJ (2003a) Expression patterns and subcellular localization of the Down Syndrome candidate protein MNB/DYRK1A suggest a role in late neuronal differentiation. Eur J Neurosci 17: 2277–2286

Hämmerle B, Elizalde C, Galceran J, Becker W, Tejedor FJ (2003b) The MNB/DYRK1A protein kinase: neurobiological functions and Down Syndrome implications. J Neural Transm [Suppl] 67 (this volume)

Himpel S, Tegge W, Leder S, Joost HG, Becker W (2000) Specificity determinants of substrate recognition by the protein kinase DYRK1A. J Biol Chem 275: 2431–2438

Himpel S, Panzer P, Eirmbter K, Czajkowska H, Sayed M, Packman LC, Blundell T, Kentrup H, Grötzinger J, Joost HG, Becker W (2001) Identification of the autophosphorylation sites and characterization of their effects in the protein kinase DYRK1A. Biochem J 359: 497–505

Kentrup H, Becker W, Heukelbach J, Wilmes A, Schurmann A, Huppertz C, Kainulainen H, Joost HG (1996) Dyrk, a dual specificity protein kinase with unique structural features whose activity is dependent on tyrosine residues between subdomains VII and VIII. J Biol Chem 271: 3488–3495

Leder S, Weber Y, Altafaj X, Estivill X, Joost HG, Becker W (1999) Cloning and characterization of DYRK1B, a novel member of the DYRK family of protein kinases. Biochem Biophys Res Commun 254: 474–479

Mao J, Maye P, Kogerman P, Tejedor FJ, Toftgard R, Xie W, Wu G, Wu D (2002) Regulation of Gli1 transcriptional activity in the nucleus by Dyrk1. J Biol Chem 277: 35156–35161

Marti E, Altafaj X, Dierssen M, de la Luna S, Fotaki V, Alvarez M, Perez-Riba M, Ferrer I, Estivill X (2003) Dyrk1A expression pattern supports specific roles of this kinase in the adult central nervous system. Brain Res 964: 250–263

Matsuo,R, Ochiai W, Nakashima K, Taga T (2001) A new expression cloning strategy for isolation of substrate-specific kinases by using phosphorylation site-specific antibody. J Immunol Methods 247: 141–151

Raich WB, Moorman C, Lacefield CO, Lehrer J, Bartsch D, Plasterk RH, Kandel ER, Hobert O (2003) Characterization of Caenorhabditis elegans homologs of the Down syndrome candidate gene DYRK1A. Genetics 163: 571–580

Smith DJ, Stevens ME, Sudanagunta SP, Bronson RT, Makhinson M, Watabe AM, O'Dell TJ, Fung J, Weier H-UG, Cheng J-F, Rubin EM (1997) Functional screening of 2 Mb of human chromosome 21q22.2 in transgenic mice implicates minibrain in learning defects associated with Down syndrome. Nat Genet 16: 28–36

Song W-J, Chung S-H, Kurnit DM (1997) The murine Dyrk protein maps to chromosome 16, localizes to the nucleus, and can form multimers. Biochem Biophys Res Commun 231: 640–644

Tejedor F, Zhu XR, Kaltenbach E, Ackermann A, Baumann A, Canal I, Heisenberg M, Fischbach KF, Pongs O (1995) minibrain: a new protein kinase family involved in postembryonic neurogenesis in Drosophila. Neuron 14: 287–301

von Groote-Bidlingmaier F, Schmoll D, Orth HM, Joost HG, Becker W, Barthel A (2003) DYRK1 is a coactivator of FKHR (FOXO1a)-dependent glucose-6-phosphatase gene expression. Biochem Biophys Res Commun 300: 764–769

Wang J, Kudoh J, Shintani A, Minoshima S, Shimizu N (1998) Identification of two novel 5′ noncoding exons in human MNB/DYRK gene and alternatively spliced transcripts. Biochem Biophys Res Commun 250: 704–710

Wiechmann S, Czajkowska H, de Graaf K, Grötzinger J, Joost HG, Becker W (2003) Unusual function of the activation loop in the protein kinase DYRK1A. Biochem Biophys Res Commun 302: 403–408

Woods YL, Rena G, Morrice N, Barthel A, Becker W, Guo S, Unterman TG, Cohen P (2001a) The kinase DYRK1A phosphorylates the transcription factor FKHR at Ser329 in vitro, a novel in vivo phosphorylation site. Biochem J 355: 597–607

Woods YL, Cohen P, Becker W, Jakes R, Goedert M, Wang X, Proud CG (2001b) The kinase DYRK phosphorylates protein-synthesis initiation factor eIF2Bepsilon at Ser539 and the microtubule-associated protein tau at Thr212: potential role for DYRK as a glycogen synthase kinase 3-priming kinase. Biochem J 355: 609–615

Yang EJ, Ahn YS, Chung KC (2001) Protein kinase Dyrk1 activates cAMP response element-binding protein during neuronal differentiation in hippocampal progenitor cells. J Biol Chem 276: 39819–39824

Zhang Z, Smith MM, Mymryk JS (2001) Interaction of the E1A oncoprotein with Yak1p, a novel regulator of yeast pseudohyphal differentiation, and related mammalian kinases. Mol Biol Cell 12: 699–710

Authors' address: W. Becker, Institut für Pharmakologie und Toxikologie, Medizinische Fakultät der RWTH Aachen, Wendlingweg 2, D-52074 Aachen, Germany, e-mail: walter.becker@post.rwth-aachen.de

Cytoskeleton derangement in brain of patients with Down Syndrome, Alzheimer's disease and Pick's disease

D. Pollak[1], **N. Cairns**[2], and **G. Lubec**[1]

[1] Department of Pediatrics, Division Basic Sciences, University of Vienna, Vienna, Austria
[2] Department of Neuropathology, King's College, London, United Kingdom

Summary. Although cytoskeleton derangement has been reported in brain of patients with neurodegenerative disorders, basic information on integral constituents forming this network including stoichiometric composition is missing. It was therefore the aim of the study to qualitatively and quantitatively evaluate individual proteins of the three major classes representing the cytoskeleton of human brain.

Cytoskeleton proteins β-actin (βA), alpha-actinin (Act), tubulin beta-III (βIII), microtubule associated protein 1 (MAP1), neurofilaments NF-L, NF-M and NF-H and neuron specific enolase (NSE), a marker for neuronal density, were determined by immunoblotting. Brain samples (frontal cortex) of controls (CO), patients with Down Syndrome (DS), Alzheimer's disease (AD) and Pick's disease (PD) were used for the study.

In DS brain βIII, NF-H and NF-M, in AD brain NF-M and NF-H and in PD brain NF-L, NF-M and NF-H were significantly reduced. Stoichiometry of cytoskeleton proteins in control brain revealed the following relations:

βA : Act : βIII : MAP1 : NF-L : NF-M : NF-H = 1.0 : 0.8 : 3.8 : 2.4 : 3.2 : 2.2.

This stoichiometrical ratios were aberrant in DS, AD and PD with the main outcome that ratios of members of the neurocytoskeleton (βIII, NF's) in relation to βA were remarkably decreased. This finding confirms data of decreased neuronal density using NSE in DS and AD.

We propose stoichiometry of cytoskeleton elements in normal brain and confirm and extend knowledge on cytoskeleton defects in neurodegenerative diseases. The finding of significantly decreased individual elements may well lead to or represent disassembly of the neurocytoskeleton observed in neurodegenerative diseases.

Introduction

The cytoskeleton of neurons comprises three interacting structural and dynamic complexes: microfilaments (MFs), microtubules (MTs) and neurofilaments (NFs) (Siegel, 1999). MFs are built from 43-kDa actin monomers arranged like two strings of pearls intertwined into fibrils. The core structure of MTs is a polymer of 50-kDa tubulin subunits with heterodimers of alpha- and beta-tubulin aligning end to end to form a protofilament in the shape of a hollow tube. Three subunit proteins represent the NF-triplet: NF high-molecular-weight-subunit (NF-H, 180–200 kDa), NF middle-molecular-weight-subunit (NF-M, 130–170 kDa) and NF low-molecular-weight-subunit (NF-L, 60–70 kDa) (Siegel, 1999).

Furthermore, a series of cytoskeletal-associated proteins serving as crosslinkers as well as contributing to several of the cytoskeleton's most essential functions, have already been identified. To name but two, alpha-actinin, a polypeptide of 100-kDa represented by several isoforms, acts as F-actin bundling protein and MAP1 belongs to the heterogenous collection of microtubule-associated proteins involved in the assembly and stabilisation of microtubules and in the interaction of microtubules with other cytoskeletal elements (Doering, 1994).

While each set of cytoskeletal elements has a distinctive spectrum of composition, stability and distribution, they interact in the crucial events of neuronal morphogenesis and establishment of the structural plasticity of the brain regulating the balance between stability and motility in neuronal structures (Kaech et al., 2001). Actin-containing MFs are associated with structural plasticity, both, during development when their dynamic activity drives the exploratory activity of growth cones and after circuit formation when actin-rich dendritic spines of excitatory synapses retain the capacity for rapid changes in morphology (Kaech et al., 2001).

Cytoskeleton elements, highly phosphorylated themselves (Sanchez et al., 2002; Veeranna et al., 2000; Avila et al., 1994), are involved in the complex pathways of signaling (Schmidt et al., 1998); Bloch and coworkers (2001) e.g. showed that disruption of cytoskeletal integrity impairs G_1-mediated signaling due to replacement of G_1 proteins and phosphorylation-regulated association of MAPs with proteins of intracellular signal-transduction pathways suggests a link between cellular signaling and neuronal cytoskeleton (Lim et al., 2000).

Although literature on cytoskeleton structure and function is abundant, information on the composition and stoichiometry of basic cytoskeleton elements is fragmentary (Cambiazo et al., 1995; Coffey et al., 1995; Frappier et al., 1987; Ishikawa et al., 1992; Kamal et al., 2000; Kirsch et al., 1991; Pedrotti et al., 1994; Scott et al., 1985; Sihag et al., 1998; Steiner et al., 1987; Terasaki et al., 2002).

It was therefore the aim of this study to provide an integrated view on basic constituents forming this protein network in quantitative terms and in respect to stoichiometry under "physiological" conditions i.e. controls in human brain. As cytoskeleton derangement in neurodegenerative disorders has been reported and may be shedding light on its tentative involvement in pathogenesis, we included analysis of cytoskeleton composition in Down Syndrome (DS), Alzheimer's disease (AD) and Pick's disease (PD).

Materials and methods

Brain samples

Frontal cortex from controls (n = 6; three females and three males with mean age of 65.7 ± 6.2 years and mean postmortem interval (PMI) of 38.5 ± 22 hours), DS patients (n = 7; two females and five males with a mean age of 57.3 ± 6.5 years and mean PMI of 26.7 ± 13.1 hours), AD patients (n = 4; one female and three male with mean age of 63.3 ± 3.4 years and mean PMI of 35.7 ± 24.4 hours) and PD patients (n = 5; only females with mean age of 55 ± 11.7 years and mean PMI of 33 ± 23.1 hours), were used for the present study. The samples were obtained from Dr.N.Cairns, King's College, Brain Bank for Neurodegenerative Diseases, Departement of Neuropathology, Institute of Psychiatry, London UK. All samples were stored at −70°C and the freezing chain was never interrupted.

Antibodies

For western blot the following antibodies were purchased:

First antibodies: Monoclonal Anti-β-Actin Clone AC-74, Anti-alpha-Actinin, Monoclonal Anti-β-Tubulin Isotype III Clone SDL.3D10, Monoclonal Anti-MAP1 Light Chain Clone E12 (Sigma-Aldrich, Inc., Missouri, USA), Rabbit Anti Neurofilament L, Rabbit Anti Human Neurofilament M 150 kD, Rabbit Anti Neurofilament H 200 kD (Serotec Ltd, Oxford, UK), Mouse Anti-Neuron Specific Enolase Monoclonal Antibody (Chemicon, CA, USA)

Secondary antibodies: Goat Anti-Mouse IgG2a, Goat Anti-Rabbit IgG (H + L), goat anti-Sheep Anti-Rabbit IgG Star 54 (Serotec Ltd, Oxford, UK)

Western blotting

Brain tissue ground under liquid nitrogen was homogenised in lysis buffer containing protease inhibitor cocktail tablets (Roche, Germany) at 4°C and centrifuged at 8,000 ×g for 10 minutes. The BCA protein assay kit (Pierce, USA) was applied to determine concentration of protein in the supernatant. Samples (10 μg) were mixed with the sample buffer (100 mM Tris-HCl, 2% SDS, 1% 2-mercaptoethanol, 2% glycerol, 0.01% bromophenol blue, pH 7.6), incubated at 95°C for 15 minutes and loaded onto a ExcelGel SDS homogenous gel (7.5% for NF-M and NF-H; 12.5% for NF-L, MAP1, βIII, βA, Act and NSE; Amersham Pharmacia Biotech, Sweden). Electrophoresis was performed with Multiphor II Electrophoresis System (Amersham Pharmacia Biotech). Proteins separated on the gel were transferred onto PVDF membrane (Millipore, USA) and membranes were incubated in blocking buffer (10 mM Tris-HCl, pH 7.5, 150 mM NaCl, 0.1% Tween 20 and 2% non-fat dry milk).

Membranes were incubated for 2 hours at room temperature with diluted primary antibodies (1:5,000 for βA; 1:2,000 for Act; 1:1,000 for βIII and NSE; 1:1,200 for MAP1; 1:3,000 for NF-L, NF-M and NF-H).

After 3 times washing for 15 minutes with blocking buffer, membranes were probed with secondary antibodies (goat anti-mouse IgG2a for βA and NSE, goat anti-rabbit IgG (H + L) for Act, goat anti-mouse IgG 2b for βIII, goat anti mouse IgG1 for MAP1, sheep anti-rabbit IgG Star 54 for NF-L, NF-M and NF-H, coupled to horseradish peroxidase for 1 hour. Membranes were washed 3 times for 15 minutes and developed with the Western Lightning™ chemiluminescence reagents (PerkinElmer Life Sciences, Inc., USA).

Statistics

Density of immunoreactive bands was measured by RFLPscan version 2.1 software program (Scanalytics, USA). Between-group differences were calculated by non-parametric Mann-Whitney U test using GraphPad Instat2 program and the level of significance was considered at $P < 0.05$.

Results

Figure 2 shows representative bands for the individual cytoskeleton elements.

As shown in Table 1 a series of cytoskeleton proteins were deranged in neurodegenerative disorders as compared to controls.

Briefly, in DS brain βIII, NF-H and NF-M were significantly reduced.

When protein levels were normalised versus NSE, a marker for neuronal density, βIII, NF-M and NF-H were comparable between controls and DS patients indicating that decreased levels from above may have been due to neuronal loss. βA in relation to NSE was significantly decreased in DS patients.

In AD NF-M and NF-H levels were reduced significantly, however when normalised versus NSE, NF-M became significantly increased while NF-H was comparable between controls and AD. In addition, βA, Act and MAP1 were significantly increased when normalised versus NSE probably due to neuronal loss.

Table 1. Results (mean and standard deviation [SD]) of cytoskeleton protein levels (abbreviations given in text) in brains of controls (CO), Down syndrome (DS), Alzheimer's disease (AD) and Pick's disease (PD). To rule out effects of neuronal loss neuron-specific-enolase (NSE) was used for correction. The density of detected bands was measured and calculated by non-parametric Mann-Whitney U test; the level of significance was considered at $p < 0.05$. An asteriks reveals statistical significance as compared to control

	CO	DS	AD	PD
Mean ± SD		Arbitrary units		
βA	0.45 ± 0.23	1.21 ± 0.84	0.96 ± 1.46	3.08 ± 0.89*
βA/NSE	0.39 ± 0.19	2.89 ± 2.83	2.70 ± 1.56*	2.91 ± 1.35*
Act	0.42 ± 0.25	0.31 ± 0.19	0.63 ± 0.15	0.39 ± 0.31
Act/NSE	3.36 ± 2.79	0.78 ± 0.72*	4.51 ± 3.82*	0.30 ± 0.26
βIII	1.85 ± 0.48	0.86 ± 0.52*	1.49 ± 0.67	1.56 ± 0.21
βIII/NSE	1.57 ± 0.35	1.73 ± 1.45	12.93 ± 12.81	1.48 ± 0.50
MAP1	1.93 ± 1.15	1.01 ± 0.70	2.02 ± 1.48	1.08 ± 0.76
MAP1/NSE	1.64 ± 0.99	2.12 ± 1.86	9.68 ± 11.56*	1.13 ± 0.86
NF-L	1.19 ± 0.42	0.99 ± 0.89	0.73 ± 0.30	0.56 ± 0.68*
NF-L/NSE	1.05 ± 0.48	1.86 ± 1.75	5.14 ± 4.06	0.05 ± 0.07*
NF-M	1.64 ± 0.53	0.74 ± 0.39*	0.79 ± 0.23*	0.24 ± 0.21*
NF-M/NSE	1.42 ± 0.50	1.72 ± 1.88	4.72 ± 3.56*	0.22 ± 0.18*
NF-H	1.14 ± 0.31	0.68 ± 0.26*	0.51 ± 0.43*	0.91 ± 0.15*
NF-H/NSE	1.00 ± 0.36	1.45 ± 1.07	4.28 ± 4.55	0.09 ± 0.15*

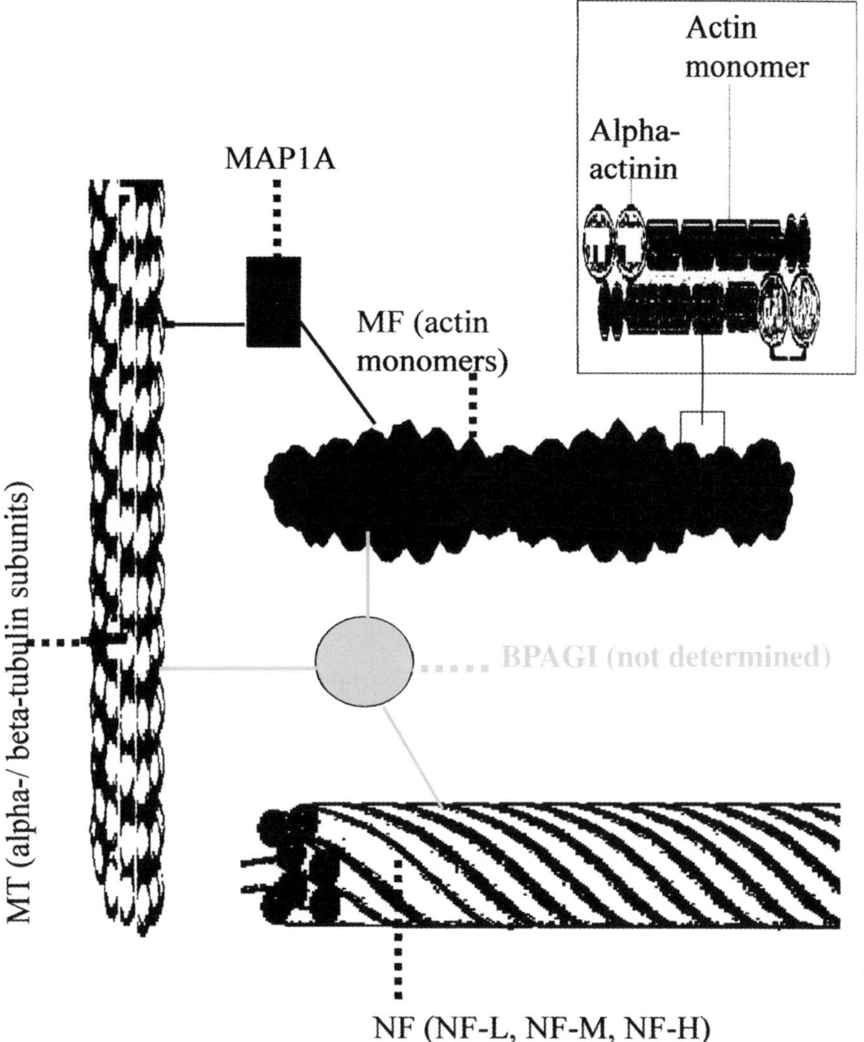

Fig. 1. Arrangement of the three basic classes of cytoskeletal elements (microtubules [MT], microfilaments [MF] and neurofilaments [NF]) connected by cytoskeletal crosslinker proteins: *Microtubule-associated protein 1* (*MAP1*): microtubule-binding protein comprised of two developmentally regulated isoforms MAP1A and MAP1B, is responsible for forming the network of bridges between microtubules as well as interconnecting cytoskeletal elements. *BPAG1* (not determined in this study): member of the spectraplakin superfamily of proteins, characterised by their ability to link different elements of the cytoskeleton. Individual isoforms have the ability to make all possible linkages between actin, microtubules and intermediate filaments. Furthermore, another protein analysed in this study is indicated: *Alpha-actinin*: F-actin binding protein mediating association of actin fiaments into bundles hereby increasing actin elasticity

In PD NF-L, NF-M, NF-H were significantly reduced and actin was significantly elcvated.

Normalisation versus NSE did not change the outcome for proteins from above.

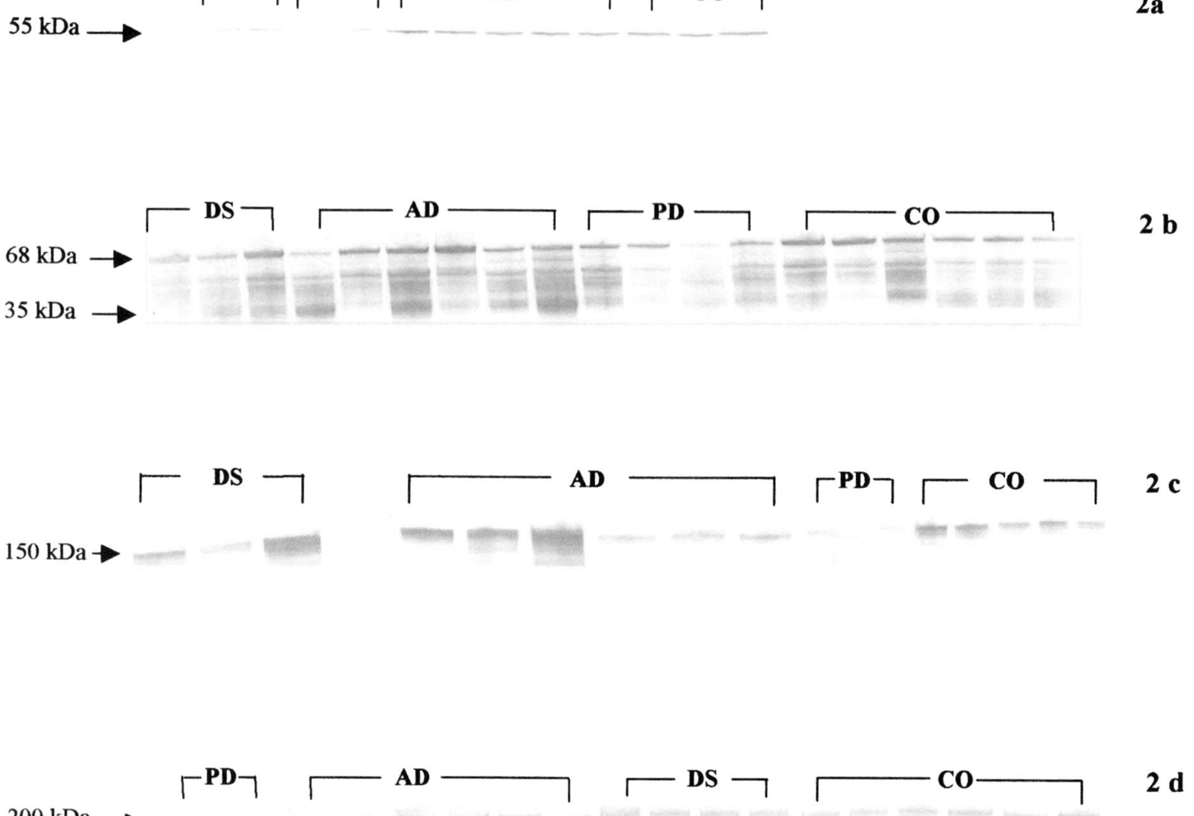

Fig. 2. **a** Results from immunoblotting using βIII antibody: a single band at 55 kDa was observed and the decrease of βIII in DS and AD was already observed by inspection. **b** As already reported in literature we found a series of NF-L immunoreactive bands with an apparent molecular weight ranging from 35 kDa to 68 kDa, with the faster mobility probably resulting from phosphorylation. The 68 kDa band was used for quantification in agreement with literature and the paternal apparent molecular weight. **c** NF-M-immunoreactivity revealed a major band at 150 kDa and **d** NF-H was represented by a major band at 200 kDa

Stoichiometrical results are presented in Table 2 and revealed aberrant relative composition of cytoskeletal elements in all neurodegenerative disorders. The main outcome shows that ratios of members of the neurocytoskeleton (βIII, NF's) in relation to βA were remarkably decreased probably confirming data of decreased neuronal density.

The ratio of Act to βA was remarkably decreased in DS and PD.

Discussion

The main findings of our study are represented by the characterisation of several cytoskeletal elements (cyt el) in stoichiometrical terms and aberrant ratios of the individual cyt el in neurodegeneration.

Table 2. **a** Stoichiometry of individual cytoskeleton proteins (abbreviations given in text) in relation to the level of βA, set at "1", in brains of controls (CO), patients with Down syndrome (DS), Alzheimer's disease (AD) and Pick's disease (PD)

	βA	:	Act	:	βIII	:	MAP1	:	NF-L	:	NF-M	:	NF-H
CO:	1	:	0.8	:	3.8	:	3.8	:	2.4	:	3.2	:	2.2
DS:	1	:	0.2	:	0.7	:	0.8	:	0.8	:	0.6	:	0.6
AD:	1	:	0.6	:	1.5	:	2.0	:	0.7	:	0.8	:	0.5
PD:	1	:	0.1	:	0.5	:	0.3	:	0.2	:	0.1	:	0.3

b Stoichiometry of neurocytoskeleton proteins in relation to the level of expression of βIII, set at "1", in brains of CO, patients with DS, AD and PD

	βIII	:	NF-L	:	NF-M	:	NF-H
CO:	1	:	0.6	:	0.8	:	0.6
DS:	1	:	1.1	:	0.9	:	0.9
AD:	1	:	0.5	:	0.5	:	0.3
PD:	1	:	0.4	:	0.2	:	0.6

The finding of reduced βIII, NF-H and NF-M levels, essential components of the neurocytoskeleton, in DS brain may be due to neuronal loss, supported by the observation that normalisation versus NSE led to comparable levels with controls, which is in agreement with stoichiometrical results revealing decreased ratios of βIII, NF-H and NF-M versus cellular βA in DS and significantly decreased NSE levels in frontal cortex of DS (Table 2a). Only taking neurocytoskeleton proteins in account, ratios were not remarkably altered thus also suggesting the mechanism given above (Table 2b). Another interpretation may be made considering that actin is a death substrate in apoptosis that unequivocally takes place in adult DS. With the exception of an earlier report describing neurofilament derangement in other brain regions of DS, no stoichiometric data on neurocytoskeleton proteins are available (Bajo et al., 2000). A series of cytoskeleton and associated proteins including drebrin, a dendritic spine protein (Shim et al., 2002; Weitzdoerfer et al., 2001), centractin and capping proteins, integral constituents of the dynactin complex (Gulesserian et al., 2002), actin-related protein complex 2/3 (Weitzdoerfer et al., 2002) and moesin, a plasma membrane-actin cytoskeleton crosslinking protein (Lubec et al., 2001) has already been derscribed altered pointing to cytokeletal deficits in DS brain.

In AD, known to present with decreased neuronal density which is confirmed by decreased NSE levels in our study, an expressional pattern similar to DS was observed, although βIII reduction did not reach statistical significance. Stoichiometrical evaluation showed a trend comparable to DS (Table 2a). Reduced NF-L has already been reported at the protein (Bajo et al., 2000) and the transcriptional level (Kittur et al., 1994) in AD pointing to decreased neuronal density. Indeed, NF-L is the major and essential determinant of the neurofilament triplet forming the structural core with NF-M and

NF-H as accessory proteins responsible for mediating interaction with other neurocytoskeleton components (Scott et al., 1985).

In PD results of decreased neurofilament proteins are not paralleling those of NSE and βIII which have been comparable between controls and PD. Stoichiometrical results show decreased ratios of ubiquitous βA versus neurofilament proteins, MAP1 and βIII in PD possibly caused by increased βA which in turn may reflect glial proliferation (Dickson et al., 1998). Alternatively, altered cytoskeleton composition may be due to aberrant posttranslational modifications shown to occur in neurodegeneration (reviewed in: Julien and Mushynski, 1998). On the other hand, we used an antibody recognising the non-phosphorylated form of NF-L at 68–70 kDa and although a series of bands ranging from approximately 35 to 68–70 kDa were detected, only the 68–70 kDa band was used for quantification. But not only phosphorylation (Doering, 1994) has to be taken into account, also glycosylation is a major confounding factor (Dong et al., 1993). These posttranslational modifications may lead to stoichiometrical changes varying with age, species and even with individual brain regions (Scott et al., 1987).

As the cytoskeleton proteins examined in this study are strongly interacting and closely assembled (Fig. 1) we aimed to quantitatively establish an interrelationship (Table 2a). Moreover, we calculated ratios of neurocytoskeletal proteins separately and observed unchanged proportions in neurodegenerative disorders (Table 2b).

Our results have to be considered in respect to post-mortem effects and protein stability (Fountoulakis, 2001), age-related and apoptotic mechanisms forming the basis for stoichiometrical analysis of cyt el in human control brain and neurodegeneration. Cytoskeletal stoichiometry, although highly variable, is a fundamental cellular characteristic that should be determined in order to understand assembly and complex structural interaction mechanisms in health and disease.

References

Avila J, Dominguez J, Diaz-Nido J (1994) Regulation of microtubule dynamics by microtubule-associated protein expression and phosphorylation during neuronal development. Int J Dev Biol 38(1): 13–25

Bajo M, Yoo BC, Cairns N, Gratzer M, Lubec G (2001) Neurofilament proteins NF-L, NF-M, NF-H in brains of patients with Down syndrome and Alzheimer's disease. Amino Acids 21: 293–301

Bloch W, Fan Y, Han J, Xue S, Schoeneberg T, Ji G, Lu ZJ, Walther M, Faessler R, Hescheler J, Addicks K, Fleischmann BK (2001) Disruption of cytoskeletal integrity impairs G_i-mediated signaling due to displacement of G_i proteins. J Cell Biol 154(4): 753–761

Cambazio V, Gonzales M, Maccioni RB (1995) DMAP-85: a tau-like protein from Drosophila larvae. J Neurochem 64(3): 1288–1297

Coffey RL, Purich DL (1995) Non-cooperative binding of the MAP-2 microtubule-binding region to microtubules. J Biol Chem 270(3): 1035–1040

Dickson DW (1998) Pick's disease: a modern approach. Brain Pathol 8: 339–354

Doering LC (1993) Probing modifications of the neuronal cytoskeleton. Mol Neurobiol 7(3–4): 265–291

Dong D, Xu ZS, Chevrier MR, Cotter RJ, Cleveland DW, Hart GW (1993) Glycosylation of mammalian neurofilaments. J Biol Chem 268(22): 16679–16687

Fountoulakis M, Hardmeier R, Hoeger H, Lubec G (2001) Postmortem changes in the level of brain proteins. Exp Neurol 167(1): 86–94

Frappier T, Regnouf F, Pradel LA (1987) Binding of brain spectrin to the 70-kDa neurofilament subunit protein. Eur J Biochem 169(3): 651–657

Gulesserian T, Kim SH, Fountoulakis M, Lubec G (2002) Aberrant expression of centractin and capping proteins, integral constituents of the dynactin complex, in fetal Down syndrome brain. Biochem Biophys Res Commun 291: 62–67

Ishikawa R, Kagami O, Hayashi C, Kohama K (1992) The binding of nonmuscle caldesmon from brain to microtubules. Regulations by calcium calmodulin and cdc2 kinase. Biochemistry 33(41): 12463–12470

Julien JP, Mushynski WF (1998) Neurofilaments in health and disease. Prog Nucl Acid Res Mol Biol 6: 1–23

Kaech S, Parmar H, Roelandse M, Bornmann C, Matus A (2001) Cytoskeletal microdifferentiation: a mechanism for organizing morphological plasticity in dendrites. Proc Natl Acad Sci USA 98(13): 7086–7092

Kamal A, Stokin GB, Yang Z, Xia CH, Goldstein LS (2000) Axonal transport of amyloid precursor protein is medaited by direct binding to the light chain subunit of kinesin-I. Neuron 28(2): 449–459

Kirsch J, Langosch D, Prior P, Littauer UZ, Kohama K (1991) The 93-kDa glycine receptor-associated protein binds to tubulin. J Biol Chem 266(33): 22242–22245

Kittur S, Hoh J, Endo H, Tourtelotte W, Weeks BS, Markesbery W, Adler W (1994) Cytoskeletal neurofilament gene expression in brain tissue from Alzheimer's disease patients. Decrease in NF-L and NF-M message. J Geriat Psychiatry Neurol 7(3): 153–158

La Monte B, Wallace KE, Holloway BA, Shelly SS, Ascano J, Tokito M, Van Winkle T, Howland DS, Holzbaur ELF (2002) Dysruption of dynein/dynactin inhibits axonal transport in motor neurons causing late-onset progressive degeneration. Neuron 34: 715–727

Lim RW, Halpain S (2000) Regulated association of microtubule-associated protein 2 (MAP2) with Src and Grb2: evidence for MAP2 as a scaffolding protein. J Biol Chem 275(27): 20578–20587

Lubec B, Weitzdoerfer R, Fountoulakis M (2001) Manifold reduction of moesin in fetal Down syndrome brain. Biochem Biophys Res Comm 286(5): 1191–4

Sanchez C, Diaz-Nido J, Avila J (2000) Phosphorylation of microtubule-associated protein 2 (MAP2) and its relevance for the regulation of neuronal cytoskeleton function. Prog Neurobiol 61(2): 133–168

Scott D, Smith KE, O'Brien BJ, Angelides KJ (1985) Characterization of mammalian neurofilament triplet proteins. Subunit stoichiometry and morphology of native and reconstituted filaments. J Biol Chem 260 (19): 10736–10747

Shim KS, Lubec G (2002) Drebrin, a dendritic spine protein, is manifold decreased in brains of patients with Alzheimer's disease and Down syndrome. Neurosci Lett 324(3): 209–212

Siegel GJ, Agranoff BW, Albers RW, Fisher SK, Uhler MD (1999) Basic neurochemistry, molecular, cellular and medical aspects, 6th edn. Lippincott Williams & Wilkins, Philadelphia

Sihak RK (1998) Brain beta-spectrin phosphorylation: phosphate analysis and identification of threonin as a heparin-sensitive protein kinase phosphorylation site. J Neurochem 71(5): 2220–2228

Steiner JP, Ling E, Bennet V (1987) Nearest neighbor analysis for brain synapsin I. Evidence from in vitro reassociation for association with membrane protein(s) and the Mr = 68.0000 neurofilament subunit. J Biol Chem 262(2): 905–914

Terasaki AG, Morikawa K, Suzuki H, Oshima H, Ohashi K (2002) Characterisation of the arp2/3 complex in chicken tissues. Cell Struct Funct 27(5): 383–391

Weitzdoerfer R, Dierssen M, Fountoulakis M, Lubec G (2001) Fetal life in Down syndrome starts with normal neuronal density but impaired dendritic spines and synaptosomal structure. J Neural Transm [Suppl] 61: 59–70

Weitzdoerfer R, Fountoulakis M, Lubec G (2002) Reduction of actin-related protein complex 2/3 in fetal Down Syndrome brain. Biochem Biophys Res Commun 293(2): 836–841

Authors' address: G. Lubec, CChem, FRSC (UK), Department of Pediatrics, University of Vienna, Währinger Gürtel 18, A-1090 Vienna, Austria, e-mail: gert.lubec@akh-wien.ac.at

The cerebral cortex in Fetal Down Syndrome

U. Unterberger[1], **G. Lubec**[2], **M. Dierssen**[3], **G. Stoltenburg-Didinger**[4], **J. C. Farreras**[5], and **H. Budka**[1]

[1] Institute of Neurology, and
[2] Department of Pediatrics, University of Vienna, Vienna, Austria
[3] Genes and Disease Program, Genomic Regulation Center, Barcelona, Spain
[4] Department of Neuropathology, University Hospital Benjamin Franklin, Free University Berlin, Germany
[5] Unidad de Patología, Corporació Sanitaria Parc Tauli, Sabadell, Barcelona, Spain

Summary. Brain histopathology of 32 fetuses with Down syndrome was compared to that of 25 age-matched normal controls and 9 brains of fetuses of HIV positive mothers. Four cases of Down syndrome and 1 HIV case showed microdysgenesia of the cerebral cortex. As the pathogenetic background of cortical irregularities is presently not known, we analyzed the neuronal expression of drebrin, an actin-binding protein of neuronal dendritic spines. This protein is thought to play a role in synaptic formation and was recently shown to be manifold reduced in brains of fetuses with Down syndrome. However, immunocytochemistry revealed no differences in drebrin expression pattern between Down patients and controls. We conclude that cerebral cortical microdysgenesia is an infrequent non-specific pathology in fetal Down syndrome.

Introduction

The postnatal pathology of the cerebral cortex in patients with Down syndrome has been well studied. Golgi impregnations showed neuronal dendrites in the visual cortex to be shorter, and the number of dendritic spines to be lower than in age-matched controls (Takashima et al., 1981). The dendritic length seems to be normal at the time of birth, but decreases steadily during the first years of life (Becker et al., 1986). These abnormalities continue into adulthood and may relate with mental retardation, or early onset of dementia, respectively, in Down syndrome (Takashima et al., 1989). On the other hand, reports on histopathological findings in fetal Down brains are scarce. Some authors found regular neuronal density and normal neuronal morphology, i.e. dendritic arborization and number of spines, in different cortical regions (Takashima et al., 1981; Vukšić et al., 2002), and gross neuropathological parameters in fetuses from 15 to 22 weeks of gestation not to differ from controls (Schmidt-Sidor et al., 1990). On the other hand, disturbances in cortical development (Golden et al., 1994) and synaptic formation (Petit et al., 1984) were described. Moreover, the number of ramified microglia in the fetal brain is significantly higher in Down syndrome than in age-related controls (Wierzba-Bobrowicz et al., 1999).

It is unlikely that neuronal development in Down syndrome becomes pathologic only after birth. Many biochemical parameters have been found to be abnormal in fetal tissues (Engidawork and Lubec, 2003). Drebrin, an actin-binding protein of neuronal dendritic spines believed to play an important role in synaptic plasticity, is manifold reduced in fetuses (and adults) with trisomy 21 (Weitzdoerfer et al., 2001; Shim and Lubec, 2002). This lack might result in inadequate formation of synapses and aberrant cortical lamination patterns. The aim of this work was to study an eventual relation between histopathology of the cortex and drebrin expression in the fetal Down syndrome brain.

Materials and methods

Formalin-fixed, paraffin-embedded material of 41 fetal brains was provided by the centers in Barcelona and Berlin. In 32 cases, trisomy 21 had been clinically and cytogenetically diagnosed. Nine brains were from fetuses of HIV positive mothers. Unstained 4 μm paraffin sections of 25 normal brains from fetuses who had been aborted for unrelated causes were provided by the Barcelona laboratory. Gestational ages ranged from 12 to 35 weeks in trisomy 21 patients, and 14 to 37 weeks in controls. In most cases, one or two coronal blocks of cerebral hemispheres, including hippocampus, and samples from cerebellum and brainstem were available. Sections of 3–5 μm thickness were mounted on glass slides and kept at 65°C for one hour. Slides were then deparaffinized and either stained with hematoxylin and eosin (H&E) or further processed for immunocytochemistry. For that, batches of 20 slides underwent antigen retrieval in a microwave oven in 200 ml citrate buffer (pH 6,0) for 2 minutes at 850 W and then 10 minutes at 250 W. Subsequently, a monoclonal anti-drebrin antibody (clone M2F6 from MBL, Nagoya, Japan) was applied at a 1:75 dilution. This immunocytochemical protocol was established to work best on control fetal brain in a previous pilot study of various pretreatments and dilutions. The DAKO ChemMate™ Detection Kit (DAKO, Glostrup, Denmark) was used for chromogene visualization.

Results

H&E sections revealed microdysgenesia of the cerebral cortex in four of the 32 Down syndrome patients and one of nine fetuses of HIV positive mothers. Four brains of trisomy 21 (ages 18, 19, and twice 23 weeks, respectively) showed widespread wart-like protrusions of cortical layer II into the molecular layer, resembling a so called "status verrucosus" (Fig. 1B). Similar dysgenetic changes of the cortex could be observed in one fetus born to an HIV infected mother (19 weeks). However, they were not present in normal controls (Fig. 1A). In addition, in one Down syndrome case aged 23 weeks, the dentate nuclei of the cerebellar hemispheres were still ungyrated. In another case, also aged 23 weeks, the configuration of the dentate nucleus was regular; in the others the dentate was not cut.

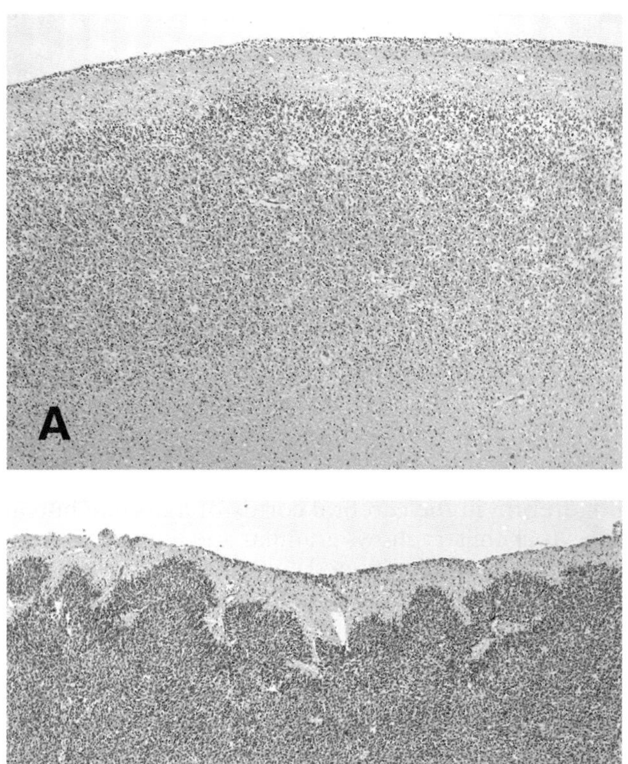

Fig. 1. Microdysgenesia of the cerebral cortex in a fetus with Down syndrome, 23rd week of gestation **(B)**. Layer II protrudes irregularly into the molecular layer, while the cortical surface is relatively smooth, with preserved outer granular layer. An age-related control with regular cortical structure is presented in **A** (H&E, ×40)

Other pathologic findings were equally present in Down and HIV fetuses and control cases, including lesions attributable to abortion, like bleedings within the germinal layer, intraventricular and leptomeningeal hemorrhages, as well as ventricular dilatation.

Immunocytochemical localization of drebrin revealed a granular staining mainly of neuronal dendrites, but also of some nerve cell somata (Fig. 2). There was no difference in expression pattern either between controls and Down syndrome patients, or between regions showing varying degrees of dysgenetic changes within the same brain.

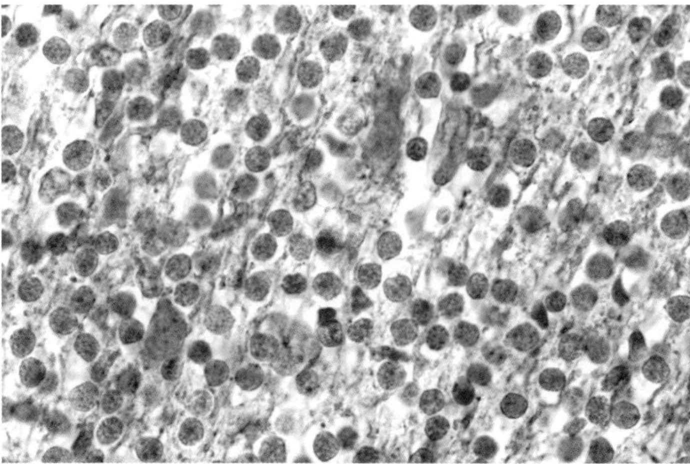

Fig. 2. Expression of drebrin in the cerebral cortex of a normal human fetus, 23rd week of gestation. Immunocytochemistry shows granular and linear staining mainly of neuronal dendrites between the nuclei (counterstained as round structures), but also of some nerve cell somata (×600)

Discussion

Brain histopathology in fetal Down syndrome has been incompletely characterized. Microdysgenesia of the cerebral cortex as observed here has not been described. Although what has been described as "status verrucosus" in fetal brains may be also due to preparational artifacts, preservation of the overlying outer granular layer and restriction of changes to Down/HIV but not control brains argues against an artifactual origin. Like other pathologies in Down syndrome, such changes are not specific, but may occur due to a variety of influences during the fetal period, including intrauterine ischemia or infection. In concordance with that is our observation of a similar pathology in the brain of a fetus born to an HIV infected mother. In this case, it is not clear whether the disturbed neuronal migration is a result of intrauterine infection or is caused by other influences like maternal drug abuse.

The clinical significance of these brain lesions remains undetermined. While polymicrogyria is generally accepted as a malformation that may manifest with varying degrees of neurologic disability, depending on the size and the localization of the lesions (Friede, 1989; Ellison and Love, 1998), the situation is less straightforward in case of the so called "status verrucosus" that is sometimes considered as a transient phase in normal development (Ellison and Love, 1998). Our study shows that there is at least a more frequent appearance of these changes in fetal Down brain. The extent of the lesion is likely to play a key role with respect to clinical outcome, as small and focal cortical irregularities may be present in otherwise normal brains.

We demonstrate drebrin as regularly expressed in cerebral cortex of Down brains. Discrepancy to biochemical results of drebrin derangement in DS may be due to protein modifications including posttranslational modifications and the presence of several isoforms of this protein. Our morphological results are in agreement with previous reports demonstrated normal development of the dendritic tree until the first postpartal months (Takashima et al., 1981; Vukšić et al., 2002). Thus the pathogenesis of fetal cortical microdysgenesia as observed here remains unclear.

Acknowledgements

The authors are indebted to the Red Bull Company, Salzburg, Austria, the Verein zur Durchführung der wissenschaftlichen Forschung auf dem Gebiet der Neonatologie „Unser Kind", the Fundación CIEN, and the Jerôme Lejeune Foundation for generous financial assistance to UU and GL. We want to thank U. Köck and M. Strohschneider for their excellent technical work.

References

Becker LE, Armstrong DL, Chan F (1986) Dendritic atrophy in children with Down's syndrome. Ann Neurol 20(4): 520–526

Ellison D, Love S (1998) Neuropathology. Mosby International, London, pp 3.26–3.37

Engidawork E, Lubec G (2003) Molecular changes in fetal Down syndrome brain. J Neurochem 84: 895–904

Friede RL (1989) Developmental neuropathology, 2nd rev and expanded edn. Springer, Berlin Heidelberg New York, pp 330–346

Golden JA, Hyman BT (1994) Development of the superior temporal neocortex is anomalous in Trisomy 21. J Neuropathol Exp Neurol 53(5): 513–520

Petit TL, LeBoutillier JC, Alfano DP, Becker LE (1984) Synaptic development in the human fetus: a morphometric analysis of normal and Down's syndrome neocortex. Exp Neurol 83(1): 13–23

Schmidt-Sidor B, Wisniewski KE, Shepard TH, Sersen EA (1990) Brain growth in Down syndrome subjects 15 to 22 weeks of gestational age and birth to 60 months. Clin Neuropathol 9(4): 181–190

Shim KS, Lubec G (2002) Drebrin, a dendritic spine protein, is manifold decreased in brains of patients with Alzheimer's disease and Down syndrome. Neurosci Lett 324(3): 209–212

Takashima S, Becker LE, Armstrong DL, Chan F (1981) Abnormal neuronal development in the visual cortex of the human fetus and infant with Down's syndrome. A quantitative and qualitative Golgi study. Brain Res 225: 1–21

Takashima S, Ieshima A, Nakamura H, Becker LE (1989) Dendrites, dementia and the Down syndrome. Brain Dev 11(2): 131–133

Vukšić M, Petanjek Z, Rašin MR, Kostović I (2002) Perinatal growth of prefrontal layer III pyramids in Down syndrome. Pediatr Neurol 27(1): 36–38

Weitzdoerfer R, Dierssen M, Fountoulakis M, Lubec G (2001) Fetal life in Down syndrome starts with normal neuronal density but impaired dendritic spines and synaptosomal structure. J Neural Transm [Suppl] 61: 59–70

Wierzba-Bobrowicz T, Lewandowska E, Schmidt-Sidor B, Gwiazda E (1999) The comparison of microglia maturation in CNS of normal human fetuses and fetuses with Down's syndrome. Folia Neuropathol 37(4): 227–234

Authors' address: Prof. H. Budka, MD, Institute of Neurology, University of Vienna, P.O.B. 48, Währinger Gürtel 18-20, A-1097 Vienna, Austria, e-mail: h.budka@akh-wien.ac.at

Polysomnography in transgenic hSOD1 mice as Down syndrome model

D. Colas[1], **J. London**[2], **R. Cespuglio**[1], and **N. Sarda**[1]

[1] INSERM Unit 480, Claude Bernard University, Lyon, and
[2] EA 3508, Paris 7 Denis-Diderot University, Paris, France

Summary. Sleep-wake homeostasis is crucial for behavioral performances and memory in the general population and in learning disability populations among them Down syndrome patients. We investigated, in a mouse model of Down syndrome, cortical EEG and sleep-wake architecture under baseline conditions and after a 4 hr sleep deprivation (SD). Young heterozygous transgenic mice (S/+) for the human Cu/Zn superoxide dismutase (hSOD-1) were obtained on FVB/N background. Baseline records for slow wave sleep (SWS) and wake (W) parameters were the same in S/+ and control mice whereas paradoxical sleep (PS) episode number decreased and PS latency increased after light off in S/+ mice. These data correlate well the polysomnographic phenotype of young DS patients.

Abbreviations

DS Down's syndrome; *hSOD1* human CuZn superoxide dismutase 1; *PS* paradoxical sleep; *SWS* slow wave sleep; *W* wakefulness

Introduction

Trisomy 21 or Down syndrome (DS)[2] is the main autosomal aneuploidy that is not lethal in fetal or early postnatal life. DS results from the triplication of the whole or distal part of human autosome 21 (Lejeune, 1959) and is the primary cause of mental retardation (Caviedes, 1990; Epstein, 1995). DS phenotypes show variable penetrance, affecting many different organs including the neural system. Among the brain abnormalities present in DS are the smaller volumes of the cerebellum, frontal cortex and hippocampus, the enlargement of the hippocampal gyrus, decrease cell density of the cerebellum granular layer and abnormal numbers and ramification of spines (Wisniewski, 1985; Head, 2001). Cognitive deficits on the part of subjects with DS have been closely related with sleep-disordered breathing (SDB) deficits (Andreou, 2002). Sleep patterns abnormalities might also be relevant for their poor cognitive efficiency and constant fatigability of DS patients.

Polysomnographic recordings obtained from DS patients have shown a reduction in the percentage of paradoxical sleep (PS), a prolonged latency to the first PS episode, an increase in undifferential sleep and a reduced ratio of oculomotor frequencies (Grubar, 1986; Diomedi, 1999; Levanon, 1999; Andreou, 2002).

Several strategies have been used for modeling DS in mice: transgenic mice overexpressing single or a combination of genes or large fragments of DNA (human Yac or murine BAC) and mouse trisomies carrying part of mouse 16 chromosome which carries numerous orthologues genes to chromosome 21 (Dierssen, 2001). Human Cu/Zn superoxide dismutase 1 gene (hSOD1) was the first chromosome 21 gene to be characterized (Groner, 1985) and it was even shown earlier to be overexpressed at the protein level in DS patients (Sinet, 1975). Transgenic mice expressing wild-type hSOD1 were the first model for Down syndrome (Epstein, 1987). SOD1 overexpression have been shown to be either protective against in vivo insults as glutamate and MPTP toxicity (Chan, 1990; Przedborski, 1992) or deleterious for aging processes (Kola, 1998). These SOD mice still remain a good model for some abnormalities present in DS patients: neuromuscular abnormalities of the tongue and the legs (Avraham, 1991), cognitive deficit and impairment in long term potentiation (LTP) (Gahtan, 1998), premature thymic involution (Nabarra, 1996) and diminished serotonin uptake in platelets (Schickler, 1989).

To explore whether sleep architecture anomalies may be related to the SOD1 gene dosage effect, we compared sleep parameters in mice overexpressing the hSOD1 transgene at 2–3 months of age. These transgenic mice and their controls were obtained on the inbred FVB/N background. Because responses to total sleep deprivation (SD) reveal differences in sleep regulation (Tobler, 2000), the effects of a 4 hr SD were analyzed.

Materials and methods

Animals

hSOD1wt mice were generated as previously described (Paris, 1996). Heterozygous (S/+) expressing the transgene were discriminated from non transgenic littermates used as controls (+/+ mice) on the inbred FVB/N background by measuring inhibition of nitroblue tetrazolium reduction by Cu/Zn-SOD1 using a Ransod kit (Randox Laboratories). All experiments followed EEC Directive (86/609/EEC) and every effort was made to minimize the number of animals used as well as to avoid any pain and discomfort. Six male (2–3 months) of each type were used ($S/^{+}$ and FVB/N). Immediately after their arrival (4 weeks), animals were housed in transparent barrels (diameter 20 cm; height 30 cm) and placed in a insulated sound-proofed recording room maintained at an ambient temperature at 22°C ± 1°C and on a 12 hr light-dark cycle (lights on at 05.00 a.m), water and food being available ad libitum. These conditions were respected throughout the study.

Surgery

Animals with a body weight of 21–30 gm, were chronically implanted, under deep sodium pentobarbital anaesthesia (60–70 mg/kg, i.p.), with two cortical electrodes (gold-plated screw, ø = 0.4 mm) and two muscle electrodes (fluorocarbon-coated gold-plates stainless steel wire, ø = 0.03 mm; Cooner Wire, Chatwork, CA) to record the electroenmyogram (EEG) and electromyogram (EMG) and to monitor the sleep-wake cycle. All electrodes were previously soldered to a multichannel electrical connector, and each was separately insulated with a covering of heat-shrinkable polyoletin-polyester tubing. The cortical electrodes were inserted into the dura through two holes (ø = 0.4 mm) made in the skull, located, respectively, in the frontal (1 mm lateral and anterior to the bregma) and parietal (1 mm lateral to the midline at the midpoint between the bregma and lambda) cortices. The muscle electrodes were inserted into the neck muscles. Finally, the electrode assembly was anchored and fixed to the skull with Super6-Bond (Sun Medical Co., Shiga, Japan) and dental cement. At least 2 weeks were allowed for recovery and adaptation to the recording situation.

Experimental protocol and data acquisition

Six mice were recorded simultaneously in one session. One S/+ and its control were included and equally distributed in each session to avoid non-specific variations among strains. Each session consisted of 3 consecutive 24 hr period recordings. Mice were subjected to 4-hr SD in the last day by gentle handling and recorded for the remaining 4 hr. SD began 4 hr after lights on. Cortical EEG and EMG signals were recorded simultaneously and continuously with an Embla-Somnologica device (Flaga, Island). All signals were amplified, analog-to-digital converted with a sampling rate of 100 Hz and digitally filtered (EEG 0.5 to 40 Hz, EMG 10 to 50 Hz). The EEG signal was subjected to a fast Fourier transformation (FFT) analysis (sampling rate 256 Hz with 50% overlap over 10 s windows), yielding power spectra between 0 and 49.6 Hz with a 0.39 Hz frequency resolution.

On the basis of EEG and EMG animal's behavioral states were classified as wakefulness (W), slow wave sleep (SWS) and paradoxical sleep (PS). States were scored for 2 consecutive 24 hr, by consecutive 30 s epochs to determine the duration and number of episodes of each state. Quantitative analysis of each state (W, SWS and PS) was calculated for different periods: 12 hr light, 12 hr dark and over 24 hr for each animal. PS latency was determined as the time spent between light off or light on and the first subsequent event scored as PS. For spectral analysis states were scored by consecutive 10 s epochs according to Colas (2003).

Statistical analysis

For the sleep-wake cycle parameters, statistical comparisons were performed using multifactorial ANOVA for repeated measures (Statgraphics software, Manugistic) involving combined parameters (genotype, experimental conditions and x-hr epochs). When ANOVA was significant at $p < 0.05$, a post hoc Fisher's least significant differences test was performed.

Results

Under basal conditions, in which S/+ mice were left undisturbed, sleep states amounts in S/+ and FVB/N are distributed differently throughout the light-dark periods, i.e. larger amounts of SWS and PS during the light period and opposite distributions for W state corresponding to a circadian distribution

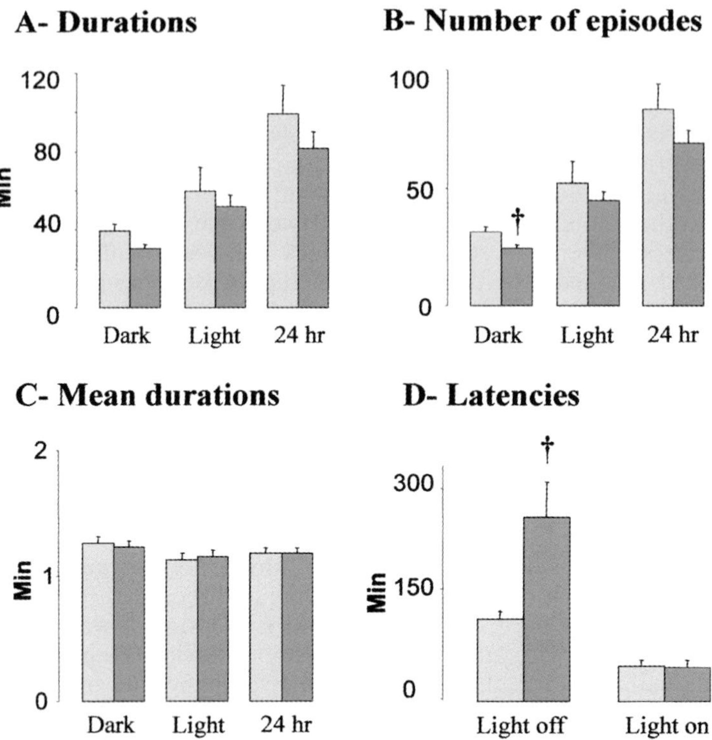

Fig. 1. Quantitative variations of paradoxical sleep (PS) parameters in SOD/+ mice. Data are the mean ± SE for each parameter (n = 6 per group) in FVB/N mice (control; light blue) and in SOD/+ mice (dark blue). **A** durations (in minutes); **B** number of PS episodes; and **C** mean duration in (minutes) are given for 12 hr dark, 12 hr light and 24 hr periods. **D** Latencies to PS (in minutes) are given after light off. Statistics: ANOVA, followed by post-hoc Fischer's test; variable: strain; † $p < 0.05$

with a low amplitude. Quantitative analysis of each sleep-wake stage showed no significant changes in the total amount of W, SWS and PS during the 12 hr light and 12 hr dark or over 24 hr periods (data not shown). Despite no major modification in the daily amount of PS (Fig. 1A), when the analysis was performed hour per hour, we found a significantly lower amount of PS (42.8%) in S/+ mice during the period running from 20:00–23:00 hr (illustrated Fig. 2). This decrease was accompanied by a significant decreased number of PS episodes (24%) within the 12 hr of the dark period (Fig. 1B) and no significant change in the mean duration of PS episode. Moreover, the latency to PS was significantly increased more than two fold after light off: 275 ± 55 min in S/+ instead of 122 ± 13 min seen in FVB/N (Fig. 1D). Cortical EEG power density during spontaneous sleep-wake states did not reveal any differences between $S/^{+}$ and control mice. All these modifications are noticed on a typical hypnogram illustrating the spontaneous sleep-wake cycle before and after light-off and sleep-wake transitions in FVB/N and S/+ mice (Fig. 2). S/+ mice did not show SWS or PS rebounds after SD but the delta-SWS power spectra was enhanced like in controls mice (data not shown).

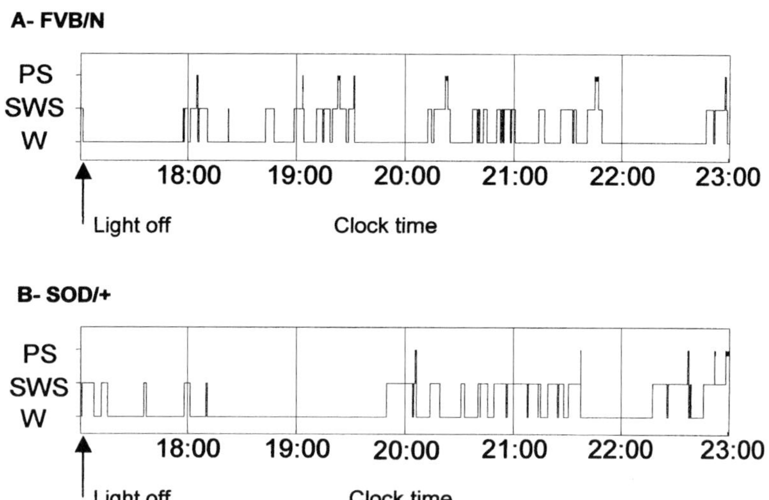

Fig. 2. Typical hypnograms illustrating the effects of the environmental change light off at 17:00 p.m on FVB/N (**A**) and SOD/+ (**B**) mice

Discussion

The aim of the present study was to investigate the sleep-wake cycle in S/+ transgenic mice for the wild-type human gene, as model of Down syndrome. Control and transgenic mice were obtained on the same FVB/N background. The inter-individual standard deviations for the sleep-wake stages were small within each genotype indicating that each group was homogenous.

Transgenic mice (S/+) for the hSOD1 gene exhibit some polysomnographic characteristics that are observed in DS patients. In this study, the most important quantitative modifications were associated to PS: a decrease in PS amount and number of episodes during the dark period and a significant increase in PS latency after light-off. Many authors have analyzed the importance of the relationship of PS to a learning ability and memory. In fact, PS deprivation in animals caused a deficit of mnemonic and learning abilities (Pearlman, 1972). On the basis of these observations, PS has been considered as a neurophysiological marker of the ability of the central nervous system to receive new information, i.e. an index of brain plasticity. Under normal conditions, S/+ mice seem to develop a DS-like phenotype. One variable not controlled for in relation to Down syndrome is the increased probability of sleep apnea. In fact, an increase of PS sleep latency, a reduction in PS sleep percentage and PS cycles number were reported in DS patients. It has also been suggested that the deficiency in PS for DS patients could be due to increased levels of sleep apnea (Grubar, 1986; Diomedi, 1999; Levanon, 1999).

With reference to sleep mechanisms, a growing body of evidence suggests that NO in cholinergic neurons play a role in the circadian and homeostatic processes of PS. We have recently demonstrated that during aging, NO maintains PS production (Clément, 2003). In S/+ mice, we have found that the metabolism of NO in brainstem was accelerated to generate nitrite/nitrate production without modification in NOS enzymatic activities (Colas, 2003). We have hypothesized that the NO decrease in brainstem of S/+ mice may explain the decrease in PS production. Additional studies are necessary to confirm this point by using mice in which the SOD1 gene has been invalidated or mice with partial trisomy 16 (Ts65Dn) in which only one extracopy of the SOD1 gene is present.

In conclusion, the modifications in sleep-wake architecture in SOD-1 transgenic mice shown in the present study confirm the characteristics previously reported in the litterature in developmental disabilities with DS behavioral phenotype. These mice are a good model which may be useful to better understand the biochemical basis of sleep abnormalities in DS patients.

Acknowledgements

This work was supported by grants from Foundation Jérôme Lejeune (Paris, France).

References

Andreou G, Galanopoulou C, Gourgoulianis K, Karapetsas A, Molyvdas P (2002) Cognitive status in Down syndrome individuals with sleep disordered breathing deficits (SDB). Brain Cogn 50: 145–149

Avrahan KB, Sugarman H, Rotshenker S, Groner Y (1991) Down's syndrome: morphological remodelling and increased complexity in the neuromuscular junction of transgenic CuZn-superoxide dismutase mice. J Neurocytol 20: 208–215

Caviedes P, Ault B, Rapoport SI (1990) The role of altered sodium currents in the action potential abnormalities of cultured dorsal root ganglion neurons from trisomy 21 (down's syndrome) human fetuses. Brain Res 510: 229–236

Chan PH, Chu L, Chen SF, Carlson EJ, Epstein CJ (1990) Attenuation of glutamate-induced neuronal swelling and toxicity in transgenic mice overexpressing human Cu/Zn-superoxide dismutase. Acta Neurochir [Suppl]51: 245–247

Clément P, Gharib A, Cespuglio R, Sarda N (2003) Changes in the sleep-wake cycle architecture and cortical nitrix oxide release during ageing in the rat. Neurosci 116: 863–870

Colas D, London J, Gharib A, Cespuglio R, Sarda N (2003) Sleep wake architecture in Down syndrome murine models. Neurobiol Dis (in press)

Dierssen M, Fillat C, Crnic L, Arbones M, Florez J, Estivill X (2001) Murine models for Down syndrome. Physiol Behav 73: 859–871

Diomedi M, Curatola P, Scalise A, Pacidi F, Caretto F, Gigli GL (1999) Sleep abnormalities in mentally retarded autistic subjects: Down's syndrome with mental retardation and normal subjects. Brain Dev 21: 548–553

Epstein CJ (1995) In: Scriver CR, Beaudet AL, Sly WS, Valle D (eds) The metabolic and molecular basis of inherited disease. McGraw Hill, New York, p 749

Epstein CJ, Avraham K, Lovett M, Smith S, Elroy-Stein O, Rotman G, Bry C, Groner Y (1987) Transgenic mice with increased CuZn superoxide dismutase activity: animal model of dosage effects in Down sydrome. Proc Natl Acad Sci USA 84: 8044–8048

Gahtan E, Auerbach JM, Groner Y, Segal M (1998) Reversible impairment of long-term potentiation in transgenic Cu/Zn-SOD mice. Eur J Neurosci 10: 538–544

Groner Y, Lieman-Hurwitz J, Dafni N, Sherman L, Levanon D, Bernstein Y, Danciger E, Elroy-Stein O (1985) Molecular structure and expression of the gene locus on chromosome 21 encoding the Cu/Zn superoxide dismutase and its relevance to Down syndrome. Ann NY Acad Sci 450: 133–156

Grubar JC, Gigli GL, Colognola RM, Ferri R, Musumeci SA, Bergonzi P (1986) Sleep patterns of Down's syndrome children: effects of butoctamide hydrogen succinate (BAHS) administration. Psychopharmacol 90: 119–122

Head E, Azizeh BY, Lott IT, Tenner AJ, Cotman CW, Cribbs DH (2001) Complement association with neurons and beta-amyloid deposition in the brains of aged individuals with Down syndrome. Neurobiol Dis 8: 252–265

Kola I, Herzog PJ (1998) Down syndrome and mouse models. Curr Opin Genet Dev 8: 316–321

Lejeune J, Gautier M, Turpin R (1959) Etude des chromosomes somatiques de neufs enfants mongoliens. CR Acad Sci 248: 1722–1724

Levanon A, Tarasiuk A, Tal A (1999) Sleep characteristics in children with Down syndrome. J Pediatr 134: 755–760

Nabarra B, Casanova M, Paris D, Nicole A, Toyama K, Sinet PM, Ceballos I, London J (1996) Transgenic mice overexpressing the human Cu/Zn-SOD gene: ultrastructure studies of a premature thymic involution model of Down's syndrome (trisomy 21). Lab Invest 74: 617–626

Paris D, Toyama K, Magarbane A, Casanova M, Sinet PM, London J (1996) Rapid fluorescence in situ hybridation on interphasic nuclei to discriminate between homozygous and heterozygous transgenic mice. Transgenic Res 5: 397–403

Pearlman C, Greenberg R (1972) Brief REM sleep deprivation impair consolidation in complex learning in rats. Psychophysiology 9: 109–110

Przedborski S, Kostic V, Jackson-Lewis V, Naini AB, Simonetti S, Fahn S, Carlson E, Epstein CJ, Cadet JL (1992) Transgenic mice with increased Cu/Zn-superoxide dismutase activity are resistant to N-methyl-4-phenyl-1,2,3, 6-tetrahydropyridine-induced neurotoxicity. J Neurosci 12: 1658–1667

Schickler M, Knobler H, Avraham KB, Elroy-Stein O, Groner Y (1989) Diminished serotonin uptake in platelets of transgenic mice with increased Cu/Zn-superoxide dismutase activity. EMBO 8: 1385–1392

Sinet PM, Lavelle F, Michelson AM, Jerome H (1975) Superoxide dismutase activities of blood platelets in trisomy 21. Biochem Biophys Res Com 67: 904–909

Tobler I (2000) Phylogeny of sleep regulation. In: Kryger MH, Roth T, Dement WC (eds) Principles and practice of sleep medicine. Saunders, Philadelphia, p 72

Wisniewski HM, Rabe A (1985) Discrepancy between Alzheimer-type neuropathology and dementia in persons with Down's syndrome. Ann NY Acad Sci 477: 247–260

Authors' address: Dr. N. Sarda, INSERM Unit 480, Claude Bernard University, 8 av. Rockefeller, F-69375 Lyon, France, e-mail: sarda@laennec.univ-lyonl.fr

Spectrum of cognitive, behavioural and emotional problems in children and young adults with Down syndrome

R. Nicham, R. Weitzdörfer, E. Hauser, M. Freidl, M. Schubert, E. Wurst, G. Lubec, and **R. Seidl**

Department of Pediatrics, University of Vienna, Austria

Summary. In comparison to most other groups with intellectual disability individuals with Down syndrome are at lower risk for significant psychopathology, although relative to their typically developing peers they have higher rates of behavioural and emotional problems. A total of 43 Down syndrome patients (21 females and 22 males), who ranged in age from 5.33 to 30.58 years, were examined for the presence of age-related changes in the spectrum of externalizing and internalizing problems. Intelligence tests included Hamburg-Wechsler-Intelligenz Test für Kinder III (HAWIK-III), Hamburg-Wechsler-Intelligenz Test für Erwachsene (HAWIE-R) and Kaufman-Assessment-Battery for Children, German Version (K-ABC). Behavioural and emotional problems were assessed by the the Strengths and Difficulties Questionnaire for Parents, German Version (SDQ) and the Clinical Assessment Scale for Child and Adolescent Psychopathology (CASCAP). IQ was significantly inversly related to the age of patients. Externalizing behaviours (dominant, opposing/refusing, impulsiveness, inattention and increased motor activity) were significantly higher in the 5–10 years old group, whereas internalizing behaviours (shy/insecure, low selfconfidence, decreased motor activity) where more prevalent in adolescents and adults (10–30 years). Possible relationships between this age-related changes and increased risks of later-onset psychopathology (depression and dementia) are discussed.

Introduction

Psychiatric disorders in young people with Down syndrome (DS) comprise disturbances of mood, thought, and behaviour that stem from different (biologic and psychologic) causes and have varying relationships to Down syndrome. Biologic factors include excess chromosomal/genetic material, psychiatric disorders independent of DS and disorders related to mental retardation. Psychosocial stressors and other environmental influences may be conflicts, losses, life changes, and the individual's increasing awareness of his or her disabilities (Myers, 1997).

Relative to their typically developing peers, children and adolescents with DS are at increased risk for behavioural and emotional problems (Rutter et al., 1976; Gath and Gumley, 1986, 1987; Cuskelly and Dads, 1992), in particular attention deficit, noncompliance, thought disorder, and social withdrawal (Coe et al., 1999). In a study by Stores et al. (1998), children with DS and children with other intellectual disabilities showed significantly higher rates of behavioural disturbance on all five of the ABC subscales (Irritability, Lethargy, Stereotypies, Hyperactivity and Inappropriate Speech) and on the Total ABC score. However, the children with other intellectual disabilities also showed significantly higher scores than the children with DS on four of the ABC subscales: Irritability, Lethargy, Stereotypies and Hyperactivity, as well as the Total ABC score. Maternal stress was significantly higher in the group with other forms of intellectual disability than in the DS group.

So interestingly, in comparison to most other groups with intellectual disability DS subjects are at lower risk for significant psychopathology (only 18–23% of children with DS) and score significantly lower on rating scales of maladaptive behaviour (for review see Cuskelly and Gunn, 1997; Myers, 1997; Dykens, 2002). Collacott et al. (1992) published their findings of a study of diagnosed mental disorders in 371 adults with DS. The individuals were matched on the basis of age, sex and residential placement with a similar number of control subjects with other developmental disabilities. People with DS were less likely to have an additional psychiatric diagnosis. In particular, people with DS appeared to be protected from conduct and personality disorder, neurotic disorders (other than depression), and schizophrenia. On the other hand, they were considerably more vulnerable to presenile dementia (×16) and depressive disorder (×2–3) (Collacott et al., 1992; Collacott, 1999). In a later study the Disability Assessment Schedule was used to characterize behaviour of 360 adults with DS compared to 1829 adults with developmental disorders due to other aetiologies. The DS group was less likely to demonstrate maladaptive behaviours (Collacott et al., 1998).

Since the first description by Langdon Down, numerous studies have focused on behaviour and personality in DS children (for review see Myers, 1997), with conflicting results. Although, relatively few children with DS have severe maladaptive behaviour, early identification and interventions may minimize possible withdrawal, nurture sociability and ameliorate personality shifts. In particular, they have more externalizing types of problems than normal controls, including stubbornness, oppositionality and inattention (Pueschel, 1991; Coe et al., 1999).

In future studies age-related patterns have to be clarified to shed some light on the developmental course of behavioural problems. The hyperactivity domain of the ABC (Aberrant Behaviour Checklist) declined in children 4–11 years to age 16–19 years (Stores et al., 1998). The internalizing domain of the CBCL (Child Behaviour Checklist) correlated with age in 4–19 year old DS subjects (Dykens and Kasari, 1997).

Recently Dykens et al. (2002) examined age related changes in the maladaptive behaviour in a large cohort of children and adolescents with DS aged 4–19 years. The 211 participants were divided into four age groups: 4–6 years,

7–9 years, 10–13 years and 14–19 years. Both community and clinical samples of people with DS showed a similar pattern of decreased externalizing symptoms in older adolescents as well as increased internalizing symptoms during these same years. Increases were found in withdrawal, seen in 63% of community-based adolescents, and 75% of adolescents regularly seen in the clinic.

In this present study the authors examined age-related changes in the behavioural (maladaptive) and emotional spectrum in a cohort of children, adolescents and adults with DS, aged 5 to 30 years.

Furthermore, it was the aim of this investigestion to start a collaboration with families and authorities in Austria who care for individuals with DS in order to set up new standards in the care and search for new, well-controlled strategies.

Subjects and methods

Inclusion and exclusion criteria

Inclusion: Patients with DS aged 5–40 years of age were recruited via announcement in an Austrian Down Syndrome magazine. Parents or a caregiver who knows the patient well (spends at least 10 hours per week with him or her), and who is in good health, had to agree to participate as well. The participant and his/her legal representative must provide informed consent.

Exclusion criteria: Inborn errors of metabolism, additional diseases leading to psychomotor retardation, and other neurological diseases. Instable medical conditions like cardiac heart failure or untreated compromised thyroid function. The study protocol was approved by the Ethics Committee of the University of Vienna.

Participants

A total of 43 DS patients (21 females and 22 males) (Table 1), who ranged in age from 5.33 to 30.58 years, were enrolled in the present study. The mean age of the participants was 13.14 ± 7.29 years. All subjects were recruited from family and never lived in an institution. All participants were accompanied by their parents who completed behavioral surveys. Medical histories were obtained and subjects had physical examination and laboratory tests to rule out unstable medical illnesses. A comprehensive checklist of medical conditions inluding cardiac disease, hypothyroidism, head injury, visual and hearing impairment was undertaken.

Procedures

Hamburg-Wechsler-Intelligenz Test für Kinder III (HAWIK-III)

Hamburg-Wechsler-Intelligenz Test für Erwachsene (HAWIE-R)

The HAWIK-III (Tewes et al., 2000) is an individual test for examining the cognitive development of children and adolescents aged from 6 to 16. Through all together 10 subtests and 3 additional tests several aspects of intelligence can be measured. Apart from measuring strengths and weaknesses in the various subtests, a general intelligence quotient as well as a verbal and action IQ is determined. The HAWIK-III was performed by 9 DS patients.

The HAWIE-R (Tewes, 1994) is an intelligence test for adolescents and adults aged from 16 to 74. The test consists of 11 subtests (6 verbal and 5 action tests). It allows both measuring total IQ as well as analysis of these subtests. 3 young adult DS patients underwent the HAWIE-R.

Kaufman-Assessment-Battery for Children, German Version (K-ABC)

The K-ABC (Melchers and Preuß, 1991) is an intelligence test for children from 2;6 to 12;5 years for measuring the intelligence and abilities. The test is based on a definition of intelligence as the ability to solve problems through cognitive processing. Intellectual abilities are measured separately from the degree of acquired abilities (learning and knowledge). 19 of the DS participants were examined with the K-ABC, out of which only 8 were able to perform the required subtests for determine an IQ.

The K-ABC also fits for mentally retarded persons (Maluck and Melchers, 1998). Therefore, its standards (12;3 to 12;5) were applied to some adult Down syndrome patients.

Degree of intellectual retardation was stratified using ICD 10 criteria into mild <70, moderate <50, severe <35 and profound <20.

Developmental Screening Test (Denver II)

The Denver II (Frankenburg and Dodds, 1990) is a developmental screening tool for children from birth to six years of age. The Denver II will assess a child's development in four areas: personal/social, fine motor/adaptive, language and gross motor. The Denver II was applied to children who were either too young or due to difficulties regarding concentration and understanding unable to perform the K-ABC or the HAWIK-III. 4 DS children were tested with the Denver II.

The Strengths and Difficulties Questionnaire for Parents, German Version (SDQ)

The SDQ (Goodman, 1997) is a questionnaire screening the strengths and weaknesses of children and adolescents aged 3 to 16. The SDQ asks about 25 attributes, some positive and others negative. These 25 items are divided between 5 scales: 1) emotional symptoms, 2) conduct problems, 3) hyperactivity/inattention, 4) peer relationship problems and 5) prosocial behaviour. Alltogether 24 parents completed the SDQ.

Clinical Assessment Scale for Child and Adolescent Psychopathology (CASCAP)

The CASCAP is a psychopathological diagnostic tool for children and adolescents. It is a semi-structured interview with the patient (if possible) and caregiver. In addition to that the investigator's evaluation of the setting is taken into account. The proband's respective behaviour is categorised as "unknown", "non-existent", "moderate", "significant" and "strong". In the present study the evaluation rests on information gained from both the parents and the investigator. 40 out of a total of 43 participants aged from 5 to 30 were evaluated with the CASCAP.

Verbaler Lern- und Merkfähigkeitstest (VLMT)

In the VLMT (Helmstaedter et al., 2001) probands are asked five times to reproduce the same series of nouns. After being distracted by a differing list and a break of 30 minutes the proband is asked to repeat the original list. Finally, the proband is confronted with random nouns and asked to recognise those being part of the original list. The test is designed for probands aged from 9 to 42. 12 DS participants were able to perform the VLMT.

Comparisons of frequencies of different questionnaire items were performed using Chi-square analysis. Correlations of continuous variables such as age with IQ were made using the nonparametric Spearman rank correlation coefficient. Statistical significance was tested at the 5% level. SPSS 10.0 package was used for all analyses.

Results

Based on the results of HAWIK-III and HAWIE-R respectively as well as K-ABC a level of intelligence according to ICD10 could be determined for a number of 20 DS patients. 11 DS participants were able to perform some subtests of the K-ABC only. 4 DS children were analysed according to Denver II development scales. 8 participants were incapable of performing any of the before mentioned tests (Table 1).

All of the tested children and young adults with DS show intellectual results below average. The lowest IQ level amounts to 40, while the highest is 72. In average the patients exhibit an IQ of 46.85. Summing up 75% show moderate intellectual retardation (defined as IQ 35 to 49), 20% a slight intellectual retardation (defined as IQ 50 to 69). 1 out of 20 probands exhibits low intelligence (IQ 70 to 84). Within the sample only the 6- to 13-year olds reached IQs of above 50. For those 20 patients who were able to complete one cognitive test, IQ showed a statistical significant inverse correlation with age ($R = -0.477$, $P = 0.03$). It is noteworthy that the probands reached higher IQs in the verbal area than in the action part of the HAWIK or HAWIE respectively. With regard to the K-ABC we want to mention that the subtest "Gestaltschließen" (comprehensive understanding) could be performed by 17 out of 19 participants. Considering the fact that only 8 out 19 patients were able to undergo all necessary subtests this number is high.

The findings of Denver II indicate that all 4 DS children tested do by far not achieve results corresponding to their chronological age. Generally speaking their developmental age is below their respective chronological age in all areas tested.

The results of the SDQ (only parental questionnaire) show that the area "peer relationship problems" is most conspicuous. Here 29.2% of the children and adolescents exhibit unequivocal problems. 16.7% show border-line results. The results in the areas of "hyperactivity/inattention" (16.7%) and "conduct problems" (12.5%) highlight additional problem areas. Emotional

Table 1. Characteristics of the 43 Down syndrome patients

Pat. nr.	Sex	Age in years	IQ-value	Test-procedure	Denver results	IQ action part of HAWIK/HAWIE	IQ verbal part of HAWIK/HAWIE
5	m	5.33		denver	pers>grm>fim>lang		
44	w	5.33					
6	w	6.17	59	k-abc			
26	m	6.25		k-abc			
16	m	6.50		k-abc			
31	m	6.58		denver	pers>grm>fim>lang		
4	w	6.67		k-abc			
88	w	7.25	52	k-abc			
36	w	7.44		denver	pers>fim>grm>lang		
13	w	7.58		k-abc			
40	w	7.58		k-abc			
11	m	8.08		k-abc			
34	m	8.08		k-abc			
33	m	8.75					
8	m	8.92		k-abc			
24	w	8.92		denver	grm>fim>pers>lang		
20	m	9.25		k-abc			
35	m	9.33	46	k-abc			
28	m	9.50	42	k-abc			
9	m	9.67		k-abc			
1	w	9.83	53	hawik		54	61
14	w	9.83	41	k-abc			

3	m	10.42		k-abc		
37	w	10.75	72	hawik	64	82
87	w	11.33	46	k-abc		
19	w	12.25	46	hawik	52	50
38	w	12.25				
42	w	13.92	50	hawik	52	57
12	m	14.00				
43	w	15.00	46	hawik	46	55
22	m	15.25	40	hawik	46	46
32	m	15.42				
21	w	15.75	44	hawik	50	48
27	w	16.92	40	k-abc		
89	m	16.92	41	k-abc		
25	w	21.75	44	hawie	39	51
10	m	22.92	43	hawik	52	46
15	w	24.17	44	hawie	39	51
30	m	26.67				
2	m	27.67				
41	m	28.17	47	hawie	48	62
23	m	30.08	41	hawik	46	47
5	w	30.58		denver		

denver results: > indicates order of decreasing developmental areas

Table 2. Percentages of CASCAP items showing age affects in 43 children with DS, chi^2 test

Scales of symptoms	CASCAP item	Items	5–10 years N = 22 [m/f]	>10 years N = 18 [m/f]	X^2	P
Externalizing						
Aggressive Symptomatology	2.1	dominant	13 [6/7] 59%	6 [3/3] 33%	6.57	**0.010**
	2.2	opposing/refusing	14 [9/5] 63%	7 [4/3] 39%	5.67	**0.017**
	2.3/2.4	aggression verbal/physical	4 [2/2] 18%	3 [1/2] 17%	0.36	0.849
Hyperkinetic symptoms	4.2	increased motor activity	9 [6/3] 41%	2 [2/0] 11%	19.78	**0.000**
	4.3	impulsive	16 [9/7] 73%	7 [5/2] 39%	10.6	**0.001**
	4.4	inattentive/ distractible	21 [12/9] 96%	12 [7/5] 67%	8.2	**0.004**
Social-emotional impulsiveness	8.2	irritable/dysphoric	1 [1/0] 5%	2 [0/2] 11%	0.96	0.327
	1.5	demonstrative	8 [5/3] 36%	4 [3/1] 22%	2.55	0.111
Internalizing						
Anxiety	1.2	shy/insecure	1 [0/1] 5%	8 [3/5] 44%	14.18	**0.000**

	1.1	overadjusted	2 [0/2] 9%	3 [2/1] 17%	0.91	0.340
	6.1–6.6	anxious	13 [8/5] 59%	9 [5/4] 50%	0.73	0.394
Depressive symptoms	8.1	depression/sadness	2 [2/0] 9%	3 [1/2] 17%	0.91	0.340
	8.4	reduced selfconfidence	0 0%	7 [2/5] 39%	10.21	**0.001**
	4.1	decreased motor activity	0	8 [4/4] 44%	14.10	**0.000**
	5.1–5.4	tics, stereotypes, abnormal habits	6 [3/3] 27%	9 [4/5] 50%	4.55	**0.033**
	9.4	increased food intake	3 [0/3] 14%	6 [1/5] 33%	3.84	**0.050**
	10.6	insomnia/sleep disturbance	8 [7/1] (36%)	4 [2/2] (22%)	2.54	0.111
Language problems	3.2	articulation problems	22 [12/10] 100%	18 [9/9] 100%	NS	
	3.3	expressive speech disorder	22 [12/10] 100%	18 [9/9] 100%	NS	
	3.4	receptive speech disorder	19 [10/9] 86%	17 [9/8] 94%	2.74	0.098

Table 3. Percentages of SDQ items showing age affects in 43 children with DS, chi² test

	5–10 years N = 14 [m/f]	10–16 years N = 10 [m/f]	X^2	P
Emotional symptoms	1 [1/0] 7%	2 [0/2] 20%	1.44	0.229
Conduct problems	4 [4/0] 29%	4 [2/2] 40%	0.76	0.383
Hyperactivity/inattention	4 [3/1] 29%	1 [1/0] 10%	5.36	0.021
Peer relationship problems	5 [4/1] 36%	6 [2/4] 60%	3.44	0.064
Prosocial	0 0%	1 [0/1] 10%	NS	
Total	2 [2/0] (14%)	3 [1/2] (30%)	1.65	0.199

problems could not be found among children and adolescents. When patients were divided into 2 age groups (children 5–10 years, and adolescents 10–16 years), parents complained more often about hyperactivity/inattention in children, whereas adolescents were considered to have more often peer relation problems (Table 3).

The findings of the CASCAP underscore that more than half of the 5-to 10-years olds tested were slightly to significantly conspicuous in more externalizing domains: "dominant" (59%), "opposing/refusing" (63%), "inattentive/distractible" (96%), "impulsive" (73%). This indicates that children and adolescents are inconsiderate of others and are rather "dominant". They often refuse meeting requests or do not respect rules and regulations ("opposing/refusing"), have difficulties concentrating and are easily distracted ("inattentive/distractible"). Moreover, they tend towards acting unreflectively and fast and do hardly await their turn ("impulsive"). In this age group males were more concerned by these behaviours than the females.

Concerning internalizing domains 59% of the 5 to 10-year-olds were found to be "anxious", suffering mainly from phobias ("anxious"). Only 9% of this age group complained about depressive symptoms. Summing up externalizing behaviours were higher in 5 to 10-year-olds (Table 2).

Among the 10–30-year-olds (n = 18) the results are slightly different. In this age group behaviours of the internalizing areas were higher: 44% were found to be "shy/insecure", 50% suffered from phobias ("anxious"), 17% complained about depressive symptoms and 39% showed reduced selfconfidence.

Concerning externalizing behavioural patterns 67% were found to be "inattentive/distractible", 33% "dominant", 39% "opposing/refusing" and "impulsive". In the >10 years group psychomotor behaviours like tics, stereotypes, abnormal habits were found by 50%.

Verbal or physical aggression was not a problem in any of the age groups.

A total of 55% of 40 DS participants or caregivers regarded anxious behaviour (phobias) as a problem. These are mainly specific fears dealing with concrete topics, e.g. agoraphobia, fear of loud noises, fear of certain situations of daily life.

Comparing the frequencies of the items between both age groups, externalizing behaviour problems (dominant, opposing/refusing, increased motor activity, impulsiveness and inattention) were significantly more frequent in children 5–10 years of age, while internalizing problems (shy/insecure, reduced selfconfidence) were more common in adolescents and adults (Table 2).

One adult male (28.2 years) and one adolescent female (16 years) were suspected to have a depressive disorder and treatment with SSRI was started by a psychiatric consultant after verification of the diagnosis.

The VLMT shows that 81.8% reached below-average results with regard to the aggregate performance of learning, 63.6% with regard to the performance of recognition. Hence, DS participants have generally memorised only little and results in terms of recognition do not meet age-appropriate levels. However, the short-term memory defined as loss of information over time (in this study a period of 30 minutes), is very low. While 54.5% achieved average results 27.3% reached results above average.

Discussion

Psychopathology in individuals with DS is distinctive from other groups with intellectual disability.

Young people with DS may be at lower risk for psychiatric disorders than those with mental retardation from other causes (Collacott et al., 1992; Myers, 1997). In the survey by Collacott (1992) 25.9% of 371 individuals with DS, compared to 37.7% of 371 people with mental retardation of other causes showed psychiatric disorders of mood, thought and behavior (Collacott et al., 1992). Myers and Pueschel (1991), in a survey of 425 outpatient children, adolescents and adults with DS, reported that 13% of children under the age of 10 revealed psychiatric disorders; 20% of those between 10 and 20 and 25% of outpatient adults (over age 20) had such disorders. In the study by McCarthy and Boyd (2001), the prevalence rate of psychiatric disorders in DS was 35%, with mood disorder being the most common problem.

It is difficult to compare the rates of behavior disorders reported in various studies of DS because of the different classifications used. A differentiation between psychiatric diagnoses and behavioral problems seems useful (Cuskelly and Gunn, 1997).

As definite diagnosis (ICD 10 or DSM IV) cannot be made with the help of the used questionnaires, only the spectrum of difficult behavior/behavioral problems and emotional problems, categorized as more externalizing or internalizing domains, can be given in the present study.

Disruptive behavior disorders (i.e. attention-deficit/hyperactivity disorder [ADHD], conduct disorder, oppositional disorder) are the most common psychiatric disorders in children with DS (Myers and Pueschel, 1991), as well as in the general population. Under age of 20, 6.1% of 264 individuals showed attentional disorders (ADHD). In a study by Pueschel (1991) statistically significant differences were observed between study and control groups on externalizing and total scores on responses using the Aschenbach Child Behavior Checklist, obtained from both parents and teachers, but not on internalizing scores. Further analysis of the data on a subgroup of 28 children (70%) between the ages of 6 and 11 years revealed a "hyperactive" profile pattern for both boys (n = 16) and girls (n = 12). About one half of the boys sometimes exhibited hyperactive behaviours (Pueschel et al., 1991). *Conduct disorder* (e.g. fighting, temper tantrums, destructiveness, stealing, running away, fire setting) and *oppositional disorder* (e.g. uncooperative behavior within the family) are no more frequent in children and adolescents with DS than in those with mental retardation of other causes (about 12–20%) (Gath and Gumley, 1986; Myers and Pueschel, 1991).

Infantile Autism and Pervasive Developmental Disorders is found in 1–2.2% of DS children (Gath and Gumley, 1986; Myers and Pueschel, 1991; Collacott et al., 1992), and there is probably no specific relation between autism and DS, as it is found more frequently in patients with other causes of mental retardation. The frequency of *stereotypic behavior* (repetitive, nonfunctional, self-stimulatory behavior) appears to be low in DS (2.7% under the age of 20, and 4.3% in adults with DS) (Myers and Pueschel, 1991).

Similar as in the study by Dykens et al. (2002) age — related patterns of behaviour were found in the present study. Decreased externalizing behaviour in adolescents and adults opposed to increased internalizing problems. Specifically low-level aggressive behaviours like dominant, opposing/refusing behaviour are prevalent in children under the age of 10 years, which seems to correspond to argumentative behavior, attention demanding or swearing among the 10–13 year old group in the study by Dykens et al. (2002). However, the low frequency of more direct verbal and physical aggression (16%) is noteworthy, and has also been found in other studies (Collacott et al., 1998; Cooper and Prasher, 1998; Dykens et al., 2002).

Above the age of 10 years, internalizing behaviours increased. Dykens et al. (2002) found that primarily the withdrawn domain (as retreat from interpersonal contact and social involvement) dominates. In fact, 66% of adolescents were described as preferring to be alone than with others, and one third were cast as secretive and not wanting to talk. This clearly corresponds to the higher frequencies of shy/insecure, not selfconfident and sometimes sad behaviour in the present study.

It is not known, if subtle increases in internalizing symptoms over the adolescent period might set the stage for later depressive disorder, or be the early changes of mood or behaviour associated with dementia (Dykens et al., 2002), as discussed later in more detail. Withdrawal (Dykens et al., 2002) and shy/insecure, low selfconfident behavior (present study) was higher in females, who are approximately 1.77 times more likely to develop dementia than men (Lai et al., 1999). Reduced motor activity and increased food intake also increased in the older DS group, again very similar to the study by Dykens et al. (2002) (domains underactive and overweight).

McCarthy and Boyd (2001) were able to interview 52 adult DS subject for presence of psychiatric disorders. These subjects were from a sample of 193 subjects examined in childhood and adolescence for psychiatric and behaviour disorders. They also found, that there may be a pattern of disorder with conduct problems of the externalizing type dominating in childhood followed by mood disorder in early adult life, but there was no evidence of any continuity between these two disorders.

Turner and Sloper (1996) found a significant decline with age in the overall frequency of behaviour problems including overactivity.

In the study by Stores et al. (1998), daytime behaviour subscale hyperactivity showed higher scores in children with DS below 11 years of age than in those 12–19 years af age, where boys were reported to have more disturbed behaviour.

The results of the SDQ total score underlines that most children with DS do not appear to experience significant behavioral problems which could impede everyday adaptive functioning. Only 20% of the community sample (N = 180) had clinically elevated CBCL scores (Dykens et al., 2002), consistent with previous studies (Myers and Pueschel, 1991), and much lower than in children with intellectual disability in general (Einfeld and Tonge, 1996).

DS patients are at an elevated risk for Alzheimer-type dementia. In this study, none of the adult patients as observed by their caregivers showed signs of a cognitive decline wich could be early manifestation of a dementia. However, IQ for patients who could be tested significantly declined with age. Examining for signs of dementia in adults with DS is hampered by poor test-taking skills on standardized measures. By definition, dementia implies a change from baseline functioning and baseline is often difficult to document in this population (Nelson at al., 2001). Due to premorbid intelligence, performance on standard cognitive instruments is poor even in the non-demented at baseline (Tyrrell et al., 2001). Low baseline functioning can mask downward changes in cognitive functioning (Dalton, 1992; Gibson et al., 1988), resulting in false negative results. In addition, cognitive delay, severe language problems, and low attention combine to negatively impact test results. As individuals with DS age, their ability to withstand the rigors of psychological testing diminish (Burt et al., 1998). Alternative methods to direct evaluation of the subject with intellectual disabilities are informant-based reports, as used in this study (CASCAP, SDQ, PAS-ADD). However this carer-rated questionnaires do not directly measure cognitive function (Tyrrell, 2001).

Virtually all DS individuals manifest progressive AD-like neuropathology beyond 35 years of age (7.5% at 10–19 yrs., 15,5% at 20–29 yrs., 80% over 30 yrs.), and clinical dementia occurs in only a part of the patients beyond 40 years of age (Wisniewski et al., 1985), increasing to 75–90% of DS individuals after 60 years of age (Haxby and Schapiro, 1992). The discrepancy between the age at which neuropathological features are present and the age at which clinical dementia is present, suggests several possibilities, such as the presence of amyloid plaques in the brain does not necessarily indicate AD, or people with DS may be protected from earlier clinical onset by virtue of their different brain anatomy (Holland at al., 1998), or the detection of clinical dementia may be difficult in people with DS, as mentioned above. Clinical prevalence increases from about 7–11% between ages 40 and 49 to 50–77% between 60 and 69 and 50–100% at age 70 and over. (Visser, 1997; Tyrrell, 2001). The average age of clinical onset in prospective studies was 51–54 years, with 8% appearing in those by 35–49 years of age, 55% in those by 50–59 years of age, and 75% in those over 60 years of age (Lai and Williams, 1989). Several reasons account for the different prevalence values between studies: different criteria to make a diagnosis of dementia, some studies excluded cases of dementia on study entry, heterogeneous sampling like population-based or people from residential homes or community (Tyrrell, 2001).

The occurrence of Alzheimer-type dementia in people with Down syndrome under the age 30 is rare. Thus, the cause of a mental and social decline in a young person with DS from ages 10–30 is unlikely to be AD, other causes of dementia or pseudodementia like depression, hyperthyroidism, folate or B12 deficiencies should be excluded (Myers, 1997).

Deterioration of speech and gait, change in adaptive behaviour are early signs of dementia, epileptic seizures and myoclonus occur, loss of cognitive function, memory loss often remain rather unrecognized. Nevertheless, irrespective of whether dementia occurs, a progressive loss of intellectual function and brain metabolism appears in many older individuals with DS: diminished long-term memory and impaired visuospatial construction in neuropsychological testing (Haxby, 1989); age dependent changes in P300 component of "event related potentials" after age of about 37 years (St.Clair and Blackwood, 1985); decreased glucose metabolism (positron emission tomography) in temporal and parietal cortical areas during "stress test" by audiovisual stimulation (Pietrini et al., 2000). This raises the question and also points to the difficulties in diagnosis of dementia in mentally retarded individuals, further complicated by occurrence of non-cognitive, behavioral symptoms mimicking functional decline in adults with DS (Geldmacher et al., 1997). Non-cognitive (psychopathological) and behavioral symptoms are considered as a cardinal feature of possible functional decline in adults with Down syndrome (DS) and typically present as mood disorder (e.g. depression), anxiety-disorder with or without challenging behavior (irritability, agitation, self-injury, hyperactivity, aggression, hostility or autistic features). In fact, occurrence of non-cognitive, behavioral symptoms may mimick functional decline, where some patients who are thought to be demented are suffering from major depression (Warren et al., 1989). On the other side, behavioral, emotional symptoms and personality changes may precede or represent the early signs of cognitive decline of developing AD (Aylward et al., 1997; Holland et al., 2000). In a study by Holland et al. (2000), 71% of 49 persons had initially deteriorated in their personality or behaviour, rather than in other areas of ability such as memory. Apathy was the change most frequently reported, followed by stubbornness. Changes in personality and behaviour were also the most common first changes in those who had the diagnosis of dementia. There was a predominant trend moving from behavioural or personality changes observed in younger adults to changes across other domains (memory, self-care skills) in the older groups. Early behavioral changes could be a consequence of early, but unrecognized, cognitive decline in specific areas of functioning or that functions served by the frontal lobes are compromised early in the course of the neuropathological progression (Holland et al., 2000; Nelson, 2001). In the study by Nelson et al. (2001), a distinct pattern of emotional functioning was found in adults with DS who showed signs of probable dementia. This pattern was characterized by symptoms of depression and indifference (i.e. anosognosia, apathy). These emotional factors may be taken as a sign of prefrontal lobe dysfunction in which symptoms of inattentiveness, apathy, heightenend threshhold to stimulation and labile mood predominate (Nelson, 2001).

Two of the 43 DS individuals were diagnosed as having a depressive disorder. There is no direct evidence for depressive disorders in children with DS, but several reports of major depressive disorder in adults have been published (for review see Myers, 1997; Collacott, 1999). It may be difficult for DS individuals to verbalize complaints of depression, self-deprecation, guilt, suicidal thoughts, or fatigue, but specific behaviors such as crying, loss of interest or pleasure, poor appetite and weight loss, insomnia, psychomotor agitation or retardation, and poor agitation can be observed to support the diagnosis of depressive disorder (Myers, 1997).

In a 10-year review Myers and Pueschel (1995) found a freqency 5.5%, Collacott (1992) published a lifetime frequency of all levels of severity of 11.3%, compared to 4.3% in mental retardation of other causes. In study by McCarthy and Boyd (2001) 13% of adults had mood disorder. This greater prevalence suggests the possibility of a specific relationsship between DS and depression. As stated earlier, a marked deterioration in mental, social and adaptive abilities in a young adult with DS may suggest Alzheimer disease, but is more likely to be major depression, even though the two disorders can occur together (Warren, 1990). Depression may be likely to respond to serotonergic antidepressants, as low levels of serotonin and norepinephrine in brains of DS may have a relationship to depression (Collacott, 1992). Depression in DS patients leads to additional mental suffering, behavioral disturbance (aggression), poor cognition, poor self-care, caregiver depression, caregiver burden, and earlier entry into a nursing home.

Following reports of serotonin deficiency in blood, cerebrospinal fluid and platelets, as well as decreased 5-HT platelet uptake in DS (see Epstein, 1995), significant reductions of 5-HT have been reported in most telencephalic brain regions (Yates et al., 1986; Godridge et al., 1987; Seidl et al., 1999; Risser et al., 1997). Neuronal loss in 5-HT projection side, midbrain dorsal raphe nuclei has also been identified (Mann et al., 1985). Reversal of hypotonia by administration of 5-hydroxytryptophan (Bazelon et al., 1967), reduction of self-injurious behavior by dietary increase in serotonin (Gedye, 1990, 1991), but also no beneficial effect on intelligence development in response to 5-hydroxytryptophan (Weise, 1974) or increased risk for infantile spasms (Coleman, 1971) have been reported. Recently, treatment with selective serotonin-reuptake inhibitor (SSRI) medication in six DS patients, aged 23 to

63 years, showed improvement in behaviors and on objective measures, such as workplace productivity. These DS patients presented with functional decline in adult life, but non-cognitive symptoms were prominent and included aggression, social withdrawal, compulsive behaviors and cognitive dysfunction at varying degrees. Authors concluded, that treatment trials with SSRIs may, therefore, be warranted in such cases (Geldmacher et al., 1997). In an earlier study, low dose antidepressant treatment combined with a serotonin-enhancing diet in a mentally handicapped adult with Down syndrome showing signs of Alzheimer-type dementia decreased aggressive behavior (Gedye, 1991). However, clinical, biochemical and/or neuropathological data supporting a serotonergic dysfunction in infant and childhood DS brain have not been reported.

Anxiety disorders include panic disorders with or without agoraphobia, specific phobias, social phobias, posttraumatic stress disorder, generalized anxiety disorder and obsessive-compulsive disorder.

Gath and Gumley (1986) found 3% of DS children exhibited anxiety and fearfullness, Myers and Pueschel (1991) noted specific phobias in 1,5% of 261 DS patients under 20, and in 0.6% over 20 years of age.

In the present study, a total of 55% of 40 DS participants or caregivers regarded anxious behaviour (phobias) as a problem. These are mainly specific fears dealing with concrete topics, e.g. agoraphobia, fear of loud noises, fear of certain situations of daily life. In addition, rates of anxious behavior, remained fairly constant across age groups.

Self-Injurious behavior was not observed in this study. Previously, 0.8% of 261 children and 1.6% of 164 adults (Myers and Pueschel, 1991) exhibited this behavior, which seems to be distinctively lower in people with DS than in people with mental retardation of other causes (Myers, 1997).

Limitations of the present study that need to be considered are the cross-sectional design, limited sample size and the lack of a comparison group. The latter is the reason why the authors concentrated on the age-related, within-group changes, giving an impression of the change of the spectrum of problems rather than data of prevalence in comparison to a normal population. However, a longitudinal study will follow which might give more concrete answers when shifts in externalizing and internalizing behaviour are likely to occur and how this relates to the later onset of more serious psychiatric disorders (Dykens et al., 2002). This was not an epidemiological or population-based study, and it is possible that parents interested in a study of behavioral problems are those whose children are more likely to have behavioural disturbances. Even with these limitations, the present study confirms previous findings of age-related changes in behavioural problems in a cohort of people with DS, which warrants future longitudinal research in Austria.

Acknowledgements

We are highly indebted to the Red Bull Company, Salzburg, Austria, for generous financial support of the study. Furthermore the authors have to thank the "Verein zur

Durchführung der wissenschaftlichen Forschung auf dem Gebiet der Neonatologie und Kinderintensivmedizin".

References

Aylward E, Burt D, Thorpe L, Lai F, Dalton A (1997) Diagnosis of dementia in individuals with intellectual disability: report of the task force for development of criteria for diagnosis of dementia in individuals with mental retardation. J Intellect Disabil Res 41: 152–164

Bazelon M, Paine RS, Coeiw VA, Hunt P, Houck JC, Mahanand D (1967) Reversal of hypotonia in infants with Down's syndrome by administration of 5-hydroxytryptophan. Lancet i: 1130–1133

Burt D, Loveland K, Primeaux-Hart S, Chen Y, Phillips N, Cleveland L, Lewis K, Lesser J, Cummings E (1998) Dementia in adults with Down Syndrome: diagnostic challenges. Am J Ment Retard 103: 130–145

Coe DA, Matson JL, Russell DW, Slifer KJ, Capone GT, Baglio C, Stallings S (1999) Behavior problems of children with Down syndrome and life events. J Autism Dev Disord 29: 149–156

Coleman M (1971) Infantile spasms associated with 5-hydroxytryptophan administered in patients with Down's syndrome. Neurol 21: 911–919

Collacott RA (1999) People with Down syndrome and mental health needs. In: Bouras N (ed) Psychiatric and behavioural disorders in developmental disabilities and mental retardation. Cambridge University Press, Cambridge, pp 200–211

Collacott RA, Cooper S, McGrother C (1992) Differential rates of psychiatric disorders in adults with Down's syndrome compared to other mentally handicapped adults. Br J Psychiatry 161: 671–674

Collacott RA, Cooper SA, Branford D, McGrother C (1998) Behaviour phenotype for Down's syndrome. Br J Psychiatry 172: 85–89

Cooper SA, Prasher VP (1998) Maladaptive behaviours and symptoms of dementia in adults with Down's syndrome compared with adults with intellectual disability of other aetiologies. J Intellect Disabil Res 42: 293–300

Cuskelly M, Dadds M (1992) Behavioural problems in children with Down's syndrome and their siblings. J Child Psychol Psychiatry 33: 749–761

Cuskelly M, Gunn P (1997) Behavior concerns. In: Pueschel SM, Sustrova M (eds) Adolescents with Down Syndrome. Paul H. Brookes, Baltimore London Toronto Sydney, pp 111–127

Dalton A (1992) Dementia in Down syndrome: methods of evaluation. In: Nadel L, Epstein C (eds) Down Syndrome and Alzheimer disease. Wiley-Liss, New York, pp 51–76

Döpfner M, Berner W, Flechtner H, Lehmkuhl G, Steinhausen H-C (1999) Clinical Assessment Scale for Child and Adolescent Psychopathology. Hogrefe, Göttingen Bern Toronto Seattle

Dykens EM, Kasari C (1997) Maladaptive behaviour in children with Prader-Willi syndrome, Down syndrome, and non-specific mental retardation. Am J Ment Retard 102: 228–237

Dykens EM, Hodapp RM, Evans DW (1994) Profiles and development of adaptive behavior in children with Down syndrome. Am J Ment Retard 98: 580–587

Dykens EM, Shah B, Sagun J, Beck T, King BH (2002) Maladaptive behaviour in children and adolescents with Down's syndrome. J Intellect Disabil Res 46: 484–492

Einfeld SL, Tonge BJ (1996) Population prevalence of psychopathology in children and adolescents with intellectual disability. II. Epidemiological findings. J Intellect Disabil Res 40: 99–109

Epstein CJ (1995) Down Syndrome (Trisomy 21). In: Scriver SR, Beaudet AL, Sly WS, Valle D (eds) The metabolic and molecular bases of inherited disease. McGraw-Hill, New York, pp 749–794

Frankenburg WK, Dodds JB (1990) Denver II developmental screening test. In: Stanhope M, Knollmueller RN (1992) Handbook of community and home health-nursing: tools for assessment intervention and education. Mosby, St. Louis, pp 317–319

Gath A, Gumley D (1986) Behaviour problems in retarded children with special reference to Down's syndrome. Br J Psychiatry 149: 156–161

Gath A, Gumley D (1987) Retarded children and their siblings. J Child Psychol Psychiatry 5: 715–730

Gedye A (1990) Dietary increase in serotonin reduces self-injurious behaviour in a Down's syndrome adult. J Ment Defic Res 34: 195–203

Gedye A (1991) Serotonergic treatment of aggression in a Down's syndrome adult showing signs of Alzheimer's disease. J Ment Defic Res 35: 247–258

Geldmacher DS, Lerner AJ, Voci JM, Noelker EA, Somple LC, Whitehouse PJ (1997) Treatment of functional decline in adults with Down syndrome using selective serotonin-reuptake inhibitor drugs. J Geriatr Psychiat Neurol 10: 99–104

Gibson D, Groenewez G, Jerry P, Harris A (1988) Age and pattern of intellectual decline among Down Syndrome and other mentally retarded adults. Int J Rehabil Res 11: 47–55

Godridge H, Reynolds GP, Czudek C, Calcutt NA, Benton M (1987) Alzheimer-like neurotransmitter deficits in adult Down's syndrome brain tissue. J Neurol Neurosurg Psychiatry 50: 775–778

Goodman R (1997) The strengths and difficulties questionnaire for parents, German version. J Child Psychol Psychiatry 38: 581–586

Haxby JV (1989) Neuropsychological evaluation of adults with Down's syndrome: patterns of selective impairment in nondemented old adults. J Ment Defic Res 33: 193–197

Haxby JV, Schapiro MB (1992) Longitudinal study of neuropsychological functions in older adults with Down syndrome. In: Epstein C, Nadel L (eds) Down syndrome and Alzheimer disease. Wiley Liss, New York, pp 35–50

Helmstaedter C, Lendt M, Lux S (2001) Verbaler Lern- und Merkfähigkeitstest. Beltz, Göttingen

Holland AJ, Hon J, Huppert FA, Stevens F, Watson P (1998) Population based study on the prevalence and presentation of dementia in adults with Down's syndrome. Br J Psychiatry 172: 493–498

Holland AJ, Hon J, Huppert FA, Stevens F (2000) Incidence and course of dementia in people with Down's syndrome: findings from a population-based study. J Intellect Disabil Res 44: 138–146

Johnson J, Head E, Kim R, Starr A, Cotman C (1999) Clinical and pathological evidence for a frontal variant of Alzheimer's disease. Arch Neurol 56: 1233–1239

Lai F, Williams R (1989) A prospective study of Alzheimer disease in Down syndrome. Arch Neurol 46: 849–853

Lai F, Kammann E, Rebeck GW, Anderson A, Chen Y, Nixon RA (1999) APOE genotype and gender effects on Alzheimer disease in 100 adults with Down syndrome. Neurol 53: 331–336

Maluck A, Melchers P (1998) Kaufman Assessment Battery for Children. Differential evaluation of (partial) intellectual ability of mentally handicapped adults. Nervenarzt 69: 1007–1014

Mann DM, Yates PO, Marcyniuk B, Rawndra CR (1985) Pathological evidence for neurotransmitter deficits in Down's syndrome of middle age. J Ment Defic Res 29: 125–135

McCarthy J, Boyd J (2001) Psychopathology and young people with Down's syndrome: childhood predictors and adult outcome of disorder. J Intellect Disabil Res 45: 99–105

Melchers P, Preuß U (1991) Kaufman-Assessment-Battery for Children, German Version. Swets & Zeitlinger, Amsterdam
Myers BA (1997) Psychiatric disorders. In: Pueschel SM, Sustrova M (eds) Adolescents with Down Syndrome. Paul H. Brookes, Baltimore London Toronto Sydney, pp 129–142
Myers BA, Pueschel SM (1991) Psychiatric disorders in population with Down syndrome. J Nerv Ment Dis 179: 609–613
Nelson LD, Orme D, Osann K, Lott IT (2001) Neurological changes and emotional functioning in adults with Down Syndrome. J Intellect Disabil Res 45: 450–456
Pietrini P, Alexander GE, Furey ML, Hampel H, Guazelli M (2000) The neurometabolic landscape of cognitive decline: in vivo studies with positron emission tomography in Alzheimer's disease. Int J Psychophysiol 37: 87–98
Pueschel SM, Bernier JC, Pezzullo JC (1991) Behavioural observations in children with Down's syndrome. J Ment Defic Res 35: 502–511
Risser D, Lubec G, Cairns N, Herrera-Marschitz M (1997) Excitatory amino acids and monoamines in parahippocampal gyrus and frontal cortical pole of adults with Down syndrome. Life Sci 60: 1231–1237
Rutter M, Tizard J, Yule W, Graham P, Whitmore K (1976) Research report: Isle of Wight Studies, 1964–1974. Psychol Med 6: 313–332
Seidl R, Kaehler ST, Prast H, Singewald N, Cairns N, Gratzer M, Lubec G (1999) Serotonin (5-HT) in brains of adult patients with Down syndrome. J Neural Transm [Suppl] 57: 221–232
St.Clair DF, Blackwood D (1985) Premature senility in Down's syndrome. Lancet 6: 34
Stores R, Stores G, Fellows B, Buckley S (1998) Daytime behaviour problems and maternal stress in children with Down's syndrome, their siblings, and non-intellectually disabled and other intellectually disabled peers. J Intellect Disabil Res 42: 228–237
Tewes U (1994) Hamburg-Wechsler-Intelligenztest für Erwachsene. Huber, Bern Göttingen Toronto Seattle
Tewes U, Rossmann P, Schallberger U (2000) Hamburg-Wechsler-Intelligenz Test für Kinder III. Huber, Bern Göttingen Toronto Seattle
Turner S, Sloper P (1996) Behaviour problems among children with Down's Syndrome: prevalence, persistence and parental appraisal. J Appl Res Intellect Disabil 9: 129–144
Tyrrell J, Cosgrave M, McCarron M, McPherson J, Calvert J, Kelly A, McLaughlin M, Gill M, Lawlor BA (2001) Dementia in people with Down's syndrome. Int J Geriatr Psychiatry 16: 1168–1174
Visser FE, Aldenkamp AP, van Huffelen AC, et al. (1997) Prospective study of the prevalence of Alzheimer-type dementia in institutionalised individuals with Down syndrome. Am J Ment Retard 101: 400–412
Warren AC, Holroyd S, Folstein P (1989) Major depression in Down's syndrome. Br J Psychiatry 155: 202–207
Warren AC, Holroyd S, Folstein MP (1990) Mayor depression in Down's syndrome. Br J Psychiatry 155: 202–205
Weise P, Koch R, Shaw KNF, Rosenfeld MJ (1974) The use of 5-HTP in the treatment of Down's syndrome. Pediatr 54: 165–167
Wisniewski KE, Wisniewski HM, Wen GY (1985) Occurrence of neuropathological changes and dementia of Alzheimer's disease in Down syndrome. Ann Neurol 17: 278–282
Yates CM, Simpson J, Gordon A (1986). Regional brain 5-hydroxytryptamine levels are reduced in senile Down's syndrome as in Alzheimer's disease. Neurosci Lett 65: 189–192

Authors' address: Prof. Dr. R. Seidl, Department of Pediatrics, University of Vienna, Währinger Gürtel 18-20, A-1090 Vienna, Austria, e-mail: rainer.seidl@akh-wien.ac.at

Overexpression of transcription factor BACH1 in fetal Down Syndrome brain

R. Ferrando-Miguel[1], **M. S. Cheon**[1], **J.-W. Yang**[1], and **G. Lubec**[1]

Department of Pediatrics, University of Vienna, Austria

Summary. There is a series of about 12 transcription factors expressed on chromosome 21. These transcription factors (TFs) are major candidates for playing a pathogenetic role for the abnormal wiring of the brain in fetal Down Syndrome (DS) as approximately 5,000 TFs are developmentally involved in the complex architecture of the human brain. TF derangement in DS has been already reported and we decided to contribute to the problem by studying four TFs encoded on chromosome 21 in fetal DS brain.

We used fetal cortex of 8 DS fetuses and 6 controls (females) from the 18–19th week of gestation. Brain homogenates were subject to immunoblotting using goat-anti-BACH1, rabbit anti-heme oxygenase 1 (HO1), rabbit anti-ERG, rabbit anti-RUNX1 and goat anti-SIM2 l. Antibodies against beta-actin were used to normalise cell loss and antibodies against neuron-specific enolase were used to compensate neuronal loss.

BACH1 was significantly overexpressed in fetal DS ($p < 0.008$) as compared to controls whereas RUNX1 and ERG proteins were comparable between groups, and SIM2 l was not detectable in any specimen. BACH1 was even significantly increased in the DS panel when normalised versus the housekeeping protein beta-actin ($p < 0.01$) or the neuron specific enolase ($p < 0.01$). HO-1 was found comparable between groups.

BACH1, a member of the family of BTB-basic leucine zipper transcription factors, regulates gene expression through the NF-E2 site. More specifically, BACH1 suppresses expression of HO1. Increased BACH1, however, did not lead to decreased HO1, which would have explained oxidative stress observed in fetal DS.

Abbreviations

DS Down Syndrome; *co* control; *BACH1* Transcription regulation protein BTB and CNC homolog 1; *HO1* heme oxygenase 1; *SIM2* Single-minded homolog 2; *RUNX1* Runt-related transcription factor 1; *NSE* Neuron specific enolase.

Introduction

Down Syndrome (DS), trisomy 21, is one of the most common causes for mental retardation. There is evidence for anatomic, histological and ultrastructural alterations including reduction of brain size, abnormal neuronal migration, differentiation and abnormal dendritic arborization (Epstein, 1995; Capone, 2001). A host of biochemical mechanisms has been reported (Engidawork and Lubec, 2003) and all these features, at least in part may be due to the impaired transcriptional machinery.

Regulation of gene expression during brain development is crucial for correct cell differentiation and histological architecture. Thousands of transcription factors (TFs) are involved in the process of gene activation (or inactivation) required for cell migration, cell-cell interaction and signaling pathways in the different phases of development and wiring of the brain, resulting in a highly regulated spatial and temporal network (He and Rosenfeld, 1991). Thus, dysregulation of this system may have serious consequences, both at the morphological and functional level (Johnston et al., 2003).

Altered expression of several TFs has been reported in DS brain. At the mRNA level Scleraxis, a helix-loop-helix type TF that regulates cell differentiation was up-regulated. Moreover, NF-kappaB, controlling genes for immune and inflammatory responses, HOXA-13 that determines development of the body axis, Ptx1, a member of the bicoid family that plays a role for development of anterior structures like face and brain, and REST, important for neuronal plasticity and synapse formation were down-regulated in fetal DS brain (Labudova et al., 1999; Bahn et al., 2002). At the protein level JunD, involved in neurogenesis, was reduced in adult DS brains (Labudova et al., 1998). ETS-2, another TF for organogenesis of brain, heart and skeletal system was unchanged (Engidawork et al., 2001), whereas protooncogens CRK and CRK-like protein (CRKL) mediating transduction of intracellular signals through Rac or related proteins were reduced in fetal brain of DS (Freidl et al., 2001).

Trying to get further insight into the role of TFs in DS we decided to study expression of BACH1, RUNX1, SIM2 1 and ERG, four TFs encoded on chromosome 21 (Fig. 1). BACH1 is an ubiquitous protein containing a BTB/POZ protein interaction domain and the CNC (Cap'n'Collar)-type bZip domain. BACH1 forms heterodimers by bZip domains with MafK, one of the small bZip protein of the NF-E2 TF, and this heterodimer recognizes the DNA sequence motif MARE coordinating transcription activation and repression (Oyake et al., 1996; Ohira et al., 1998). These binding sites are present in different genes as e.g. heme oxigenase 1 (HO-1), an inducible enzyme that converts toxic heme into antioxidants and is repressed by Bach1/Mafk heterodimer (Sun et al., 2002; Kitamuro et al., 2003).

ERG is a member of the Ets-transcription factor family and was first described in the v-ets oncogen from the avian acute leukemia virus (Reddy et al., 1987). In humans it has been described in the fusion proteins formed in myeloid leukemia and Erwing sarcoma (Ichikawa et al., 1994; Delattre et al.,

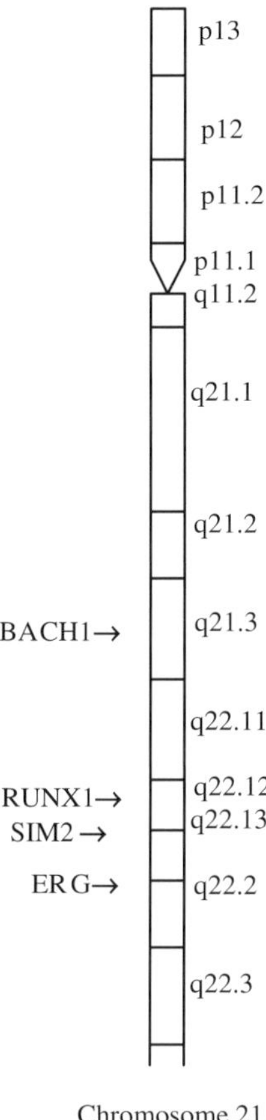

Fig. 1. Giemsa banding (G-bands) of human chromosome 21. Arrows indicate four gene products encoded on chromosome 21, examined in this study

1994). ERG is involved in epithelial-mesenchymal interactions through regulation of extracellular matrix proteins (Maroulakou et al., 2000). During mouse brain development Erg expression was found in migrating neural crest cells and expression ceased with settlement and formation of dorsal root ganglia (Valeminck-Guillem et al., 2000). Erg mRNA was detected in human fetal brain and in cortex from adult human brain at a low expression level (Su et al., 2002).

RUNX1 is the Runt-related transcription factor 1, previously termed Acute Myeloid Leukemia 1 (AML-1) since it was primarily found as a fusion protein in stage M2 acute myeloid leukemia (Miyoshi et al., 1991; Erickson et al., 1992). RUNX1 encoded the α subunit of heterodimeric TF polyomavirus enhancer binding protein 2 (PEBP2), also known as core binding factor (CBF). The lack of hematopoiesis in AML-1-deficient embryos supports the role of AML-1 as a critical factor for the development of the hematopoietic system (Lutterbach, 2000). In human brain Runx1 expression in fetal and adult tissues was weak as detected at the mRNA level (Su et al., 2002).

SIM2 is the human homolog of *Drosophila* single-minded (*sim*) gene that acts as a master regulator of central nervous system midline development of the fly. The human SIM2 protein is located in the nucleus and presents with a basic helix-loop-helix (bHLH) domain which functions by forming heterodimers through helix-loop-helix domains and by binding to specific DNA sequences, two Per-Arnt-Sim (PAS) dimerization domains and H1F1α-SIM-TRH (HST) domain (Fan et al., 1996; Chrast et al., 1997). Different transgenic mice have been constructed. Mice overexpressing mSim2 under the control of the β-actin promoter exhibited a moderate defect in context-dependent fear conditioning and mild defect in the Morris water maze test, tasks that require the integrated neural circuit of the hippocampus and amygdala, where mSim2 was particularly overexpressed (Ema et al., 1999). mSim2 gene was overexpressed by approximately 50% in the zona limitans region in E12.5 fetuses from Ts1Cje segmental trisomy 16, a model for human trisomy 21 (Vilard et al., 2000). Another transgenic mouse with a bacterial artificial chromosome and with only one or two additional copies of mSim2, overexpressed (0.2–0.5 times) mSim2 gene only in tissues where the endogenous gene was expressed, and also presented anxiety-related/reduced exploratory behaviour and sensitivity to pain (Chrast et al., 2000). Thus, SIM2 has been suggested to be involved in the pathogenesis of brain deficits in DS.

Quantification of the four chromosome 21 encoded TFs, tentative candidates for involvement in abnormal brain development in trisomy 21 not only addressed the gene dosage effect but also identified BACH1 as a possible pathogenetic factor in DS.

Materials and methods

Fetal brain samples

Fetal brain tissues (cerebral cortex) of DS (8 females with 19.4 ± 1.1 weeks of gestational age) and controls (6 females with 19.1 ± 1.6 weeks of gestational age) were used in this study.

Brain samples were obtained from Drs. JC Farreras Unidad de Patología, Corporació Sanitaria Parc Tauli, Sabadell, Barcelona, Spain and Mara Dierssen, Genes and Disease Program, Genomic Regulation Center, Passeig Marítim 37-49, 08003 Barcelona, Spain.

All samples had a postmortem time of less than 6 hours, and were stored at −70°C with a freezing chain never interrupted until use.

Antibodies

The antibody against RUNX1 was obtained from Prof. Y. Groner, Dpt of Molecular Genetics, The Weizmann Institute of Science, Israel. Details of the preparation and characterization of the rabbit anti-RUNX1 antibody have been described previously in Ben Aziz-Aloya et al. (1998). Five antibodies for BACH1 (goat polyclonal antibody, Santa Cruz Biotechnology, USA), HO1 (rabbit polyclonal antibody, Affinity Bioreagents), ERG (rabbit polylonal antibody, Santa Cruz Biotechnology, USA), SIM2 1 (Santa Cruz Biotechnology, USA), β-actin (mouse monoclonal antibody (IgG2a), Sigma, USA) and neuron specific enolase (NSE, rabbit polyclonal antibody, Chemicon, UK) were purchased.

Western blotting

Fetal brain tissues ground under liquid nitrogen were homogenized in lysis buffer 1%SDS containing protease inhibitor cocktail tablets (Roche, Germany), incubed 10 minutes at 37°C, boiled 10 minutes at 95°C and centrifuged at 8,000 ×g for 10 minutes. The BCA protein assay kit (Pierce, USA) was applied to determine the concentration of protein in the supernatant. Samples (10 μg; 15 μg for HO1) were mixed with the sample buffer (100 mM Tris-HCl, 2% SDS, 1% 2-mercaptoethanol, 2% glycerol, 0.01% bromophenol blue, pH 7.6), incubated at 95°C for 15 minutes and loaded onto an 12.5% ExcelGel SDS homogenous gel (Amersham Pharmacia Biotech, Sweden). Electrophoresis was performed with Multiphor II Electrophoresis System (Amersham Pharmacia Biotech). Proteins separated on the gel were transferred onto PVDF membrane (Millipore, USA) and membranes were incubated in blocking buffer (10 mM Tris-HCl, pH 7.5, 150 mM NaCl, 0.1% Tween 20 and 2% non-fat dry milk). Membranes were incubated for 2 hours at room temperature with diluted primary antibodies (1: 700 for Erg and Bach1; 1:1,000 for HO1; 1:1,000 for Runx1; 1:500 for SIM2 l; 1:5,000 for β-actin and 1:2,000 for NSE). After 3 times washing for 15 minutes with blocking buffer, membranes were probed with secondary antibodies (bovine anti-goat for BACH1 (1:1,400) and SIM2 l (1:1,000), (Santa Cruz Biotechnology, USA), goat anti-rabbit IgG (H+L) for ERG (1:1,400), HO1 (1:2,000), RUNX1 (1:2,000) and NSE (1:4,000), and goat anti-mouse IgG2a for β-actin (1:5,000) coupled to horseradish peroxidase (Southern Biotechnology Associates, Inc., USA)) for 1 hour. Membranes were washed 3 times for 15 minutes and developed with the Western Lightning™ chemiluminescence reagents (PerkinElmer Life Sciences, Inc., USA) (Cheon et al., 2003).

Two-dimensional gel electrophoresis (2-DE)

Fetal brain tissues were suspended in 1 ml of sample buffer consisting of 7 M urea, 2 M thiourea, 4% CHAPS, 10 mM 1,4-dithioerythritol (DTT), 1 mM EDTA, 1 mM phenylmethylsulfonyl fluoride (PMSF) and a mixture of protease inhibitors. After sonication for approximately 15 sec, the suspension was left at room temperature for 1 hour and centrifuged at 14,000 × g for 60 min at 12°C. Desalting was done with Ultrafree-4 centrifugal filter unit (Millipore, Bedford, MA). The protein content of the supernatant was determined by the Coomassie blue method (Bradford, 1976). 2-DE was performed essentially as reported (Weitzdoerfer et al., 2002). Samples of 100 μg protein were applied on immobilized pI 3–10 nonlinear gradient strips in sample cups at their basic and acidic ends. Focusing started at 200 V and the voltage was gradually increased to 8,000 V at 4 V/min and kept constant for a further 3 hrs (approximately 150,000 Vhr totally). The second-dimensional separation was performed on 9–16% gradient sodium dodecyl sulfate polyacrylamide gel (180 × 200 × 1.5 mm).

Two-dimensional western blotting

Proteins separated on the gel were transferred onto PVDF membrane (Millipore, USA) at 400 mA constant for 2 hrs and the membrane was incubated in blocking buffer (10 mM Tris-HCl, pH 7.5, 150 mM NaCl, 0.1% Tween 20 and 2% non-fat dry milk). Subsequent immunodetection was performed as described in the western blotting method.

Statistics

The density of immunoreactive bands was measured by RFLPscan version 2.1 software program (Scanalytics, USA). Between group differences were calculated by non-parametric Mann-Whitney U test using GraphPad Instat2 program and the level of significance was considered at $P < 0.05$.

Results

We evaluated expression levels of four transcription factors encoded on chromosome 21, BACH1, the dependent protein HO1, ERG, RUNX1 and SIM2 1 in fetal brain with DS compared to controls by Western blot analysis.

Two proteins, β-actin and NSE, were used as reference proteins for total cell — and neuronal density, respectively and their expression levels were comparable in controls and DS (Fig. 2A).

We observed two bands for immunoreactive BACH1 at 82 kDa and 60 kDa. The band at 82 kDa was comparable between DS and control whereas the optical density of the 60 kDa band was significantly increased in DS ($P = 0.008$) and even when normalised with both, β-actin and NSE, a significant elevation was detected ($P = 0.01$) (Fig. 2A, 3).

In the two-dimensional western blot using a representative control and DS sample we detected seven different BACH1 isoforms at 60 kDa for DS and four isoforms for controls (Fig. 2B).

We detected several HO1-immunoreactive bands in the range between 32 kDa and 250 kDa with HO1, with a major band ~60 kDa. The band with an expected molecular weight 32 kDa was quantified (Morse et al., 2002) and this band was not significantly different between DS and controls (Fig. 2A, 3).

We found one ERG-immunoreactive band at 52 kDa and no significant differences were observed between DS and controls, even when we normalised with β-actin and NSE (Fig. 2A, 3).

One RUNX1-immunoreactive band was observed at 58 kDa, without significant differences between groups (Fig. 2A).

We did not detect any band using a commercially available antibody for SIM2 1.

Discussion

The main finding of our study represents a statistically significant, approximately threefold increase of the 60 kDa BACH1-immunoreactive band and an increase of the 82 kDa band that did not reach statistical significance. Using two-dimensional immunoblotting seven spots were detectable in fetal DS brain and four in control brain and indeed, different forms of BACH1 have been already described: One alternatively spliced isoform called BACH1t with an apparent molecular weight of approx. 32 kDa has been described lacking the leucine zipper domain and accumulating in the nucleus. BACH1t recruits BACH1 to the nucleus through BTB domain mediated interaction (Kanezaki et al., 2001). We did not detect this splicing form in fetal brain either by western blotting or 2D-immunoblotting but are clearly showing the presence of several spots in fetal brain that may represent either isoforms or

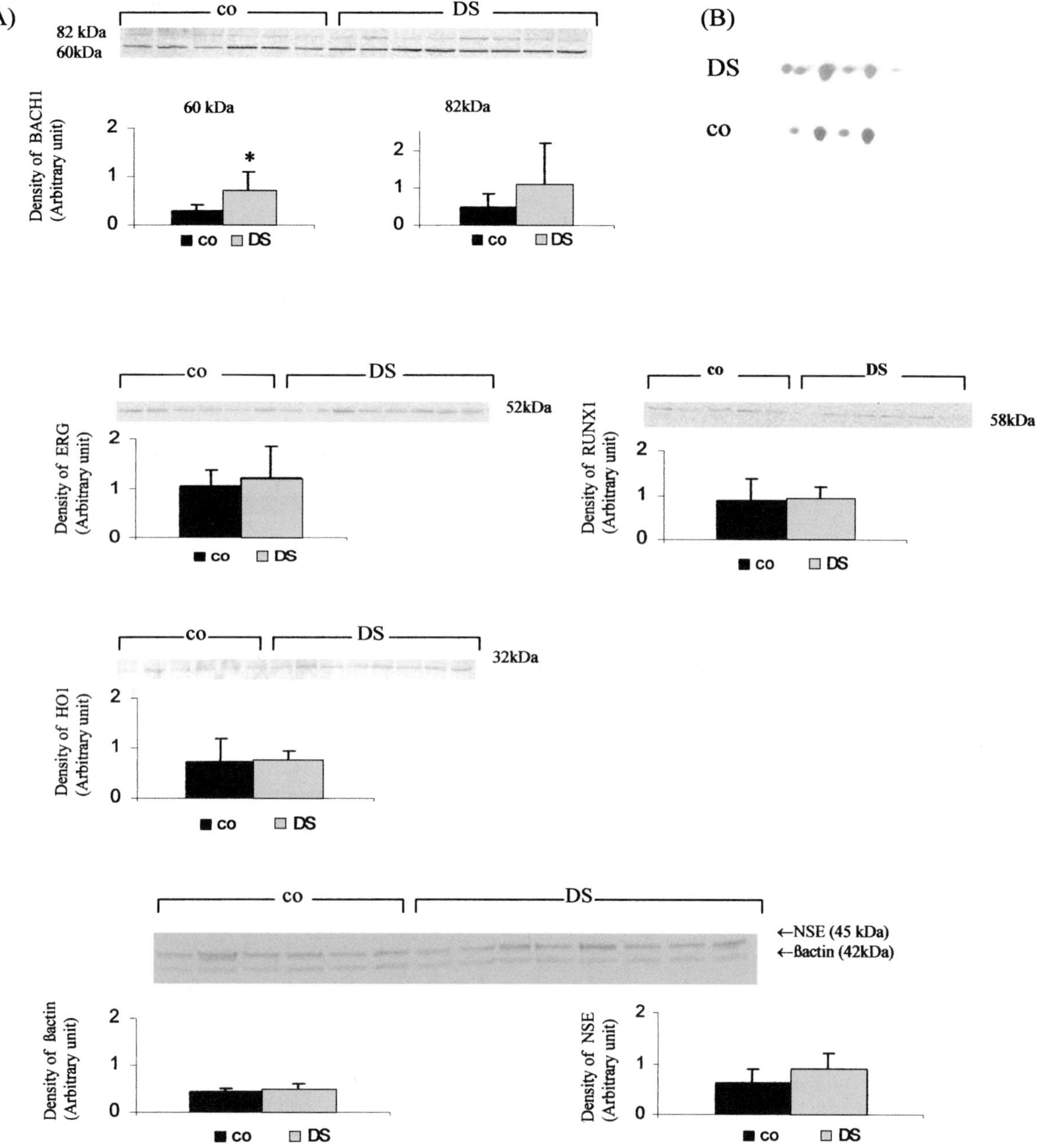

Fig. 2. **A** Western blot analysis for three TFs whose genes are encoded on chromosome 21 and related protein HO1 in cerebral cortex from fetal brain with DS and controls. Denatured proteins (10 μg) were loaded, separated on a homogeneous gel and transferred onto PVDF membrane. As described in "Materials and methods", the membranes were incubated with primary and secondary antibodies, and immunoreactive bands (BACH1, 82 kDa and 60 kDa; ERG, 52 kDa; RUNX1, 58 kDa; HO1, 32 kDa; NSE, 45 kDa; actin, 42 kDa) were detected using chemiluminescence reagents. The density of detected bands was measured and calculated by non-parametric Mann-Whitney U test, and the level of significance was considered at $P < 0.05$. **B** Two dimensional western blot using BACH1 antibody. *Significant differences between DS and control with $p < 0.05$

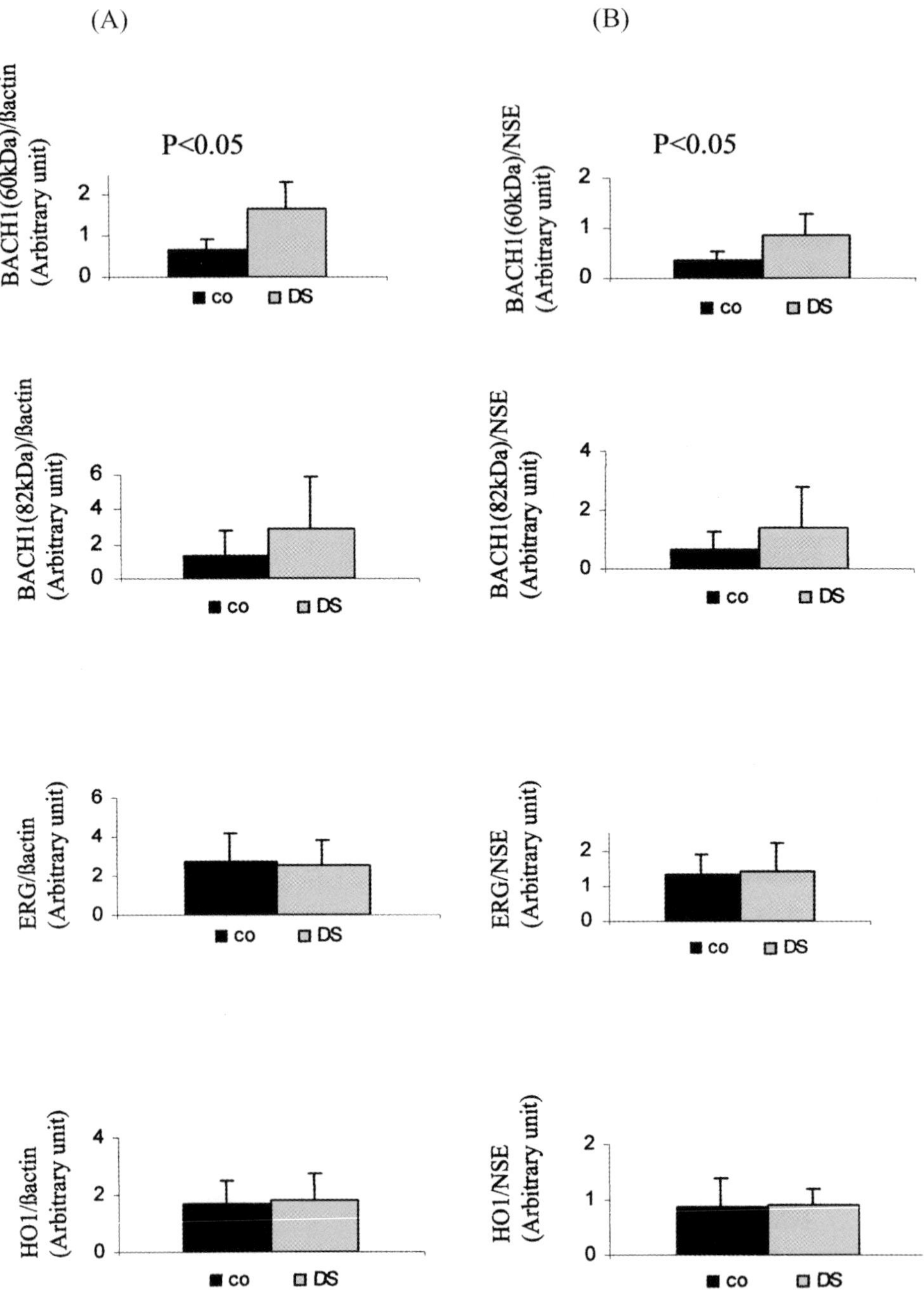

Fig. 3. Expression levels of three TFs and HO1 protein normalised with those of β-actin (**A**) or NSE (**B**). The density of immunoreactive band for each protein was normalised with that of actin or NSE, used as reference proteins

posttranslational modifications, which is highly likely as these spots with a comparable apparent molecular weight showed different horizontal migration in a pI range of 6.0–6.3 (controls) and 5.9–7.2 (in DS).

BACH1 is known to control beta-globin and the antioxidant protein heme oxygenase 1 (HO1) expression (Sun et al., 2002). Since oxidative stress is considered a main pathogenetic factor in DS (Epstein, 1995; Ianello et al., 1999; Busciglio et al., 1995) we focused work on HO1. This microsomal enzyme encoded on chromosome 22 is the ubiquitous inducible isoform of HO. HO1 accomplishes the heme break into carbon monoxide (CO), free iron, inducing ferritin, and biliverdin, which is quickly transformed into bilirubin, products with important cytoprotective and antioxidant effects (Barañano et al., 2001). The role of HO1 in oxidative stress has been described by several groups indicating induction of HO1 by heavy metal, hydrogen peroxide, etc. (Morse et al., 2002; Poss et al., 1997; Chen et al., 2000).

BACH1 represses HO1 expression by complex formation with MafK under physiological conditions. We showed comparable levels of the representative (Keyse et al., 1989) 32 kDa HO1- immunoreactive-band, an observation that does not support the notion of regulation of HO1 expression by BACH1. Under circumstances of oxidative stress (shown to exist in DS fetal brain; Gulesserian et al., 2001), however, this effect could have been masked by a concomitant hydrogen peroxide — induced increase of HO1 (Ishii et al., 2000) as increased hydrogen peroxide levels have been described in DS brain (Busciglio et al., 1995). Although BACH1 was also significantly and about threefold increased when normalised versus beta-actin and neuron specific enolase to rule out cell or neuronal loss, the biological meaning of increased BACH1 expression therefore remains elusive.

ERG expression has been reported in several organ systems and species presenting with different isoforms (Dhordain et al., 1995; Vlaemnick-Guillem et al., 2000; Hewett et al., 2001) including Erg mRNA levels in fetal and adult brain (Su et al., 2002). ERG protein levels have not been reported so far in fetal brain and we here show comparable expression of a single band at 52 kDa in fetal DS brain. Unchanged expression of ERG in fetal DS brain makes an involvement in the pathogenesis of the brain defect in DS unlikely.

RUNX1 is essential for generation and differentiation of definitive hematopoietic stem cell and is expressed in various cell lines and tissues (Otto et al., 2003; Perry et al., 2002; Yamashiro et al., 2002) and about twelve splicing forms have been reported (Tsuji et al., 2000; Asou, 2003). In human fetal and adult brain RUNX1 was expressed at the mRNA steady state level and we reported for the first time protein expression of a RUNX1-immunoreactive 58 kDa band in fetal brain that was comparable between DS and controls. Again, no pathogenetic role for the development of the brain deficit in DS may be assigned to this structure.

Sim2 is encoded in the Down Syndrome Chromosomal Region (DSCR) triplicated in DS and Sim2 mRNA is expressed during early fetal life (E8-E16) in the central nervous system (Fan et al., 1996; Dahmane et al., 1995). In humans, SIM2 is expressed in the fetal neuroepithelium, with weaker and more diffuse labeling observed in certain cortical layers in sections from brains at 16 weeks of gestation (Dahmane et al., 1995). It was reported that Northern blot analysis failed to detect any band in fetal brain (Dahmane et al., 1995; Chrast et al., 1997; Yamaki et al., 2001). In agreement with transcriptional results in reported literature we could not find any band using immunoblotting in fetal brain albeit in adult brain (frontal cortex) a weak band was detected (unpublished own data) probably reflecting temporal SIM-2 expression.

We conclude that BACH1 is one of several aberrant transcription factors in DS brain that may contribute to abnormal brain development or at least to the already described defective transcription machinery in DS. No effect of BACH1 on the antioxidant protein heme oxygenase 1 could be shown, most probably as oxidative stress in DS brain may have been inducing this enzyme directly thus masking repressive regulation by this TF. About threefold overexpression of BACH1 in DS brain may not be due to the gene dosage effect proposed in literature (Epstein, 1995) and in turn, comparative levels of other TFs encoded on chromosome 21 as ERG and RUNX1 in DS do not support the gene dosage hypothesis. We are about to confirm immunochemical findings in our laboratory by extending protein chemical studies using proteomic methodology.

Acknowledgements

We are highly indebted to the Red Bull Company (Salzburg, Austria) for generous support of the study. We thank Dr. Y. Groner (Department of Molecular Genetics, The Weizmann Institute of Science, Israel) for kindly providing the antibody against RUNX1).

References

Asou N (2003) The role of Runt domain transcription factor AML1/RUNX1 in leukemogenesis and its clinical implications. Crit Rev Oncol/Hematol 45: 129–150

Aziz-Aloya RB, Levanon D, Karn H, Kidron D, Goldenberg D, Lotem J, Polak-Chaklon S, Groner Y (1998) Expression of AML1-d, a short human AML1 isoform, in embryonic stem cells suppress in vivo tumor growth and differentation. Cell Death Differ 5: 765–773

Bahn S, Mimmack M, Ryan M, Caldwell MA, Jauniaux E, Starkey M, Svendsen CN, Emson P (2002) Neuronal target genes of the neuron-restrictive silencer factor in neurospheres derived from fetuses with Down's syndrome: a gene expression study. Lancet 359: 301–315

Barañano DE, Snyder SH (2001) Neural roles for heme oxygenase: contrast to nitric oxide synthase. PNAS 98: 10996–11002

Bradford MM (1976) A rapid and sensitive method for the quantitation of microgram quantities of protein utilizing the principle of protein-dye binding. Anal Biochem 72: 248–254

Busciglio J, Yanker BA (1995) Apoptosis and increased generation of reactive oxygen species in Down's syndrome neurons in vivo. Nature 378(6559): 776–779

Capone GT (2001) Down Syndrome: advances in molecular biology and the neurosciences. Dev Behav Pediatr 22: 40–59

Chen K, Gunter K, Maines MD (2000) Neurons overexpressing heme oxygenase-1 resist ocidative stress-mediated cell death. J Neurochem 75: 304–313

Cheon MS, Kim SH, Yaspo ML, Blasi F, Aoki Y, Melen K, Lubec G (2003) Protein levels of genes encoded on chromosome 21 in fetal Down syndrome brain: challenging the gene dosage effect hypothesis, part I. Amino Acids 24: 111–117

Chrast R, Scott HS, Chen H, Kudoh J, Rossier C, Minoshima S, Wang Y, Shimizu N, Antonarakis SE (1997) Cloning of two human homologs of the Drosophila single-minded gene SIM1 on chromosome 6q and SIM2 on 21q within the Down Syndrome Chromosomal Region. Genome Res 7: 615–624

Chrast R, Scott HS, Madani R, Huber L, Wolfer DP, Prinz M, Aguzzi A, Lipp HP, Antonarakis SE (2000) Mice trisomic for bacterial artificial chromosome with the single-minded 2 gene (Sim2) show phenotypes similar to some of those present in the partial trisomy 16 mouse models of Down Syndrome. Hum Mol Gen 9: 1853–1864

Dahmane N, Charron G, Lopes C, Yaspo ML, Maunoury C, Decorte L, Sinet PM, Bloch B, Delabar JM (1995) Down syndrome-critical region contains a gene homologous to Drosophila sim expressed during rat and human central nervous system development. Proc Natl Acad Sci USA 92: 9191–9195

Delattre O, Zucman J, Melot T, Garau XS, Zucker JM, Lenoir GM, Ambros PF, Sheer D, Turc-Carel C, Triche TJ, Aurias A, Thomas G (1994) The Ewing family of tumors: a subgroup of small-round-cell tumors defined by specific chimeric transcripts. N Engl J Med 331: 294–299

Dhordain P, Dewitte F, Desbiens X, Stehelin D, Duterque-Coquillaud M (1995) Mesodermal expression of the chicken erg gene associated with precartilaginous condensation and cartilage differentiation. Mech Dev 50: 17–28

Ema M, Ikegami S, Hosoya T, Mimura J, Ohtani H, Nakao K, Inokuchi K, Katasuki M, Fuji-Kuriyama Y (1999) Mild impairment of learning and memory in mice overexpressing the mSim2 gene located on chromosome 16: an animal model of Down's syndrome. Hum Mol Gen 8: 1409–1415

Engidawork E, Lubec G (2003) Molecular changes in fetal Down Syndrome brain. J Neurochem 84: 895–904

Engidawork E, Balic N, Fountoulakis M, Dierssen M, Greber-Platzer S, Lubec G (2001) β-Amyloid precursor protein, ETS-2 and collagen alpha 1 (VI) chain precursor, encoded in chromosome 21, are not overexpressed in fetal Down syndrome: further evidence against gene dosage effect. J Neural Transm [Suppl] 61: 335–346

Epstein CJ (1995) Down Syndrome. In: Scriver CR, Beaudet AL, Sly WS, Valle D (eds) The metabolic and molecular bases of inherited diseade, 7th edn, vol 1. McGraw Hill, New York, pp 749–794

Erickson P, Gao J, Chang KS, Look T, Whisenant E, Raimondi S, Lasher R, Trujillo J, Rowely J, Drabkin H (1992) Identification of breakpoints in t(8;21) acute myelogenous leukaemia and isolation of a fusion transcript, AML1/ETO, with similarity to Drosophila segmentation gene runt. Blood 80: 1825–1831

Fan C-M, Kuwana E, Bulfone A, Fletcher CF, Copeland NG, Jenkins NA, Crews S, Martinez S, Puelles L, Rubenstein JLR, Tessier-Lavigne M (1996) Expression patterns of two murine homologs of Drosophila single-minded suggest possible roles in embryonic patterning and in the pathogenesis of Down syndrome. Mol Cell Neurosci 7: 1–16

Freidl M, Gulesserian T, Lubec G, Fountoulakis M, Lubec B (2001) Deterioration of the transcriptional, splicing and elongation machinery in brain of fetal Down syndrome. J Neural Transm [Suppl] 61: 47–57

Gulesserian T, Engidawork E, Fountoulakis M, Lubec G (2001) Antioxidant proteins in fetal brain: superoxide dismutase-1 (SOD1) protein is not overexpressed in fetal Down syndrome. J Neural Transm [Suppl] 61: 71–84

He X, Rosenfeld MG (1991) Mechanisms of complex transcriptional regulation: implications for brain development. Neuron 7: 183–196

Hewett PW, Nishi K, Daft EL, Clifford Murray J (2001) Selective expression of erg isoforms in human endothelial cells. IJBCB 33: 347–355

Ianello RC, Crack PJ, de Haan JB, Kola I (1999) Oxidative stress and neural dysfunction in Down Syndrome. J Neural Transm [Suppl] 57: 257–267

Ichikawa H, Shimizu K, Hayashi Y, Ohki M (1994) An RNA-binding protein gene, TLS/FUS, is fused to ERG in human myeloid leukaemia with t(16;21) chromosomal translocation. Cancer Res 54: 2865–2868

Ishii T, Itoh K, Takahashi S, Sato H, Yanagawa T, Katoh Y, Bannai S, Yamamoto M (2000) Transcription factor Nrf2 coordinately regulates a group of oxidative stress-inducible genes in macrophages. J Biol Chem 275: 16023–16029

Johnston MV, Alemi L, Harum KH (2003) Learning, memory, and transcription factors. Pediatr Res 53, 3: 369–374

Kanezaki R, Toki T, Yokoyama M, Yomogida K, Sugiyama K, Yamamoto M, Igarashi K, Ito E (2001) Transcription factor BACH1 is recruited to the nucleus by its novel alternative spliced isoform. J Biol Chem 276: 7278–7284

Keyse SM, Tyrrell RM (1989) Heme oxygenase is the major 32-kDa stress protein induced in human skin fibroblast by UVA radiation, hydrogen peroxide, and sodium arsenite. Proc Natl Acad Sci USA 86: 99–103

Kitamuro T, Takahashi K, Ogawa K, Udono RF, Takeda K, Furuyama K, Nakayama M, Sun J, Fujita H, Hida W, Hattori T, Shirato K, Igarashi K, Shibahara S (2003) Bach1 functions as a hypoxia-inducible repressor for heme oxygenase-1 gene in human cells. J Biol Chem 278: 9125–9133

Labudova O, Krapfenbauer K, Moenkemann H, Rink H, Kitzmuller E, Cairns N, Lubec G (1998) Decreased transcription factor junD in brains of patients with Down syndrome. Neurosci Lett 252: 159–62

Labudova O, Kitzmueller E, Rink H, Cairns N, Lubec G (1999) Gene expression in fetal Down syndrome brain as revealed by substractive hybridization. J Neural Transm [Suppl] 59: 125–136

Maroulakou IG, Bowe DB (2000) Expression and function of Ets transcription factors in mammalian development: a regulatory network. Oncogene 19: 6432–6442

Miyoshi H, Shimizu K, Kozu T, Maseki N, Kaneko Y, Ohki M (1991) t(8;21) breakpoints on chromosome 21 in acute myeloid leukaemia are clustered within a limited region of a single gene AML1. Proc Natl Acad Sci USA 88: 10431–10434

Morse D, Choi AMK (2002) Heme oxygenase-1, the "emerging molecule" has arrived. J Respir Cell Mol Biol 27: 8–16

Ohira M, Seki N, Nagase T, Ishikawa K, Nomura N, Ohara O (1998) Characterization of a human homolog (BACH1) of the mouse Bach1 gene encoding a BTB-Basic leucine zipper transcription factor and its mapping to chromosome 21q22.1. Genomics 47: 300–306

Otto F, Lübbert M, Stock M (2003) Upstream and downstream targets of RUNX proteins. J Cell Biochem 89: 9–18

Oyake T, Itoh K, Motohashi N, Hayashi N, Hoshino H, Nishizawa M, Yamamoto M, Igarashi K (1996) Bach proteins belong to a novel family of BTB-Basic leucine zipper transcription factors that interact with MafK and regulate transcription through the NF-E2 site. Mol Cell Biol 16: 6083–6095

Perry Ch, Sklan EH, Birikh K, Shapira M, Trejo L, Eldor A, Soreq H (2002) Complex regulation of acetylcholinesterase gene expression in human brain tumors. Oncogene 21: 8428–8441

Poss KD, Tonegawa S (1997) Reduced stress defense in heme oxygenase 1-decficient cells. Proc Natl Acad Sci USA 94: 10925–10930

Reddy ES, Rao VN, Papas TS (1987) The erg gene: a human gene related to the ets oncogene. Proc Natl Acad Sci USA 84: 6131–6135

Su AI, Cooke MP, Ching KA, Hakak Y, Walker JR, Wiltshire T, Orth AP, Vega RG, Sapinoso LM, Morqrich A, Patapoutian A, Hampton GM, Schultz PG, Hogenesch JB (2002) Large-scale analysis of the human and mouse transcriptomes. Proc Natl Acad Sci USA 99: 4465–70

Sun J, Hoshino H, Takaku K, Nakajima O, Muto A, Suzuki H, Tashiro S, Takahashi S, Shibahara S, Alam J, Taketo MM, Yamamoto M, Igarashi K (2002) Hemoprotein

Bach1 regulates enhancer availability of heme oxygenase-1 gene. EMBO 21: 5216–5224

Tsuji K, Noda M (2000) Identification and expression of a novel 3′-exon of mouse Runx1/Pebp2αB/Cbfa2/AML1 gene. Biochem Biophys Res Com 274: 171–176

Vialard F, Toyama K, Vernoux S, Carlson EJ, Epstein CJ, Sinet PM, Rahmani Z (2000) Overexpression of mSim2 gene in the zona limitans of the diencephalons of segmental trisomy 16 Ts1Cje fetuses, a mouse model for trisomy 21: a novel whole-mount based RNA hybridisation study. Dev Brain Res 121: 73–78

Vlaeminck-Guillem V, Carrere S, Dewitte F, Stehelin D, Desbians X, Duterque-Coquillaud M (2000) The ets family member erg gene is expressed in mesodermal tissues and neural crests at fundamental steps during mouse embryogenesis. Mech Dev 91: 331–335

Weitzdoerfer R, Fountoulakis M, Lubec G (2002) Reduction of actin-related protein complex 2/3 in fetal Down Syndrome brain. Biochem Biophys Res Commun 293: 836–41

Yamaki A, Tochigi J, Kudoh J, Minoshima S, Shimizu N, Shimizu Y (2001) Molecular mechanisms of human single-minded 2 (SIM2) gene expression: identification of a promoter site in the SIM2 genomic sequence. Gene 270: 265–275

Yamashiro t, Aberg T, Levanon D, Groner Y, Thesleff I (2002) Expression of Runx-1, -2 and -3 during tooth, palate and craniofacial bone development. GEP 2: 109–112

Authors' address: Prof. Dr. G. Lubec, CChem, FRSC (UK), Department of Pediatrics, University of Vienna, Währinger Gürtel 18, A-1090 Vienna, Austria, e-mail: gert.lubec@akh-wien.ac.at

Down syndrome and associated congenital malformations

B. L. Shapiro

Department of Oral Sciences and Institute of Human Genetics, University of Minnesota, Minneapolis, MN, U.S.A.

Summary. Congenital malformations are many times more common in individuals with Down syndrome (DS) than in the general population. The scope of these defects is as broad in DS as it is in the general population. A positive correlation exists between the prevalence of these defects in both groups, but the incidence of each is many times greater in DS. Two examples, Brushfield spots and anorectal abnormalities are noted in which racial/ethnic prevalence differences exist. The incidence of each condition in the subpopulation with DS is proportional to but many times greater than the incidence of that condition in the general population from which the subpopulation with DS was derived. Findings presented in this review support the notion that the autosomal trisomic state amplifies expression of exposure to teratogens.

Structures that will subsequently develop and comprise the normally functioning organism are established in the human embryo between about 3 and 8 prenatal weeks. It is during this time that disturbances of development can lead to structural defects present at birth — congenital malformations. About 4–6% of all newborns are affected by maldevelopment of one or more organ systems that severely threatens normal function or survival of the organism. Individual major congenital malformations are not part of the classical Down syndrome (DS) phenotype; not one appears on any list of clinical diagnostic criteria for the condition. Nevertheless, with a few possible exceptions, all major congenital malformations observed in the general population are more prevalent in DS (as they are also in other major autosomal chromosomal abnormalities). More than 50% of infants with DS have associated defects that cause considerable morbidity and mortality (Torfs and Christianson, 1999). These do not include minor anomalies such as small or misshapen ears, pigmented spots or short palpebral fissures often found in association with DS that occur also in about 15% of nontrisomic newborns. A confluence of many minor anomalies permits the clinical diagnosis of DS (Jackson et al., 1976). The following text reiterates (Shapiro, 1983) and updates that trisomy 21, the cause of almost all cases of DS, is a condition that predisposes to the same kinds of congenital defects that occur in the general population.

Congenital defects and Down syndrome

Diseases and abnormalities associated with DS are the same as those observed in the general population. Every type of malformation has been observed in DS. Their prevalence is greater in DS and in individuals with DS clinical problems frequently are characterized by their multiple and severe expression. With the exception of the chromosomal abnormality (which almost always exists) and the variable mental retardation, no condition is observed in all individuals with DS and all pathology seen in association with DS occurs in the general population. The most common site of congenital malformation in the general population is the heart and this group of developmental defects also is most common in DS as well as other autosomal trisomies. Congenital heart disease occurs in about one half of newborns with DS. Clinical series as well as necropsy reports show that the abnormal hearts of these individuals display as wide a variety of malformations as those of subjects without DS (Freeman et al., 1999). Besides congenital heart disease which is about 100 times more prevalent in DS than in the general population, the incidence of other malformations is from 10 to 300 times greater in DS. There appears to be a rough parallel between the incidences of congenital malformations in DS and in the general population: the most common defects in the general population are the most common in DS (Fig. 1). In all cases, the expression of the malformation in DS is similar to that in the general population and there exists a marked overlap in type and similarly increased incidences with other autosomal trisomic conditions. Minor malformations also occur with greater frequency in DS (Shapiro, 1983).

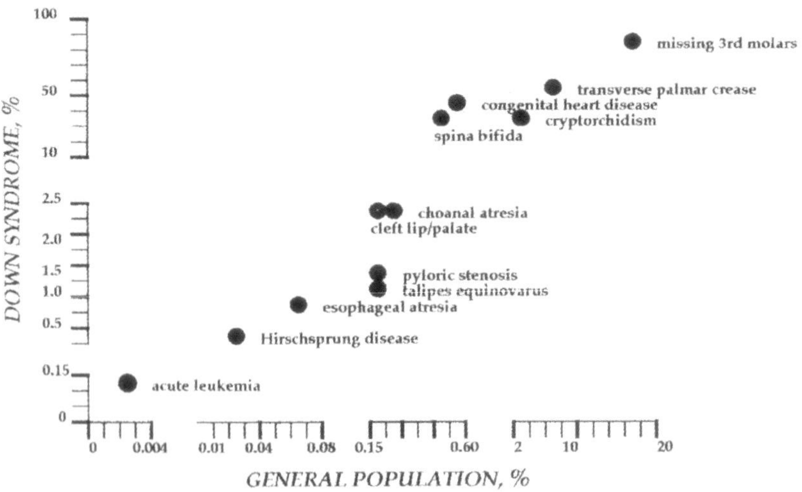

Fig. 1. Comparison of various malformations and conditions in the general population and in individuals with Down syndrome. Note the different scales within and between the two axes to accommodate the great differences among prevalence of the different conditions and between the general population and Down syndrome. There is a positive correlation of the percentages for the various conditions in the two groups (based on unpublished data)

Major congenital malformations clearly are associated with DS. This association is characterized by i) a lack of constancy; ii) a parallel with the general population in types, although incidences are far greater in DS; iii) a similarity of expression in DS and the general population; and iv) a marked overlap in type and similarly increased incidences with other autosomal trisomic conditions (Shapiro, 1983). Khoury and Erickson (1992) found that cardiac, gastrointestinal and other defects, all of which occur in the general population and with other chromosomal abnormalities, occur in infants with DS much more than expected by chance alone. In their search for environmental factors and maternal exposures in the genesis of increased congenital defects in DS they determined the presence of cardiac (22.8%) and gastrointestinal (4.1%) defects and oral clefts (1.8%) in 219 infants with DS born between 1968 and 1980 in Atlanta. While lower than many other estimates these percentages were far higher than that in the general population from which the subjects were derived. The study of Kallen et al. (1996), using 3 European birth defects registries, supported the proposal that most congenital malformations occur with increased incidence in DS. They studied major malformations in more than 5500 infants with DS. The prevalence at birth of 23 different malformations was compared with the program-specific rates for each malformation in non-DS infants. An about 300 times risk was seen for annular pancreas, cataracts and duodenal atresia and an about 100 times risk was seen for megacolon and choanal atresia. Esophageal, anal and small bowel atresia, preaxial polydactyly and omphalocele all showed risk increases between 10 and 30 times. Statistically elevated risk ratios around 3–5 times were seen for cleft palate, cleft lip/palate and limb deficiencies. In another study, among 61 defects uniformly ascertained in affected and unaffected infants, 45 were significantly more common in DS, with atrioventricular canal, duodenal atresia and annular pancreas being the most common (Torfs and Christianson, 1998).

Kallen et al. (1996) found no increased risk for neural tube defects, hydrocephaly, microtia, renal agenesis, hypospadias or polydactyly other than preaxial. In contrast, in other studies evidence exists of an increased risk in DS of hypospadias (Lang et al., 1987), spina bifida (Mautner et al., 1950) hydrocephaly (ventriculomegaly) (Nicolaides et al., 1990; Chervenak et al., 1985; Dombrowski et al., 1993) and perhaps renal development (Dombrowski et al., 1993; Nicolaides et al., 1993). Kallen et al. (1996) suggested also that some malformations, such as anencephaly, probably prevent the identification of DS and, for reasons given by them, that the recorded rates of malformations in DS are probably minimal. The statistical effect of this possible reduced incidence of birth defects in full term infants with DS would be to diminish detection of amplification of incidence of malformations that might have occurred. That is, if the presence of a severe birth defect would differentially lead to loss of fetuses with DS (a severe birth defect would presumably have greater adverse effects on an individual with DS whose viability is already reduced in comparison with a euploid fetus) the ability to detect the potential amplification in incidence of birth defects would be lessened. Despite this probable dilution of the effect of trisomy on birth defects incidence in the data available, the increase in incidence of birth defects in association with DS is dramatic.

Although there exists a nearly universal increase in specific congenital malformations in association with DS, several conditions have been found by some investigators to be no more prevalent in DS than in the general population. This may merely reflect chance and other factors discussed below. In any case, in nearly all instances the finding of no increase in incidence has been countered by an increase in other studies. For example, omphalocele was found to be no more frequent in DS (Torfs et al., 1997) and 10 to 30 times more frequent in DS (Kallen et al., 1996). No increased risk was found for microtia (Kallen et al., 1996) although a small and malformed pinna is often seen in individuals with DS (Lee and Jackson, 1972). Neural tube defects were not increased in the study of Kallen et al. (1996). These authors suggested themselves that anencephaly might prevent the diagnosis of DS; spina bifida occulta (Mautner et al., 1950) and hydrocephalus (Smith and Berg, 1976) were found to be increased in other studies.

Underestimates of congenital defects in Down syndrome

Consistent and strong data demonstrate that the prevalence of most congenital malformations is increased in DS. These increases are probably underestimates (Khoury and Erickson, 1992; Kallen et al., 1996). In most studies the number of subjects was small. As with all survey data, reported incidences are variable and dependent on many changeable factors including inconsistencies in detection and recording among studies, investigators, etc. Comparison of different registries is fraught with serious consequences such as observer differences and variable criteria for diagnoses. Differential prenatal survival probably occurs, i.e. it is probable that more severely affected subjects with DS are less likely to survive to be recorded. With more than one defect in an individual the most serious might be recorded to the neglect of another. Additional reasons for underestimates include: reports from cytogenetic laboratories where clinical data may not be supplied; defects may be regarded as part of DS and therefore not reported; and severe malformation may prevent the identification of DS. For a variety of these and other reasons, underascertainment of congenital defects in DS probably occurs. The rates recorded for congenital defects in association with DS probably represent a minimum.

Why are associated congenital defects common in Down syndrome?

Why are almost all congenital malformations that occur in the general population more prevalent in samples of individuals with DS? And why are these same malformations always found to be common in other chromosome imbalance syndromes? No evidence exists to support the presumption that clinical findings in DS are secondary to individual chromosome 21 loci (Shapiro, 1997,

1999). The common occurrence of each of the findings in DS also in the general population as well as in numerous other abnormal conditions demonstrates that these malformations are not specific to DS. A number of years ago I proposed that the superfluous gene products in trisomic cells disrupt genetic balance with consequent decreased efficiency of evolved buffering systems (Shapiro, 1983). The result is that physiological and developmental homeostasis are diminished leading to increased liability to all malformations and diseases in DS, as has been documented. Numerous investigators have supported this explanation. While acknowledging the obvious contribution of loci on chromosome 21 to the DS phenotype I indicated the indirect effect of their gene products on the phenotype and emphasized the role of environmental factors in the pathogenesis of findings in DS (Shapiro, 1975, 1983, 1994). Kallen et al. (1996) also suggested "that trisomy 21 increases the susceptibility of the embryo to teratogens." Torfs and Christianson (1999) agreed "that environmental factors can modify the occurrence of associated anomalies in the embryo with Down syndrome."

Two independent tests of the prediction concerning genetic unbalance and increased liability to pathogenic influences in DS involve the correlation between the prevalence of a trait in different populations and its parallel but amplified incidence in subpopulations of subjects with DS in the different populations. The first concerns speckling of the iris and the second anorectal abnormalities.

Race/ethnicity and birth defects in Down syndrome

Anterior uveal tissues are among those commonly involved in DS (Catalano, 1990). Some believe that speckling of the iris (Brushfield spots) is characteristic of DS. Actually, the frequency of Brushfield spots associated with DS varies among studies: as often as 80–85% or as infrequently as 20–25%. A comparable iridic difference in "normal" individuals was described twenty years earlier than Brushfield by Wolff in 1902 and these occur in 10–24% of the general population (Catalano, 1990). These spots almost never occur in Asians (as they do not in dark eyed individuals in general). In the study of Kwong and Wong (1996) only one of 127 Chinese subjects with DS had Brushfield spots and that subject was half Chinese. Brushfield spots were not found in 123 Korean children with DS (Kim et al., 2002). That is, in a racial group (Asian) where iridic speckling (Brushfield/Wolfflian spots) almost never occurs in the general population it occurred in less than 1% of subjects with DS or not at all.

A second and even more dramatic parallel prevalence of a trait between euploid and trisomic individuals involves anorectal malformations (Zlotogora et al., 1989). Imperforate anus is a relatively frequent malformation found in about 1/1,500 to 1/5,000 live-born children. The incidence in the Jewish population of Jerusalem was 1/1,500 (0.065%). DS was diagnosed in 4.6% of the children born with imperforate anus. Among 89 children with DS born between 1980 and 1986 2 had imperforate anus and one had anal stenosis (3.37%

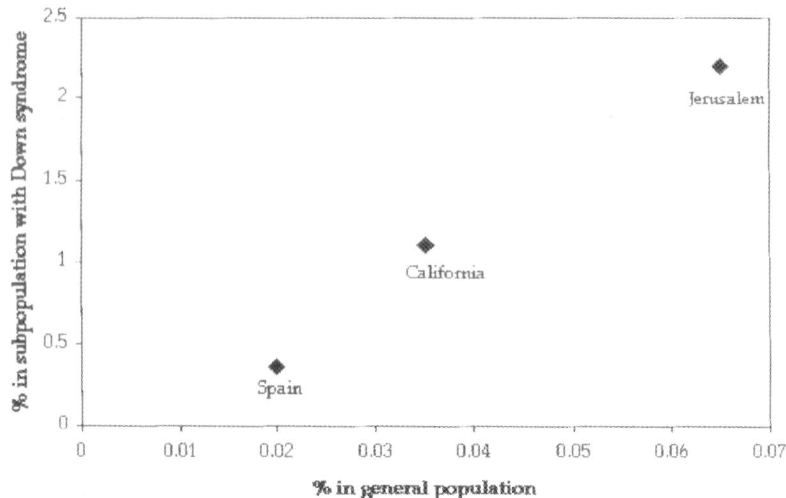

Fig. 2. Relationship between the prevalence of anorectal malformations in three different general populations (Spain, California and Jerusalem) and in subpopulations with Down syndrome derived from each. Each symbol represents the % prevalence in the general population and its subpopulation of subjects with Down syndrome. A strong positive correlation exists between the % of anorectal malformations in the three populations and in the subpopulations with Down syndrome from each. Note the more than 30 times greater % on the y (Down syndrome) axis. Graph based on data compiled by Zlotogora (1992)

with anorectal malformations). The relatively high frequency of anorectal anomalies in the Jerusalem population coincided with the relatively higher incidence in the Jewish children with DS. Subsequently, Zlotogora (1992) found that the frequency of anorectal malformations in children with DS was directly related to its prevalence in the population from which the children were derived. The highest prevalence of anorectal malformations was found in Jerusalem in both normal children (0.065%) and in children with DS (2.2%) (only anal atresia in the general population was recorded so the case of stenosis in the group with DS was not included). The lowest prevalence in both groups was found in Spain (0.020% and 0.36%, respectively). In California anorectal malformations were found in 0.035% of the general population and in 1.1% of those with DS (Fig. 2). Zlotogora concluded that the decreased buffering model of DS as proposed in 1983 predicted correctly that in populations in which the incidence of a malformation is high, the same malformation will be observed in individuals with chromosome imbalance with a relatively higher incidence than that observed in children with the same syndrome originating from a population in which the malformation is more rare. "This prediction is clearly confirmed by the observation on the prevalence of anorectal malformations in Down syndrome as compared to their prevalence in normal individuals in three different populations" (Zlotogora, 1992).

Conclusion

As previously noted (Shapiro, 1983) birth defects in general are more frequent in individuals with DS than in the general population (a possible exception for which I have no explanation might be neural tube defects). This was confirmed in several cited reports. Implicit in the model I proposed were consequences such that there would be a parallel but increased frequency between the occurrence of a particular defect in DS and euploid subjects that would extend to racial/ethnic differences. Examples in the literature support this contention. The observations described in this brief review would tend to militate against a specific teratogenic potential of particular extra loci on chromosome 21. It seems more likely that the trisomic state itself produces a generally increased susceptibility in the organism and that the instability of developmental pathways is amplified.

Strong evidence exists that: (i) birth defects associated with DS are those that occur in the euploid newborn population; (ii) most if not all birth defects are more frequent in DS; and (iii) racial/ethnic incidence differences of congenital defects which occur in the general population are parallel but amplified in DS. The basis for most birth defects in the general population is thought to be a heterogeneous, complex maldevelopment with no singular or simple cause across a population. As demonstrated by the relatively high prevalence in DS of congenital defects common in the euploid population, malformations occur in morphogenetic fields highly susceptible to developmental disruption. Those developmental systems most vulnerable to malformation in the general population are those that, because of the unbalanced genome, are even more susceptible to disruption in autosomal aneuploid conditions such as DS. The data presented provide further reason to question the commonly held notion that triplication of chromosome 21 loci accounts directly for phenotypes that permit the clinical diagnosis of DS.

References

Catalano RA (1990) Down syndrome. Surv Ophthalmol 34: 385–398

Chervenak FA, Berkowitz RL, Tortora M, Hobbins JC (1985) The management of fetal hydrocephalus. Am J Obstet Gynecol 151: 933–942

Dombrowski MP, Berry SM, Isada NB, Jones TB, Evans MI (1993) Abnormal second trimester ultrasounds: an indication for karyotype. Fet Diag Ther 8: 10–14

Freeman SB, Taft LF, Dooley KJ, Allran K, Sherman SL, Hassold TJ, Khoury MJ, Saker, DM (1999) Population-based study of congenital heart defects in Down syndrome. Am J Med Genet 87: 195–196

Jackson JF, North ER 3d, Thomas JG (1976) Clinical diagnosis of Down's syndrome. Clin Genet 9: 483–487

Kallen B, Mastroiacovo P, Robert E (1996) Major congenital malformations in Down syndrome. Am J Med Genet 65: 160–166

Khoury MJ, Erickson JD (1992) Can maternal risk factors influence the presence of major birth defects in infants with Down syndrome? Am J Med Genet 43: 1016–1022

Kim JH, Hwang J-M, Kim HJ, Yu YS (2002) Characteristic ocular findings in Asian children with Down syndrome. Eye 16: 710–714

Kwong KL, Wong G (1966) Neurodevelopment profile of Down syndrome in Chinese children. J Paediatr Child Health 32: 153–157
Lang DJ, Van Dyke DC, Heide F, Lowe PL (1987) Hypospadias and urethral abnormalities in Down Syndrome. Clin Pediat 26: 40–42
Lee LG, Jackson JF (1972) Diagnosis of Down's syndrome: clinical vs. laboratory. Clin Pediat 11: 353–356
Mautner H, Barnes A, Curtis G (1950) Abnormal findings in the spine in mongoloids. Am J Ment Defic 55: 105–107
Nicolaides KH, Berry S, Snijders RJ, Thorpe-Beeston JG, Gosden C (1966) Fetal lateral cerebral ventriculomegaly: associated malformation and chromosomal defects. Fet Diag Ther 5: 5–14
Nicolaides KH, Salvesen DR, Snijders RJ, Gosden C (1993) Fetal facial defects: associated malformations and chromosomal abnormalities. Fet Diag Ther 8: 1–9
Shapiro BL (1975) Amplified developmental instability in Down's syndrome. Ann Hum Genet Lond 38: 429–437
Shapiro BL (1983) Down syndrome — a disruption of homeostasis. Am J Med Genet 14: 241–269
Shapiro BL (1994) The environmental basis of the Down syndrome phenotype. Dev Med Child Neurol 36: 84–90
Shapiro BL (1997) Whither Down syndrome critical regions? Hum Genet 99: 421–423
Shapiro BL (1999) The Down syndrome critical region. J Neural Transm [Suppl] 57: 41–60
Smith GF, Berg JM (1976) Down's anomaly, 2nd edn. Churchill Livingstone, Edinburgh London New York, p 36
Torfs CP, Christianson RE (1998) Anomalies in Down syndrome individuals in a large population-based registry. Am J Med Genet 77: 431–438
Torfs CP, Christianson RE (1999) Maternal risk factors and major associated defects in infants with Down syndrome. Epidemiology 10: 264–270
Torfs CP, Honore LH, Curry CJR (1997) Is there an association of Down-syndrome and omphalocele? Am J Med Genet 73: 400–403
Zlotogora J (1992) Relation between the prevalence of anorectal malformations in children with Down syndrome and their prevalence in the general population. Am J Med Genet 44: 850
Zlotogora J, Abu-Dalu K, Lernau O, Sagi M, Voss R, Cohen T (1989) Anorectal malformations and Down syndrome. Am J Med Genet 34: 330–331

Author's address: Dr. B. L. Shapiro, Department of Oral Sciences and Institute of Human Genetics, 17-220 Moos Tower, 515 Delaware Street SE, Minneapolis, MN 55455, U.S.A., e-mail: burt@umn.edu

RNA Microarray analysis of channels and transporters in normal and fetal Down Syndrome (trisomy 21) brain

G. Lubec and **S. Y. Sohn**

Department of Pediatrics, University of Vienna, Austria

Summary. A couple of transporters and channels has been proposed as candidate genes involved in the pathomechanisms leading to the neuro-developmental abnormalities and the phenotype of Down Syndrome (DS, trisomy 21). No systematic study, however, has been carried out showing the concomitant expression of several candidate RNAs during fetal life.

It was therefore the aim of the study to apply an array of 96 brain RNAs mainly consisting of channels and transporters to show their expressional levels in fetal DS brain at the early second trimester.

Brain RNA was extracted from fetal cortex of the 18–19th week of gestation of controls and DS individuals and used for the GEArray Q Series Human Neuroscience-1 / Ion Channels & Transporters analysis.

15 out of 96 RNAs of the array were observed on the films in both groups during this gestational period consisting of genes for potassium, sodium, calcium channels and transporters (ASIC3, ATP1B1, CACNA1B, KCNB2, KCNC1, KCND2, KCNF1, KCNN1, KCNN3, hKCa4, KCNQ2, lipid transfer protein II, SCN2B, acetyl choline transporter, glutamate transporter3).

There was no statistically significant difference between the control and the DS group.

We provide information on the developmental expression of the aforementioned 15 RNAs and the absence of the residual examined 81 RNAs at the 18th/19th week of gestation in fetal cortex that was never reported before and show that channels and transporters present with unchanged expression in fetal DS brain.

Introduction

Information on the expression of ion channels and transporters in fetal life is limited (Stubbs et al., 1994; Thiery et al., 2000) although this knowledge would form the basis for the understanding of neurophysiology and neuropathology including epilepsies and neurological disorders and indeed a series of neurogenetic diseases is linked to aberrant ion channel or transporter expression or mutations thereof. The advent of microarrays allows the concomitant determination of RNA levels using low amounts of RNA, which is a prerequisite for fetal studies when only small tissue samples are available. The Rationale to study ion channels and transporters in fetal brain of controls and Down Syndrome (DS) individuals was to show which corresponding genes were expressed at the period of early second trimester of gestation and whether DS individuals who inevitably develop neural transmission deficits from postnatal life (Kish et al., 1989; Risser et al., 1997; Schneider et al., 1997; Gulesserian et al., 2000) show aberrant RNA expressional levels. At the protein level DS fetuses have been shown to present with comparable transporter protein expression: In a recent paper we have shown that DS individuals at the second trimester of gestation presented with normal brain protein levels for the vesicular monoamine transporter (vMAT), the vesicular acetylcholine transporter (vChAT) and the serotonin transporter (SERT) (Lubec et al., 2001). In adult DS there have been some reports showing abnormal transporter expression as reduced phospholipid transfer protein in cerebellum at the RNA level (Krapfenbauer et al., 2001) and derangement of the multidrug resistance transporter protein Pgp and MRP1 at the protein level (Engidawork et al., 2001).

The importance of ion channel functions for a wide range of developmental processes justifies the study of this system in addition: Functional studies on DS (trisomy 21) neurons have revealed that infant dorsal root ganglia neurons present with increased depolarisation and repolarisation and decreased duration of action potential (Scott et al., 1981). Moreover, it was postulated that the greater rate of depolarization in trisomy 21 neurons observed at resting potentials was primarily due to activation of residual fast sodium channels (Caviedes et al., 1990). Ion channel function and regulation can influence brain development including neuronal phenotype, neurite outgrowth, cell-cell adhesion, axon fasciculation, pathfinding, axon-glia interaction, myelination, synaptic plasticity and neuronal survival (Adamson et al., 2002; Benn et al., 2001; Fields, 1998). Precise spatial and temporal expression of each potassium channel subunit is critical for brain development (Tinel et al., 1998; Kawaschima et al., 1997; Karschin and Karschin, 1997) and overexpression of kv1.1 mRNA in amphibian embryos leads to larger delayed-rectifier potassium channel currents, shorter action potentials and a reduction in the number of morphologically differentiated neurons in culture (Gurantz et al., 1996).

Materials and methods

Brain samples

Fetal brain tissues (cerebral cortex) of controls with no obvious abnormalities (n = 5) were used for the experiment (Table 1). The brain samples were karyotyped and obtained from Dr. M. Dierssen and Dr. J. C. Ferreres of The Medical and Molecular Genetics Center-IRO, Hospital Duran I Reynals, Barcelona, Spain. The fresh brain was dissected, coronal slices snap frozen and stored at −70°C until required.

Table 1. Genes tested in microarray

Calcium channels	CACNA1A, CACNA1B, CACNB1, CACNB2, CACNB3, CACNB4, CACNG2
Potassium channels	KCNA1 (Kv1.1), KCNA2 (Kv1.2), KCNA3 (HGK5/HLK3), KCNA4 (HK1), KCNB1 (DRK1), KCNB2, KCNC1, KCNC4, KCND1, KCND2, KCNF1, KCNG1, KCNH1, KCNH2, KCNJ1 (ROMK), KCNJ3 (Kir3.1), KCNJ4 (Kir2.3), KCNJ5 (Kir3.4), KCNMB1, KCNN1, KCNN2, KCNN3, KCNN4 (hKCa4), KCNQ1, KCNQ2, KCNQ3, KCNQ4, KCNS1, KCNS2, KCNS3 (Kv9.3)
Sodium channels	ACCN1 (hBNaC1), ACCN2 (hBNaC2), ACCN3 (ASIC3), SCN1A, SCN2A2, SCN3A, SCN4A (SkM1), SCN5A, SCN6A, SCN9A, SCN1B, SCN2B
Chloride channels	CLCA1, CLCN1, CLCN2, CLCN3, CLCN4, CLCN5, CLIC1, CLIC2
Other channels	CFTR, SAS, VDAC1
Transporters and pumps	ACATN (acetyl-coenzyme A transporter), APOE, ATP1B1, ATP2B1, ATP2B4, ATP4A, ATP6A2, ATP6B2, CETP (cholesteryl ester transfer), GLUT11 (glucose transporter11) LBP, PLTP (lipid transfer protein II), SCAMP2 (secretory carrier), SCP2 (PDC-E2), SLC15A2 (H+/peptide transporter), SLC18A1 (monoamine transporter1), SLC18A2 (monoamine transporter2), SLC18A3 (acetylcholine transporter), SLC1A1 (glutamate transporter1), SLC1A2 (glutamate transporter2), SLC1A3 (glutamate transporter3), SLC25A3 (mitochondrial phosphate carrier), SLC25A4 (adenine nucleotide translocator4), SLC25A5 (adenine nucleotide translocator 5), SLC29A1 (nucleoside transporter), SLC6A11 (neurotransmitter transporter), SLC6A12 (betaine transporter), SLC6A2 (noradrenaline transporter), SLC6A3 (dopamine transporter), SLC6A4 (serotonin transporter), SLC6A6 (taurine transporter), SLC6A8 (creatine transporter), SLC6A9, TIM23, TLOC1 (Sec62), UGTREL1

Total RNA

The total RNA was extracted from each brain sample by first pulverising the frozen tissue in liquid N_2 with a mortar and pestle to a fine powder. About 0.1 g of the brain powder was homogenised in ice-cold Trizol reagent (cat. 15596-018, Invitrogen/Life Technologies). The samples were processed by the guanidinium thiocyanate method (Kim et al., 2002) according to the procedure recommended for Trizol extraction, i.e., 0.2 vol. chloroform was added to the Trizol-brain homogenate, samples hand shaken for 30 sec vigorously, and centrifuged at $12{,}000 \times g$ for 20 min at 4°C. The supernatant was transferred to a fresh tube and the RNA precipitated with 0.5 vol. isopropyl alcohol, centrifuged as above, and the pellet was extracted with 75% ethanol. The ethanol mixture was centrifuged at 7,500 $\times g$ for 10 min at 4°C. The resulting pellet was left slightly wet and partially dried at room temperature. The pellet was resuspended in DEPC-treated water and store at $-70°C$ until require.

Microarray

The relative mRNA expression of ion channels was analysed with Human Neuroscience-1 Ion Channel & Transporter GEArray™ Q series kit (HS-013, SuperArray Inc., Bethesda, MD), according to the manufacturer's protocol. Briefly, $1\,\mu g$ of Total RNA from each sample was reverse transcribed into cDNA with MMLV reverse transcriptase (M170A, Promega) in the presence of $[\alpha\text{-}^{32}P]$ dCTP ($10\,\mu Ci/\mu l$; 3,000 Ci/mmol, Amersham Pharmacia BioTech, cat. AA0005). The resulting cDNA probes were hybridised with GEArray membranes, on which ion channel gene-specific cDNA fragments including calcium channels, potassium channels, sodium channels, chloride channels, and transporters were spotted. The signals from each membrane were recorded to X-ray film and then scanner was used to convert the image into a raw image file. The raw data was extracted from the image file by using *ScanAlyze* version 2.44 by Michael Eisen, image analysis software and analysed using the GEArrayAnalyzer software version 2.1. To eliminate the potential variation in RNA quantification between samples, the expression of each ion channel and transporter was calibrated to that of GAPDH (as 1 arbitrary unit) on the same membrane and expressed as arbitrary units. These were calculated with the following formula.

$$\text{Ion channel mRNA arbitrary units} = \frac{(\text{Ion channel signal} - \text{background signal})}{(\text{GAPDH signal} - \text{background signal})}$$

Results

The 15 genes out of 96 genes related in ion channel and transporter genes — 1 gene (CACNA1B) out of 7 genes in calcium channels, 8 genes (KCNB2, KCNC1, KCND2, KCNF1, KCNN1, KCNN3, KCNN4, KCNQ2) out of 30 genes in potassium channels, 2 genes (ACCN3, SCN2B) out of 12 genes in sodium channels, and 4 genes (ATP1B1, PLTP, SLC18A3, SLC1A3) out of 36 genes in transporter and pumps — appeared to be expressed abundantly from both fetal DS brain and controls and these genes were presented in Table 2. As shown in Fig. 1, the mRNA expression of 15 genes was comparable between DS and controls. A typical microarray pattern is shown in Fig. 2.

Discussion

This is the first report on the concomitant expression of a series of ion channel and transporter genes in the early second trimester fetal brain, showing that the vast majority of these structures are not expressed at this time point of human brain development or are expressed at low abundance. 15 Genes out of 96 from the array are demonstrated and are representing calcium, potassium, sodium channels, transporters or pumps.

Also information on these 15 genes is limited: The amiloride-sensitive cation channel ASIC3, the voltage-gated potassium channel KCNC1, the

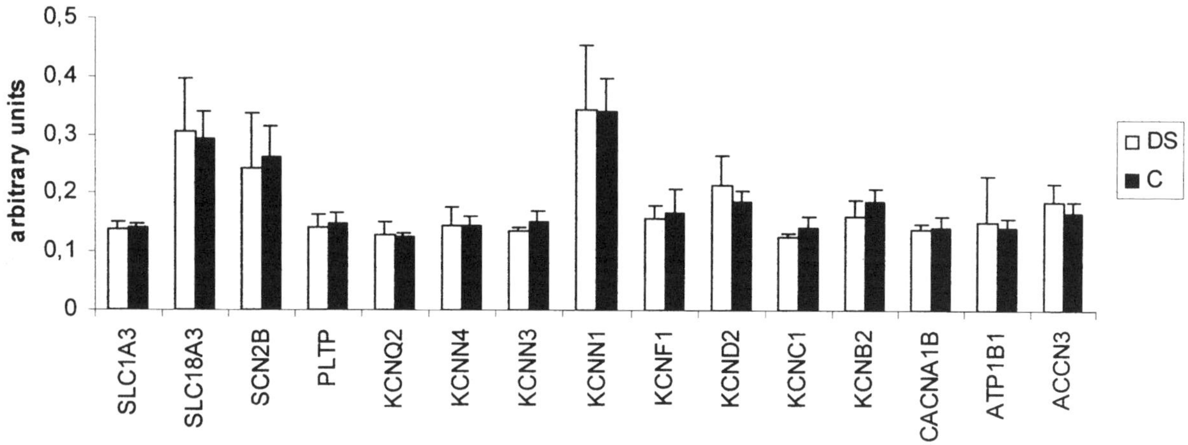

Fig. 1. The mRNA expression of 15 genes was comparable between DS and controls

potassium intermediate / small conductance calcium activated channel subfamily N and the solute carrier family 1 / glial high affinity glutamate transporter / member 3, have not been described in human brain at all, but in the mammalian system (Chen et al., 2002; Gastaldi et al., 1998; Wymore et al., 1994; Schmuckler et al., 2001; Hagiwara et al., 1996).

The Na/K-ATPase beta-subunit ATP1B1 and the intermediate conductance calcium activated potassium channel KCNN4 were never reported in brain of any animal (Shull et al., 1990; Cowley and Linsdell, 2002). For a couple of these structures only genomic data from other tissues exist including N-type calcium channel alpha-1 subunit CACNA1B, potassium voltage-gated channel, subfamily F, KCNF1, as well as potassium voltage-gated channel, KQT-like subfamily KCNQ2, linked to neonatal convulsions and shown in rat brain and sodium channel beta-2 subunit SCN2B, that has been associated with epilepsy in humans (gene bank Accession number M94172; gene bank Accession number NM_002236) (Castaldo et al., 2002; Haug et al., 2000). In human brain potassium voltage-gated channel Shaw-related subfamily member 2 KCNB2, potassium voltage-gated channel, Shal-related subfamily

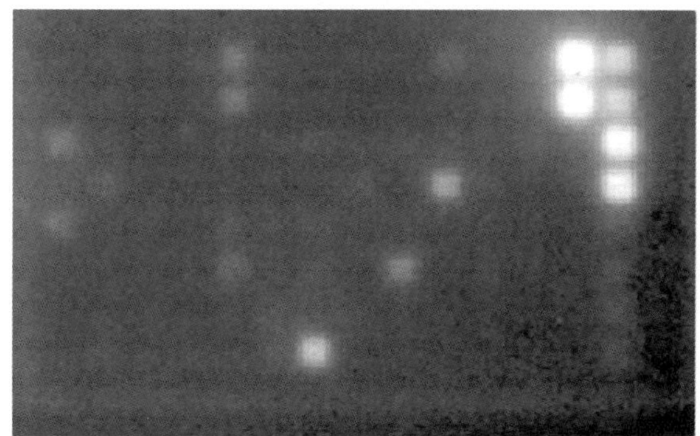

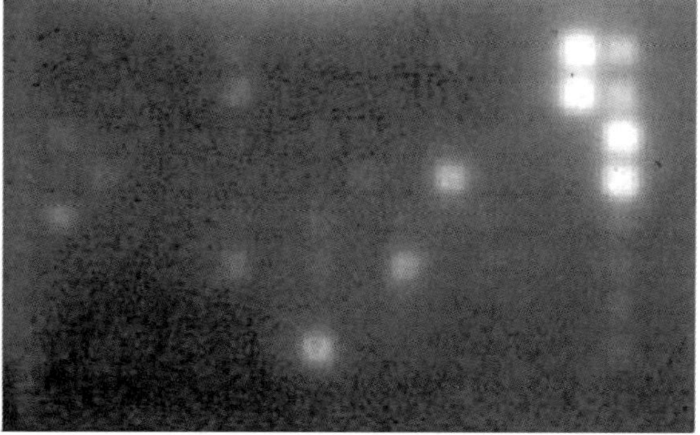

Fig. 2. A typical microarray pattern is shown

Table 2. List of genes detected in fetal brain using Human Neuroscience-1 Ion Channel & Transporters GEArray Q Series

Symbol	Gene name	Genebank	Description
ACCN3	ASIC3	NM-020321	Homo sapiens amiloride-sensitive cation channel 3, testis (ACCN3), transcript variant 1
ATP1B1	ATP1B1	X03747	MRNA for Na/K-ATPase beta subunit
CACNA1B	CACNA1B	M94172	N-type calcium channel alpha-1 subunit
KCNB2	KCNB2	NM_004770	Homo sapiens potassium voltage gated channel, Shab-related subfamily, member 2 (KCNB2)
KCNC1	KCNC1	NM_004976	Homo sapiens potassium voltage.gated channel, Shaw-related subfamily, member 1 (KCNC1)
KCND2	KCND2	NM_012181	Homo sapiens potassium voltage-gated channel, Shal-related subfamily, member 2 (KCND2)
KCNF1	KCNF1	NM_002236	Homo sapiens potassium voltage-gated channel, subfamily F, member 1 (KCNF1)
KCNN1	KCNN1	NM_002248	Homo sapiens potassium intermediate/small conductance calcium-activated channel, subfamily N, member
KCNN3	KCNN3	NM_002249	Homo sapiens potassium intermediate/small conductance calcium-activated channel, subfamily N, member
KCNN4	HKCa4	AF022797	Intermediate conductance calcium-activated potassium channel (hKCa4)
KCNQ2	KCNQ2	NM_004518	Potassium voltage-gated channel, KQT-like subfamily, member 2
PLTP	Lipid transfer protein II	L26232	Phospholipid transfer protein (PLTP); lipid transfer protein II; Extracellular Transporters & Carrier Proteins
SCN2B	SCN2B	AF049498	Sodium channel beta 2 subunit (SCN2B)
SLC18A3	Acetylcholine transporter	U10554	Vesicular acetylcholine transporter (VACHT)
SLC1A	Glutamate transporter3	U03504	SLC1A3 Solute carrier family 1 (glial high affinity glutamate transporter), member 3

member KCND2 and potassium intermediate / small conductance calcium-activated channel subfamily N member KCNN3, related to bipolar affective disorders in Man have been described already (Zhu et al., 1999; Ritsner et al., 2002).

Neither in fetal control nor in fetal DS brain these structures have been described so far.

There are, however, reports on vAChT (here SLC18A3, U10554) in fetal brain (Lubec et al., 2001) at the protein level in the identical brain region, i.e. cerebral cortex, and we can confirm this earlier report at the transcriptional level indicating that early life starts with apparently normal cholinergic innervation. Monoamine transporters SLC18A1 (monoamine transporter 1), SLC18A2 (monoamine transporter 2), SLC6A2 (noradrenaline transporter), SLC6A3 (dopamine transporter) and SLC6A4 (serotonin transporter) were not detectable in DS and controls as well. The finding of comparable expression of the serotonin transporter SERT in fetal controls and DS brain at the protein level is not contradicting the present result of absence of the serotonin transporter but was determined at a later time point within the second trimester. Also other markers for neurotransmission systems were undetectable at this developmental period: SLC1A1 (glutamate transporter 1), SLC1A2 (glutamate transporter 2), SLC1A3 (glutamate transporter 3), SLC6A11 (neurotransmitter transporter) and the taurine transporter SLC6A6 were not expressed or undetectably low. Glial high affinity glutamate transporter, a member of the solute carrier family 1, member 3 (SLC1A3) was, however, clearly expressed and one may propose that this non-specific glutamate transporter may be the first to appear in early human life.

The phospholipid transfer protein (PLTP) was shown to be decreased in adult DS brain (Krapfenbauer et al., 2001). In the present study we could detect the message comparable between groups but as Alzheimer like neuropathology inevitably develops in adult DS brain, reduced mRNA may well be due to evolving neurodegeneration rather than DS per se.

Systematically quantifying all proteins whose genes are encoded on chromosome 21, we are in the process of determining other channel and transporter proteins (Gosset et al., 1997; Shuck et al., 1997; Ohira et al., 1997; Tanizawa et al., 1996; Malo et al., 1995; Sakura et al., 1995; Tsaur et al., 1995; Yamakawa et al., 1995; Nagamine et al., 1998) that could be relevant for DS pathomechanisms in fetal brain, work that will be complementing the study presented herein.

Acknowledgements

We are highly indebted to the Red Bull Company, Salzburg, Austria for generous financial assistance. We thank Mrs. C. Avramovic for her excellent secretarial hand in the preparation of this manuscript.

References

Adamson CL, Reid MA, Mo ZL, Bowne-English J, Davis RL (2002) Firing features and potassium channel content of murine spiral ganglion neurons vary with cochlear location. J Comp Neurol 447: 331–350

Benn SC, Costigan M, Tate S, Fitzgerald M, Woolf CJ (2001) Developmental expression of the TTX-resistant voltage-gated sodium channels Nav1.8 (SNS) and Nav1.9 (SNS2) in primary sensory neurons. J Neurosci 21: 6077–6085

Castaldo P, del Giudice EM, Coppola G, Pascotto A, Annunziato L, Taglialatela M (2002) Benign familial neonatal convulsions caused by altered gating of KCNQ2/KCNQ3 potassium channels. J Neurosci 22: RC199

Caviedes P, Ault B, Rapoport SI (1990) The role of altered sodium currents in action potential abnormalities of cultured dorsal root ganglion neurons from trisomy 21 (Down syndrome) human fetuses. Brain Res 510: 229–236

Chen CC, Zimmer A, Sun WH, Hall J, Brownstein MJ, Zimmer A (2002) A role for ASIC3 in the modulation of high-intensity pain stimuli. Proc Natl Acad Sci USA 99: 8992–8997

Cowley EA, Linsdell P (2002) Characterization of basolateral K+ channels underlying anion secretion in the human airway cell line Calu-3. J Physiol 538: 747–757

Engidawork E, Roberts JC, Hardmeier R, Scheper RJ, Lubec G (2002) Expression of the multidrug resistance P glycoprotein (Pgp) and multidrug resistance associated protein (MRP1) in Down syndrome brains. J Neural Transm [Suppl] 61: 35–45

Fields RD (1998) Effects of ion channel activity on development of dorsal root ganglion Neurons. J Neurobiol 37: 158–170

Gastaldi M, Robaglia-Schlupp A, Massacrier A, Planells R, Cau P (1998) mRNA coding for voltage-gated sodium channel beta2 subunit in rat central nervous system: cellular distribution and changes following kainate-induced seizures. Neurosci Lett 249: 53–56

Gosset P, Ghezala GA, Korn B, Yaspo ML, Poutska A, Lehrach H, Sinet PM, Creau N (1997) A new inward rectifier potassium channel gene (KCNJ15) localized on chromosome 21 in the Down syndrome chromosome region 1 (DCR1). Genomics 44: 237–241

Gulesserian T, Engidawork E, Cairns N, Lubec G (2000) Increased protein levels of serotonin transporter in frontal cortex of patients with Down syndrome. Neurosci Lett 296: 53–57

Gurantz D, Ribera AB, Spitzer NC (1996) Temporal regulation of Shaker- and Shab-like potassium channel gene expression in single embryonic spinal neurons during K+ current development. J Neurosci 16: 3287–3295

Hagiwara T, Tanaka K, Takai S, Maeno-Hikichi Y, Mukainaka Y, Wada K (1996) Genomic organization, promoter analysis, and chromosomal localization of the gene for the mouse glial high-affinity glutamate transporter Slc1a3. Genomics 33: 508–515

Haug K, Sander T, Hallmann K, Rau B, Dullinger JS, Elger CE, Propping P, Heils A (2000) The voltage-gated sodium channel beta2-subunit gene and idiopathic generalized epilepsy. Neuroreport 11: 2687–2689

Karschin C, Karschin A (1997) Ontogeny of gene expression of Kir channel subunits in the rat. Mol Cell Neurosci 10: 131–148

Kawashima N, Takamiya K, Sun J, Kitabatake A, Sobue K (1997) Differential expression of isoforms of PSD-95 binding protein (GKAP/SAPAP1) during rat brain development. FEBS Lett 418: 301–304

Kim SH, Fountoulakis M, Cairns NJ, Lubec G (2002) Human brain nucleoside diphosphate kinase activity is decreased in Alzheimer's disease and Down syndrome. Biochem Biophys Res Commun 296: 970–975

Kish S, Karlinsky H, Becker L, Becker J, Gilbert J, Rebbetoy M, Chang LJ, DiStefano L, Hornykiewicz O (1989) Down's syndrome individuals begin life with normal levels of brain cholinergic markers. J Neurochem 52: 1183–1187

Krapfenbauer K, Yoo BC, Kim SH, Cairns N, Lubec G (2001) Differential display reveals downregulation of the phospholipid transfer protein (PLTP) at the mRNA level in brains of patients with Down syndrome. Life Sci 68: 2169–2179

Lubec B, Yoo BC, Dierssen M, Balic N, Lubec G (2001) Down syndrome patients start early prenatal life with normal cholinergic, monoaminergic and serotoninergic innervation. J Neural Transm [Suppl] 61: 303–310

Malo MS, Srivastava K, Ingram VM (1995) Gene assignment by polymerase chain reaction: localization of the human potassium channel IsK gene to the Down's syndrome region of chromosome 21q22.1–q22.2. Gene 159: 273–275

Nagamine K, Kudoh J, Minoshima S, Kawasaki K, Asakawa S, Ito F, Shimizu N (1998) Molecular cloning of a novel putative Ca2+ channel protein (TRPC7) highly expressed in brain. Genomics 54: 124–131

Ohira M, Seki N, Nagase T, Suzuki E, Nomura N, Ohara O, Hattori M, Sakaki Y, Eki T, Murakami Y, Saito T, Ichikawa H, Ohki M (1997) Gene identification in 1.6-Mb region of the Down syndrome region on chromosome 21. Genome Res 7: 47–58

Risser D, Lubec G, Cairns N, Herrera-Marschitz M (1997) Excitatory amino acids and monoamines in parahippocampal gyrus and frontal cortical pole of adults with Down syndrome. Life Sci 60: 1231–1237

Ritsner M, Modai I, Ziv H, Amir S, Halperin T, Weizman A, Navon R (2002) An association of CAG repeats at the KCNN3 locus with symptom dimensions of schizophrenia. Biol Psychiatry 51: 788–794

Sakura H, Bond C, Warren-Perry M, Horsley S, Kearney L, Tucker S, Adelman J, Turner R, Ashcroft FM (1995) Characterization and variation of a human inwardly-rectifying-K-channel gene (KCNJ6): a putative ATP-sensitive K-channel subunit. FEBS Lett 367: 193–197

Schmukler BE, Bond CT, Wilhelm S, Bruening-Wright A, Maylie J, Adelman JP, Alper SL (2001) Structure and complex transcription pattern of the mouse SK1 K(Ca) channel gene, KCNN1. Biochim Biophys Acta 1518: 36–46

Schneider C, Risser D, Kirchner L, Kitzmueller E, Cairns N, Prast H, Singewald N, Lubec G (1997) Similar deficits of central histaminergic system in patients with Down syndrome and Alzheimer disease. Neurosci Lett 222: 183–186

Scott BS, Petit TL, Becker LE, Edwards BA (1981) Abnormal electric membrane properties of Down's syndrome DRG neurons in cell culture. Brain Res 254: 257–270

Shuck ME, Piser TM, Bock JH, Slightom JL, Lee KS, Bienkowski MJ (1997) Cloning and characterization of two K+ inward rectifier (Kir) 1.1 potassium channel homologs from human kidney (Kir1.2 and Kir1.3). J Biol Chem 272: 586–593

Shull MM, Pugh DG, Lane LK, Lingrel JB (1990) MspI and PvuII polymorphisms in the Na,K-ATPase beta subunit gene ATP1B1. Nucl Acids Res 18: 1087

Stubbs L, Richik EM, Goldberg E, Rudy B, Handel MA, Johnson D (1994) Clustering of six human 11p15 gene homologs within a 500-kb interval of proximal mouse chromosome 7. Genomics 24: 324–332

Tanizawa Y, Matsubara A, Ueda K, Katagiri H, Kuwano A, Ferrer J, Permutt MA, Oka Y (1996) A human pancreatic islet inwardly rectifying potassium channel: cDNA cloning, determination of the genomic structure and genetic variations in Japanese NIDDM patients. Diabetologia 39: 447–452

Thiery E, Gosset P, Damotte D, Delezoide AL, de Saint Sauveur N, Vayssettes C, Creau N (2000) Developmentally regulated expression of the murine ortholog of the potassium channel KIR4.2 (KCNJ15). Mech Dev 95: 313–316

Tinel N, Lauritzen I, Chouabe C, Lazdunski M, Borsotto M (1998) The KCNQ2 potassium channel: splice variants, functional and developmental expression. Brain localization and comparison with KCNQ3. FEBS Lett 438: 171–176

Tsaur ML, Menzel S, Lai FP, Espinosa 3rd R, Concannon P, Spielman RS, Hanis CL, Cox NJ, Le Beau MM, German MS, et al. (1995) Isolation of a cDNA clone encoding a KATP channel-like protein expressed in insulin-secreting cells, localization of the

human gene to chromosome band 21q22.1, and linkage studies with NIDDM. Diabetes 44: 592–596
Wymore RS, Korenberg JR, Kinoshita KD, Aiyar J, Coyne C, Chen XN, Hustad CM, Copeland NG, Gutman GA, Jenkins NA, Chandy KG (1994) Genomic organization, nucleotide sequence, biophysical properties, and localization of the voltage-gated K+ channel gene KCNA4/Kv1.4 to mouse chromosome 2/human 11p14 and mapping of KCNC1/Kv3.1 to mouse 7/human 11p14.3–p15.2 and KCNA1/Kv1.1 to human 12p13. Genomics 20: 191–202
Yamakawa K, Mitchell S, Hubert R, Chen XN, Colbern S, Huo YK, Gadomski C, Kim UJ, Korenberg JR (1995) Isolation and characterization of a candidate gene for progressive myoclonus epilepsy on 21q22.3. Hum Mol Genet 4: 709–716
Zhu XR, Wulf A, Schwarz M, Isbrandt D, Pongs O (1999) Characterization of human Kv4.2 mediating a rapidly-inactivating transient voltage-sensitive K+ current. Receptors Channels 6: 387–400

Authors' address: Prof. Dr. G. Lubec, CChem, FRSC(UK), University of Vienna, Department of Pediatrics, Währinger Gürtel 18, A-1090 Vienna, Austria, e-mail: gert.lubec@akh-wien.ac.at

Heart type fatty acid binding protein (H-FABP) is decreased in brains of patients with Down syndrome and Alzheimer's disease

M. S. Cheon[1], **S. H. Kim**[1], **M. Fountoulakis**[2], and **G. Lubec**[1]

[1] Department of Pediatrics, University of Vienna, Austria
[2] F. Hoffmann-La Roche, Ltd., Basel, Switzerland

Summary. Fatty acid binding proteins (FABPs) are thought to play a role in the binding, targeting and transport of long-chain fatty acids, and at least three types of FABPs are found in human brain; heart type (H)-FABP, brain type (B)-FABP and epidermal type (E)-FABP. Although all three FABPs could be involved in normal brain function in prenatal and postnatal life, a neurobiological role of FABPs in neurodegenerative diseases has not been reported yet. These made us evaluate the protein levels of FABPs in brains from patients with Down syndrome (DS) and Alzheimer's disease (AD) and fetal cerebral cortex with DS using two-dimensional (2-D) gel electrophoresis with subsequent matrix-assisted laser desorption ionization mass spectroscopy (MALDI-MS) identification and specific software for quantification of proteins. In adult brain, B-FABP was significantly increased in occipital cortex of DS, and H-FABP was significantly decreased in DS (frontal, occipital and parietal cortices) and AD (frontal, temporal, occipital and parietal cortices). In fetal brain, B-FABP and epidermal E-FABP levels were comparable in controls and DS. We conclude that aberrant expression of FABPs, especially H-FABP may alter membrane fluidity and signal transduction, and consequently could be involved in cellular dysfunction in neurodegenerative disorders.

Abbreviations

AD Alzheimer's disease; *DS* Down syndrome; *FA* fatty acid; *FABP* fatty acid binding protein; *PUFA* polyunsaturated fatty acid; *2-D* two dimensional

Introduction

Fatty-acid binding proteins (FABPs), 15-kDa cytosolic proteins, are involved in uptake, transport and targeting of long-chain, unsaturated fatty acids (FAs) to regulate various cellular functions. FABPs constitute a multigene family showing expression in a variety of tissues, and human brain contains at least three different types of FABPs; heart type (H)-FABP, brain type (B)-FABP and epidermal type (E)-FABP.

H-FABP located on chromosome 1 is widely distributed and present in several tissues including brain (Veerkamp et al., 1990). Since brain H-FABP became evident after birth, with a gradual increase in gray matter, the expression of H-FABP was suggested to be neuron-specific in postnatal brain (Owada et al., 1996). Interestingly, brain H-FABP was highly concentrated in synaptosomes, neuronal nerve endings and its expression was significantly decreased in old mice (Pu et al., 1999).

B-FABP is located on chromosome 6 and has been reported to be abundantly expressed in human fetal brain (Shimizu et al., 1997), especially enriched in glial cells (Owada et al., 1996) and could be involved in the transport of hydrophobic ligands with potential morphogenic activity during central nervous system (CNS) development. The elevated levels of B-FABP at the immature stage and its subsequent decline in adult brain (Owada et al., 1996; Kurz et al., 1994) indicate that B-FABP is spatially and temporally correlated with development of the brain. E-FABP located on chromosome 15 is also found at high levels in prenatal and early postnatal brain similar to B-FABP and expressed at high levels during neurogenesis, neuronal migration and differentiation (Liu et al., 2000).

Although intracellular FABPs are present in brain and thought to be involved in various cellular functions including FAs metabolism, gene expression and cellular differentiation, almost nothing is known regarding the relationship between expression of FABPs and neurodegeneration. We therefore applied proteomics to evaluate the protein levels of FABP in adult brain with AD and DS as well as fetal DS brain.

Materials and methods

Brain sample preparation

Postmortem brain samples were obtained from Dr. N. J. Cairns in the Brain Bank, Department of Psychiatry, King's College, London, United Kingdom. All DS patients were laryotyped and possessed trisomy 21. A formal cognitive assessment of dementia in DS was not performed. In all DS brains there were abundant and extensive beta-amyloid deposits, neurofibrillary tangles and neuritic plaques. AD patients were selected prospectively and examined clinically and fulfilled the National Institute of Neurological Disorders and Stroke and Alzheimer's disease and Related Disorders Association criteria for probable AD (Tierney et al., 1988). The neuropathological diagnosis of "definite AD" was confirmed using the CERAD criteria (Mirra et al., 1991). Normal control brains were obtained from individuals with no history of neurological or psychiatric illness. The major cause of death was bronchopneumonia in AD and DS and heart disease in controls. Brain samples were dissected, coronal slices snap frozen and stored at −70°C until used. Seven brain regions (cerebellum, frontal, temporal, occipital, parietal cortex, thalamus and caudate nucleus) of patients with AD (n = 13, 58.54 ± 7.57 years old), DS (n = 9, 55.67 ± 7.48 years old) and controls (n = 18, 50.00 ± 16.94 years old) were used. The postmortem interval of brain dissection in AD, DS and controls was 30.31 ± 26.17, 31.44 ± 19.56 and 38.78 ± 18.84 hours, respectively.

Brain tissues of controls and fetus with DS were obtained from Dr. M. Dierssen in the Medical and Molecular Genetics Center-IRO, Hospital Duran i Reynals, Barcelona, Spain and Dr. J. C. Ferreres in the Department of Pathology UDIAT-CD, Corporacis Sanit'ria Parc Tauli, Sabadell, Barcelona, Spain. Cerebral cortex of controls (n = 7, male/female = 6/1, 18.79 ± 2.23 weeks) and fetuses with DS (n = 8, male/female = 6/2, 19.81 ± 2.00 weeks) was used.

Brain tissue was suspended in 0.5 ml of sample buffer consisting of 40 mM Tris, 5 M urea (Merck, Darmstadt, Germany), 2 M thiourea (Sigma, St. Louis, MO, USA), 4% CHAPS (Sigma), 10 mM 1,4-dithioerythritol (Merck), 1 mM EDTA (Merck), 1 mM PMSF (Sigma) and 1 μg/ml of each pepstatin A, chymostatin, leupeptin and antipain. The suspension was sonicated for approximately 30 seconds and centrifuged at 10,000 × g for 10 minutes and the supernatant was centrifuged further at 150,000 × g for 45 minutes to sediment undissolved material. The protein concentration of the supernatant was determined by the Coomassie blue method (Bradford, 1976).

Two-dimensional (2-D) gel electrophoresis

The 2-D gel electrophoresis was performed essentially as reported (Langen et al., 1997). Adult samples of approximately 1.5 mg were applied on immobilized pH 3–10 nonlinear gradient strips (IPG, immobilized pH-gradient strips, Pharmacia Biotechnology, Uppsala, Sweden), at both, the basic and acidic ends of the strips. The proteins were focused at 300 V for 1 hour, after which the voltage was gradually increased to 3,500 V within 6 hours. Focusing was continued at 3,500 V for 12 hours and at 5,000 V for 48 hours. The second-dimensional separation was performed on 9–16% linear gradient polyacrylamide gels. After protein fixation with 40% methanol containing 5% phosphoric acid for 12 hours, the gels were stained with colloidal Coomassie blue for 48 hours. The molecular mass was determined by running standard protein markers at the right side of selected gels. The size markers (Gibco, Basel, Switzerland) covered the range 10–200 kDa. Isoelectric point (pI) values were used as given by the supplier of the IPG strips. The gels were destained with H_2O and scanned in a Personal Densitometer (Molecular Dynamics, Sunnyvale, CA, USA). The images were processed using PhotoShop (Adobe) and PowerPoint (Microsoft) software.

For fetal brains, samples of 2 mg were applied, and focusing stated at 200 V and the voltage was gradually increased to 5000 V at 3 V/min and kept constant for a further 24 hours (approximately 180000 kVh totally). The second-dimensional separation was performed on 9–16% SDS gradient polyacrylamide gels, and the gels (180 × 200 × 1.5 mm) were run at 40 mA/gel. The gels were scanned in an Agfa DUOSCAN densitometer (resolution 200).

Matrix-assisted laser desorption ionization mass spectroscopy (MALDI-MS)

MALDI-MS analysis was performed as described with minor modification (Fountoulakis and Langen, 1997). Briefly, spots from gels for adult brain were excised, destained with 50% acetonitrile in 0.1 M ammonium bicarbonate and dried in a speedvac evaporator. The dried gel pieces were reswollen with 3 μl of 3 mM Tris-HCl, pH 8.8, containing 50 ng trypsin (Promega, Madison, WI, USA). After 15 minutes, 3 μl of H_2O were added and left at room temperature for 12 hours. Two microliters of 30% acetonitrile, containing 0.1% trifluoroacetic acid were added, the content was vortexed, centrifuged for 3 minutes and sonicated for 5 minutes. One microliter was applied onto the dried matrix spot. The matrix consisted of 15 mg nitrocellulose (Bio-Rad) and 20 mg α-cyano 4 hydroxycinnamic acid (Sigma) dissolved in 1 ml acetone: isopropanol (1:1, v/v). A 0.5 μl portion of matrix solution was applied on the sample target. Specimen were analyzed in a time-of-flight

Voyager Elite mass spectrometer (PerSeptive Biosystems, Cambridge, MA, USA) equipped with a reflectron. An accelerating voltage of 20 kV was used. Calibration was against total spots and background. The peptide masses were matched with the theoretical peptide masses of all proteins from all species of the SWISS-PROT database. For protein search, monoisotopic masses were used and a mass tolerance of 0.0075% was allowed.

For fetal brains, the spots were excised with a spot picker and placed into 96-well microtiter plates. Each spot was destained with 100 μl of 30% acetonitrile in 50 mM ammonium bicarbonate. Each dried gel pieces was rehydrated with 4 μl of 3 mM Tris-HCl, pH 9.0, containing 50 ng trypsin. After 16 hours at room temperature, 7 μl of H_2O were added to each gel piece and the samples were shaken in for 10 minutes. Four μl of 50% acetonitrile, containing 0.3% trifluoroacetic acid, the standard peptides des-Arg-bradykinin (Sigma, 904.4681 Da) and adrenocorticotropic hormone fragment 18–39 (Sigma, 2465.1989 Da), in H_2O were added to each gel piece and shaken for 10 minutes. The application of the samples was performed with a SymBiot I sample processor (PE Biosystems, Framingham, MA, USA). 1.5 μl of the peptide mixture was simultaneously applied with 1 μl of matrix, consisting of a saturated solution of α-cyano-4-hydroxycinnamic acid (Sigma) in 50% acetonitrile, containing 0.1% trifluoroacetic acid Samples were analyzed in a time-of-flight mass spectrometer (Reflex 3, Bruker Analytics, Bremen, Germany). Peptide matching and protein searches were performed automatically. Monoisotopic masses were used and a mass tolerance of 0.0025% was allowed. The algorithm used for determining the probability of a false positive match with a given MS-spectrum is described elsewhere (Bendt et al., 1999).

Quantification of proteins

In partial 2D-gel images including each protein corresponding to FABPs and the neighbouring proteins, the percentage of the volume of the spot representing each protein was quantified as compared to the total proteins present using the ImageMaster 2D Elite software (Amersham Pharmacia Biotechnology).

Statistics

Results were expressed as means ± standard deviation. Between-group differences were investigated by non-parametric Mann-Whitney U test. The significance was set at $P < 0.05$. All statistical analyses were performed by GraphPad Instat2 software, version 2.05.

Results and discussion

As shown in Table 1, FABPs were identified in human adult (Langen et al., 1999) and fetal brain based upon molecular weight, pI value and peptide matches by 2-D gel electrophoresis and MALDI-MS. In adult brain, significantly decreased protein levels of H-FABP were observed in DS (frontal, occipital and parietal cortices) and AD (frontal, temporal, occipital and parietal cortices) as shown in Table 2 and Fig. 1. Although there was no statistical significance, H-FABP was also decreased in other regions of both disorders. B-FABP was significantly increased in occipital cortices of DS. However, E-FABP was not changed in any region of adult brain from both disorders.

Table 1. Identification of FABPs by MALDI-MS. Proteins separated by 2-D gel electrophoresis were identified by MALDI-MS, following digestion with trypsin. After the peptide masses were matched with the theoretical peptide masses of all proteins from all species of the SWISS-PROT database, each protein was identified to FABP

	FABPs (accession number)	pI	MW	Matches
Adult brain	B-FABP (O15540)	5.20	14862	8
	E-FABP (Q01469)	7.00	15496	5
	H-FABP (P05413)	6.80	14905	5
Fetal brain	B-FABP (O15540)	5.34	14862	4
	E-FABP (Q01469)	6.96	15496	5

Table 2. FABP protein levels in adult brains with AD, DS and controls (arbitrary units)

	Control	AD	DS
B-FABP			
Cerebellum	4.01 ± 2.53 (n = 8)	2.37 ± 0.90 (n = 6)	4.74 ± 1.13 (n = 2)
Frontal cortex	2.61 ± 1.78 (n = 5)	2.57 ± 0.41 (n = 5)	2.58 ± 1.37 (n = 4)
Temporal cortex	1.72 ± 1.02 (n = 9)	2.21 ± 0.89 (n = 9)	2.09 ± 1.19 (n = 5)
Occipital cortex	1.25 ± 0.46 (n = 10)	1.90 ± 0.81 (n = 9)	2.16 ± 0.88* (n = 6)
Parietal cortex	1.83 ± 1.69 (n = 5)	2.13 ± 1.25 (n = 4)	2.84 ± 0.48 (n = 3)
Thalamus	1.62 ± 0.99 (n = 7)	1.89 ± 1.28 (n = 6)	2.12 ± 1.05 (n = 5)
Caudate nucleus	1.08 (n = 1)	1.07 ± 0.52 (n = 7)	1.33 ± 0.30 (n = 3)
E-FABP			
Cerebellum	3.40 ± 1.98 (n = 10)	2.94 ± 1.40 (n = 11)	3.70 ± 1.59 (n = 7)
Frontal cortex	1.87 ± 1.13 (n = 7)	1.66 ± 1.15 (n = 5)	1.20 ± 0.42 (n = 4)
Temporal cortex	1.74 ± 1.00 (n = 9)	2.78 ± 1.14 (n = 10)	1.69 ± 0.82 (n = 9)
Occipital cortex	2.77 ± 1.46 (n = 11)	2.62 ± 1.54 (n = 9)	1.90 ± 0.81 (n = 7)
Parietal cortex	1.76 ± 0.33 (n = 5)	2.58 ± 1.51 (n = 4)	2.19 ± 0.91 (n = 6)
Thalamus	1.98 ± 1.14 (n = 10)	3.30 ± 2.07 (n = 5)	2.06 ± 0.64 (n = 5)
Caudate nucleus	4.47 ± 3.96 (n = 6)	2.30 ± 2.08 (n = 8)	1.99 ± 0.42 (n = 6)
H-FABP			
Cerebellum	3.19 ± 2.14 (n = 11)	2.19 ± 1.10 (n = 11)	1.96 ± 1.05 (n = 7)
Frontal cortex	4.36 ± 1.66 (n = 8)	2.26 ± 0.49** (n = 6)	1.97 ± 0.94** (n = 6)
Temporal cortex	4.87 ± 2.10 (n = 13)	3.21 ± 0.59* (n = 9)	4.48 ± 1.76 (n = 9)
Occipital cortex	4.42 ± 1.72 (n = 14)	2.75 ± 1.05* (n = 11)	2.85 ± 1.12* (n = 8)
Parietal cortex	4.20 ± 1.05 (n = 6)	2.09 ± 1.09** (n = 6)	3.18 ± 1.22* (n = 7)
Thalamus	3.77 ± 2.37 (n = 10)	2.90 ± 2.65 (n = 6)	3.08 ± 0.86 (n = 6)
Caudate nucleus	5.47 ± 3.16 (n = 7)	3.02 ± 0.87 (n = 7)	3.42 ± 1.33 (n = 8)

The proteins from seven brain regions of patients with DS and AD, and the control group were separated on 2-D gels and visualized following stain with colloidal Coomassie blue. In partial gel images, including FABPs and the neighboring proteins, the percentage of the volume of the spots representing FABPs were quantified, compared to the total proteins present. The quantification was performed using the ImageMaster 2D software. The statistical analyses were performed by Graphpad Instat2 software. Significant differences are indicated as follow: $^{**}P < 0.01$; $^{*}P < 0.05$

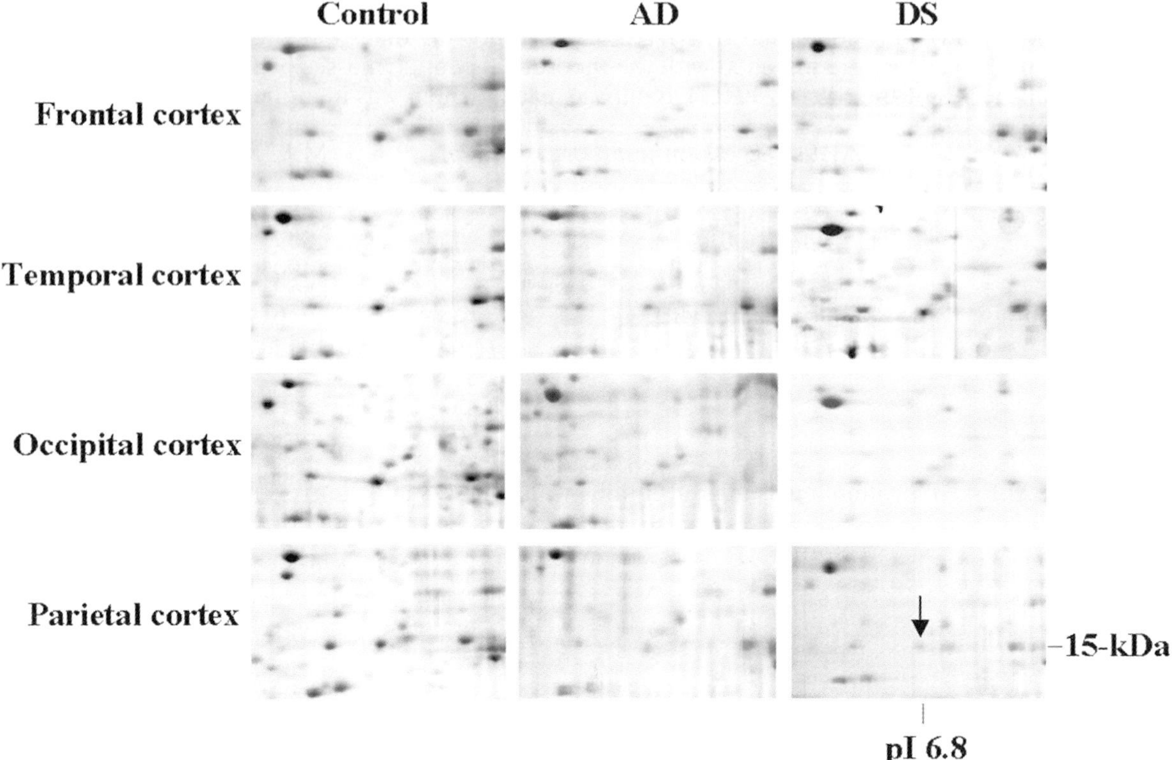

Fig. 1. Two-dimensional (2-D) gel images of H-FABP in adult brain regions with significant changes. The human adult brain proteins were extracted and separated on an immobilized pH 3–10 nonlinear gradient strip, followed by a 9–16% linear gradient polyacrylamide gel, as stated under experimental procedures. The gel was stained with Coomassie blue and the spots were analyzed by MALDI-MS. The partial 2-D gel images are shown and an arrow indicates the protein spot corresponding to H-FABP with significant changes

Considering that both H-FABP and E-FABP are distributed in neurons, the decrease of H-FABP could be specific, not resulting from neuronal loss. The levels of B-FABP and E-FABP were comparable between fetal DS and control brain (Table 3). The deranged protein level of FABPs in adult brain from DS and AD could be explained by impaired transcriptional machinery in both diseases as reported in our and other previous work (Labudova et al., 1998; Yeghiazaryan et al., 1999; Greber-Platzer et al., 1999; Sikora et al., 1993).

Among the cytosolic lipid binding proteins, FABPs bind to polyunsaturated fatty acid (PUFA) including arachidonic acid and docosahexaenoic acid (Myers-Payne et al., 1996; Richieri et al., 1994), stimulate cellular FA uptake, enhance the intracellular diffusion of FAs, and affect the metabolic targeting of FAs. PUFAs are not only involved in membrane lipid metabolism, but equally important in signal transduction, gene regulation, and cellular differentiation. PUFAs are also essential for normal cell membrane functioning

Table 3. FABPs levels in fetal brain with DS and controls (arbitrary units)

	Control	DS
B-FABP	7.42 ± 1.92 (7)	9.31 ± 1.45 (8)
E-FABP	2.15 ± 0.43 (7)	1.85 ± 0.66 (8)

The gels were prepared and the intensities of the spots representing each protein were quantified and analysed as described in legend of Table 2

because many membrane properties, such as fluidity and permeability, are closely related to the presence of unsaturated and polyunsaturated side chains (Yehuda et al., 1999).

The decrease of brain H-FABP expression may alter lipid transport, which in turn may modify lipid composition and structure of brain membranes. It has been indicated that phospholipid composition of plasma membranes was altered in DS and AD brain (Brooksbank and Martinez, 1989; Wells et al., 1995). These findings could be supported by our recent study showing downregulated mRNA levels of phospholipid transfer protein in DS brain (Krapfenbauer et al., 2001). Changes in membrane lipid composition of neurodegenerative brain could be associated with decreased membrane bulk fluidity and diminished membrane transbilayer fluidity gradients leading to the neurotoxicity and cell death seen in neurodegenerative diseases. In addition, significant decreases of PUFAs such as arachidonic acid and docosahexaenoic acid have been found in brain membrane of AD (Guan et al., 1999; Prasad et al., 1998). It has been already known that PUFA deficiency cause hypomyelination. Our recent study also demonstrated deceased levels of 2′, 3′-cyclic nucleotide-3′-phosphodiesterase, a marker for oligodendroglia reflecting deteriorated myelination in DS and AD (Vlkolinsky et al., 2001). Since nervous tissue has a high requirement for PUFAs, and H-FABP binds such PUFAs, decreases in the levels of H-FABP and PUFAs may play a significant role for cell dysfunction in the brain with neurodegeneration.

Interestingly, by sequestering arachidonic acids and other FAs, brain H-FABP could modulate membrane receptor activity. Moreover, a 33% identity of amino acids between the NMDA receptor and bovine H-FABP has been found (Petrou et al., 1993), suggesting that binding of arachidonic acid and other FAs to this domain may regulate NMDA receptor activity. Na^+-dependent uptake of amino acids is highly sensitive to free FAs, and brain FABPs markedly stimulated synaptosomal amino acid uptake by sequestering inhibitory FAs (Bass et al., 1984). Decreased levels of H-FABP in synaptosomes of AD and DS brain may contribute to exacerbate the toxic effects of released FAs or undesirable effects of excess levels of FAs that are precursors of signaling molecules in the brain.

On the other hand, the protein levels of B-FABP and E-FABP were comparable between fetal DS brain and controls. Along with reports showing that there were no major abnormalities in lipid composition (Brooksbank et al., 1989), nor alterations in the concentration of total combined PUFAs in fetal DS brain (Brooksbank and Balazs, 1984), the normal expression of FABPs found in DS fetal brain suggests that the impaired FA metabolism of adult brain may not be due to initial defects in prenatal life.

In conclusion, the present data suggest that decreased H-FABP levels may alter lipid composition and/or fluidity in membrane of the brains with neurodegeneration. Reduced brain H-FABP levels could modify the FA metabolism, modulate sensitivity of signaling molecules to bind receptors and contribute or reflect impaired synaptosomal structure and functions already reported in our previous work (Yoo et al., 2001). Therefore, modifications of FABPs in brain may be an important factor involved in cellular dysfunction in both neurodegenerative disorders.

Acknowledgements

We would like to thank Dr. N. J. Cairns (Brain Bank, Department of Psychiatry, King's College, London, United Kingdom), Dr. M. Dierssen (Medical and Molecular Genetics Center-IRO, Hospital Duran i Reynals, Barcelona, Spain) and Dr. J. C. Ferreres (Department of Pathology UDIAT-CD, Corporacis Sanit'ria Parc Tauli, Sabadell, Barcelona, Spain) for providing tissue samples.

References

Bass NM, Raghupathy E, Rhoads DE, Manning JA, Ockner RK (1984) Partial purification of molecular weight 12,000 fatty acid binding proteins from rat brain and their effect on synaptosomal Na+ dependent amino acid uptake. Biochem 23: 6539–6544

Bendt P, Hobohm U, Langen H (1999) Reliable automatic protein identification from matrix-associated laser desorption/ionization mass spectrometric peptide fingerprints. Electrophoresis 20: 3521–3526

Bradford MM (1976) A rapid and sensitive method for the quantitation of microgram quantities of protein utilizing the principle of protein-dye binding. Anal Biochem 72: 248–254

Brooksbank BW, Balazs R (1984) Superoxide dismutase, glutathione peroxidase and lipoperoxidation in Down's syndrome fetal brain. Brain Res 318: 37–44

Brooksbank BW, Martinez M (1989) Lipid abnormalities in the brain in adult Down's syndrome and Alzheimer's disease. Mol Chem Neuropathol 11: 157–185

Brooksbank BW, Walker D, Balazs R, Jorgensen OS (1989) Neuronal maturation in the foetal brain in Down's syndrome. Early Hum Dev 18: 237–246

Fountoulakis M, Langen H (1997) Identification of protein by matrix-assisted laser desorption ionization-mass spectroscopy following in-gel digestion in low-salt nonvolatile buffer and simplified peptide recovery. Anal Biochem 250: 153–156

Greber-Platzer S, Schatzmann-Turhani D, Cairns N, Balcz B, Lubec G (1999) Expression of the transcription factor ETS2 in brain of patients with Down syndrome- evidence against the overexpression-gene dosage hypothesis. J Neural Transm [Suppl] 57: 269–281

Guan Z, Wang Y, Cairns NJ, Lantos PL, Dallner G, Sindelar PJ (1999) Decrease and structural modifications of phosphatidylethanolamine plasmalogen in the brain with Alzheimer disease. J Neuropathol Exp Neurol 58: 740–747

Krapfenbauer K, Yoo BC, Kim SH, Cairns N, Lubec G (2001) Differential display reveals downregulation of the phospholipid transfer protein (PLTP) at the mRNA level in brains of patients with Down syndrome. Life Sci 68: 2169–2179

Kurz A, Zimmer A, Schnutgen F, Bruning G, Spener F, Muller T (1994) The expression pattern of a novel gene encoding brain-fatty acid binding protein correlates with neuronal and glial cell development. Development 120: 2637–2649

Labudova O, Krapfenbauer K, Moenkemann H, Rink H, Kitzmuller E, Cairns N, Lubec G (1998) Decreased transcription factor junD in brains of patients with Down syndrome. Neurosci Lett 252: 159–162

Langen H, Roeder D, Juranville JF, Fountoulakis M (1997) Effect of the protein application mode and the acrylamide concentration on the resolution of protein spots separated by two-dimensional gel electrophoresis. Electrophoresis 18: 2085–2090

Langen H, Berndt P, Roder D, Cairns N, Lubec G, Fountoulakis M (1999) Two-dimensional map of human brain proteins. Electrophoresis 20: 907–916

Liu Y, Long LD, De Leon M (2000) In situ and immunocytochemical localization of E-FABP mRNA and protein during neuronal migration and differentiation in the rat brain. Brain Res 852: 16–27

Mirra SS, Heyman A, McKeel D, Sumi S, Crain BJ (1991) The consortium to establish a registry for Alzheimer's disease. Neurology 41: 479–486

Myers-Payne SC, Hubbell T, Pu L, Schnutgen F, Borchers T, Wood WG, Spener F, Schroeder F (1996) Isolation and characterization of two fatty acid binding proteins from mouse brain. J Neurochem 66: 1648–1656

Owada Y, Yoshimoto T, Kondo H (1996) Spatio-temporally differential expression of genes for three members of fatty acid binding proteins in developing and mature rat brains. J Chem Neuroanat 12: 113–122

Petrou S, Ordway RW, Singer JJ, Walsh JVJr (1993) A putative fatty acid-binding domain of the NMDA receptor. Trends Biochem Sci 18: 41–42

Prasad MR, Lovell MA, Yatin M, Dhillon H, Markesbery WR (1998) Regional membrane phospholipid alterations in Alzheimer's disease. Neurochem Res 23: 81–88

Pu L, Igbavboa U, Wood WG, Roths JB, Kier AB, Spener F, Schroeder F (1999) Expression of fatty acid binding proteins is altered in aged mouse brain. Mol Cell Biochem 198: 69–78

Richieri GV, Ogata RT, Kleinfeld AM (1994) Equilibrium constants for the binding of fatty acids with fatty acid binding proteins from adipocytes, intestine, heart, and liver measured with the fluorescent probe ADIFAB. J Biol Chem 269: 23918–23930

Shimizu F, Watanabe TK, Shinomiya H, Nakamura Y, Fujiwara, T (1997) Isolation and expression of a cDNA for human brain fatty acid-binding protein (B-FABP). Biochim Biophys Acta 1354: 24–28

Sikora E, Radziszewska E, Kmiec T, Maslinska D (1993) The impaired transcription factor AP-1 DNA binding activity in lymphocytes derived from subjects with some symptoms of premature aging. Acta Biochim Pol 40: 269–272

Tierney MC, Fisher RH, Lewis AJ, Torzitto ML, Snow WG, Reid DW, Nieuwstraaten P, Van Rooijen LAA, Derks HJGM, Van Wijk R, Bischop A (1988) The NINCDA-ADRDA work group criteria for the clinical diagnosis of probable Alzheimer's disease. Neurology 38: 359–364

Veerkamp JH, Paulussen RJ, Peeters RA, Maatman RG, van Moerkerk HT, van Kuppevelt TH (1990) Detection, tissue distribution and (sub)cellular localization of fatty acid-binding protein types. Mol Cell Biochem 98: 11–18

Vlkolinsky R, Cairns N, Fountoulakis M, Lubec G (2001) Decreased brain levels of 2′, 3′-cylic nucleotide-3′-phosphodiesterase in Down syndrome and Alzheimer's disease. Neurobiol Aging 22: 547–553

Wells K, Farooqui AA, Liss L, Horrocks LA (1995) Neural membrane phospholipids in Alzheimer disease. Neurochem Res 20: 1329–1333

Yeghiazaryan K, Turhani-Schatzmann D, Labudova O, Schuller E, Olson EN, Cairns N, Lubec G (1999) Downregulation of the transcription factor scleraxis in brain of patients with Down syndrome. J Neural Transm [Suppl] 57: 305–314

Yehuda S, Rabinovitz S, Mostofsky DI (1999) Essential fatty acids are mediators of brain biochemistry and cognitive functions. J Neurosci Res 56: 565–570

Yoo BC, Cairns N, Fountoulakis M, Lubec G (2001) Synaptosomal proteins, beta-soluble N-ethylmaleimide-sensitive factor attachment protein (beta-SNAP), gamma-SNAP and synaptotagmin I in brain of patients with Down syndrome and Alzheimer's disease. Dement Geriatr Cogn Disord 12: 219–225

Authors' address: Prof. Dr. G. Lubec, CChem, FRSC (UK), Department of Pediatrics, University of Vienna, Währinger Gürtel 18, A-1090 Vienna, Austria, e-mail: gert.lubec@akh-wien.ac.at

Stem cell marker expression in human trisomy 21 amniotic fluid cells and trophoblasts

A.-R. Prusa, E. Marton, M. Rosner, A. Freilinger, G. Bernaschek,
and **M. Hengstschläger**

Obstetrics and Gynecology, Prenatal Diagnosis and Therapy, University of Vienna, Vienna, Austria

Summary. Down Syndrome is the most frequent genetic cause of mental retardation. Deregulation of specific differentiation processes is a major cause for the neuropathological cell features typical for this syndrome. The molecular mechanisms leading to Down Syndrome are likely to be operative from the very earliest time of embryonic/fetal development. We therefore analysed human amniotic fluid cell samples and cytotrophoblastic cells from placental biopsies, both with normal karyotypes and with trisomy 21, for the mRNA expression of stem cell marker genes. Here we describe for the first time that these human primary cell sources contain cells that express telomerase reverse transcriptase, leukemia inhibitory factor receptor, and bone morphogenetic protein receptor II. A specific difference between aneuploid and normal cells could not be detected. These data provide evidence that human amniotic fluid and cytotrophoblastic cell cultures might provide a new source for research on primary cell systems expressing these stem cell markers. In addition, it is suggested that early deregulation of the expression of these genes in the here analysed cell sources does not contribute to the molecular development of Down Syndrome.

Introduction

Down Syndrome is the most frequent genetic cause of mental retardation, affecting up to 1 in 700 livebirths. 95% of the patients have complete trisomy of chromosome 21 due to nondisjunction during gamete formation, while in the remaining patients the syndrome is due to chromosomal translocation (Antonarakis, 1993; Yoon, 1996). This syndrome is characterized by a specific phenotype featured by the classic facial appearance, mental retardation and other major congenital malformations such as those of the heart and of the gastrointestinal tract (Hernandez, 1996). Furthermore, Down Syndrome patients develop Alzheimer's disease-type neuropathology (β-amyloid plaques, neurofibrillary tangels and neuronal loss) by age 30–40 years (Engidawork, 2003). The neurohistological features of this syndrome are complex and variable. Brains of Down Syndrome patients have been reported to exhibit two hallmark lesions, senile plaques and neurofibrillary tangles, and delayed myelination, fewer neurons, lower neuronal density and distribution, and abnormal synaptic density and length (reviewed in Cairns, 1999; Lubec, 2002; Engidawork, 2003). Despite a widespread interest in Down Syndrome research, there has been no clear explanation as to how the typical cognitive and cellular changes are produced.

The molecular mechanisms that lead to Down Syndrome are likely to be operative from the very earliest time of embryonic/fetal development. It therefore makes sense to take the most primitive cells from embryos/fetuses and compare levels of gene expression in those from cells with normal karyotype with those from embryos/fetuses with trisomy 21 (Miller, 2002). Very recently, PCR mRNA expression analyses were performed on neuronal precursor cells derived from the cortex of a fetus with Down Syndrome, and findings were compared with those of two control samples. It was demonstrated that genes regulated by the neuron-restrictive silencer factor (REST) were selectively repressed in Down Syndrome neuronal precursor cells (Bahn, 2002). These data suggest that indeed the additional chromosome 21 triggers gene deregulations already at the very early levels of cell lineage differentiation.

Despite the wide and well established usage of human amniotic fluid cells in routine prenatal testing of fetal abnormalities caused by genetic alterations, the current knowledge about origin and properties of these cells is limited. For the mixture of cells within the amniotic fluid a wide variety of different origins has been suggested. Even in case of pregnancies with normal fetal development, the cells in the amniotic fluid are heterogeneous, what makes it tempting to speculate that they could represent an important source for isolation and characterization of different lineages of primary human cell types (Milunsky, 1979; Hoehn, 1982; Gosden, 1983; Prusa, 2002).

We decided to chose human amniotic fluid cells together with human cytotrophoblastic cells as biological systems for the here presented investigations because of the following reasons: 1) These are primary cells, which are neither immortalised nor transformed, making them at least to some extent more representative for *in vivo* situations. 2) The mixture of cells in human amniotic fluid is speculated to origin from all three germ layers, ectoderm, mesoderm and endoderm (Prusa, 2002). Accordingly, the drawn conclusions from amniotic fluid cells and from cytotrophoblastic cells are not restricted to only one specific cell type. 3) These cells exhibit natural occuring proliferative potential and need not to be artificially stimulated to proliferate.

In the here presented study we investigated two aspects: 1) Do early human fetal cell samples, such as amniotic fluid samples and placental biopsies, contain cells that express genes characteristic for human stem cells? 2) Is there any difference in the expression of stem cell marker genes between cells with a normal karyotype and trisomy 21 cells?

Materials and methods

Cells

Prenatal diagnosis was offered 1) to pregnant women with an abnormality in the prenatal ultrasound like a nuchal translucency thickness of 7.5 mm, a golf ball phenomen, a double bubble or a cleft lip, 2) to women with a positive Triple-Test, and 3) to women aged ≥35 years (compare e.g. Benn, 2002) (for details see Tables 1 and 2). Human cytotrophoblastic cells were isolated from placentae, obtained after uncomplicated placental biopsies, performed between the 13th and 22nd week of gestational age. Amniotic fluid cell samples were obtained from amniocentesis, performed between the 16th and 34th week of gestational age. This project has been reviewed and accepted by the ethics commission of the University of Vienna, Austria (project number: 036/2002) and each patient signed a written informed consent.

Cell cultivation and cytogenetics

The cells were plated on 100-mm culture dishes and maintained and expanded in standard medium: Nutrient Mixture Ham's F10 medium (Gibco, Austria) supplemented with 10% fetal calf serum (Gibco, Austria), Ultroser G (BioSepra, France), Gentamycin (Biochrom, Germany), L-Glutamin (Biochrom Germany) in a 95% humidified, 5% CO_2 chamber at 37°C. Cytogenetic analysis was performed according to standard protocols and chromosome banding was produced by means of the Trypsin-Giemsa method. An up-to-date summary of standard cytogenetic protocols, which were used for this study, can be found in (Wegner, 1999).

Table 1. Indications for amniocentesis

Amniotic fluid cells	F1	F2	F3	F4
Karyotype	47,XX,+21	46,XX	47,XY,+21	46,XY
Gestational age (weeks)	16	16	34	34
Maternal age	+	+	–	–
Gastrointestinal anomalies	–	–	+	–
Cleft lip	–	–	–	+

Table 2. Indications for placental biopsies

Cytotrophoblastic cells	T1	T2	T3	T4
Karyotype	47,XX,+21	46,XX	47,XY,+21	46,XX
Gestational age (weeks)	21	22	13	22
Maternal age	+	–	–	–
Positive Triple-Test	+	–	–	–
Golf ball phenomen	+	+	–	–
Cleft lip	–	–	–	+
Increased fetal NT[a]	–	–	+	–

[a] *Fetal NT* fetal nuchal translucency thickness

Detection of mRNA expression

Total RNA was prepared from cultivated human amniotic fluid cells and cytotrophoblastic cells by the TriReagent method (Molecular Research Centre, Inc., Cincinnati, USA). First-strand cDNA synthesis and PCR was performed using the RT-for-PCR-Kit (Clontech, Palo Alto, USA) following the protocol provided by the manufacturer. Expression of 18S-rRNA (VBC Genomics, Vienna) was analysed as a control. PCR primers and conditions:

- human telomerase reverse transcriptase (hTERT): for primer sequence see (Xu, 2001), annealing at 60°C, 40 PCR cycles
- leukemia inhibitory factor receptor (LIFR): primers
  5′ TGGAGGGACTGCACAGGTTATTTA 3′ and
  5′ CGGTGACACTGTTAATCGTTTGGT 3′, annealing at 54°C, 30 PCR cycles
- bone morphogenetic protein receptor II (BMPRII): primers
  5′ CCACTAGAAGGTGGCCGAACTAAT 3′ and
  5′ AGTGCCTCCTTCTGCAAGGTAAAC 3′, annealing at 54°C, 30 PCR cycles
- 18s-rRNA: primers 5′ AGGAATTGACGGAAGGGCACC 3′ and
  5′ GGACATCTAAGGGCATCACAG 3′, annealing at 54°C, 30 PCR cycles

Products were separated on 1% or 2% agarose gels and visualized by ethidium bromide staining.

Results

Telomerase activity is detectable in most immortalised cell lines and in tumour samples, but it is also a marker for pluripotent human stem cells (for a detailed description of such markers see also http://www.nih.gov/news/stemcell/scireport.htm). We performed RT-PCR analyses of the expression of human telomerase reverse transcriptase (hTERT) in human amniotic fluid cell samples and in cytotrophoblastic cell samples (Figs. 1 and 2; Tables 1 and 2). Although the expression levels differed from undetectable (F1, T1, T2, T4), very low (F2, T3) to highly expressed (F3, F4), specific differences between normal and trisomy 21 cells could not be detected. Expression of

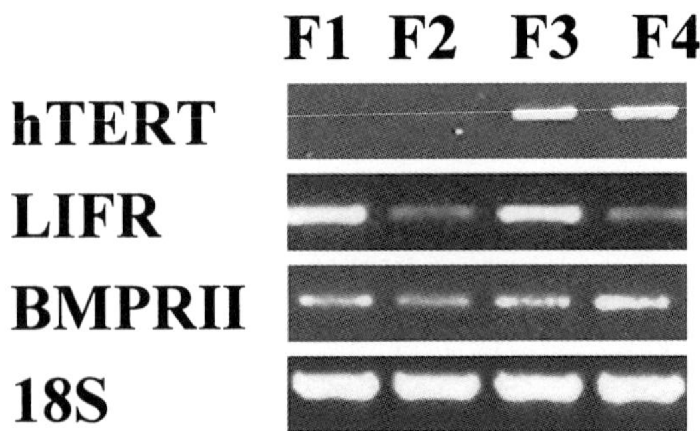

Fig. 1. Expression of stem cell marker genes in cultivated human amniotic fluid cell samples. RT-PCR analyses of mRNA expression of human telomerase reverse transcriptase (hTERT), leukemia inhibitory factor receptor (LIFR), and bone morphogenetic protein receptor II (BMPRII). As internal standard the expression of 18S-rRNA (18S) was analysed. F1, F3: trisomy 21; F2, F4: normal karyotype. For details compare Table 1

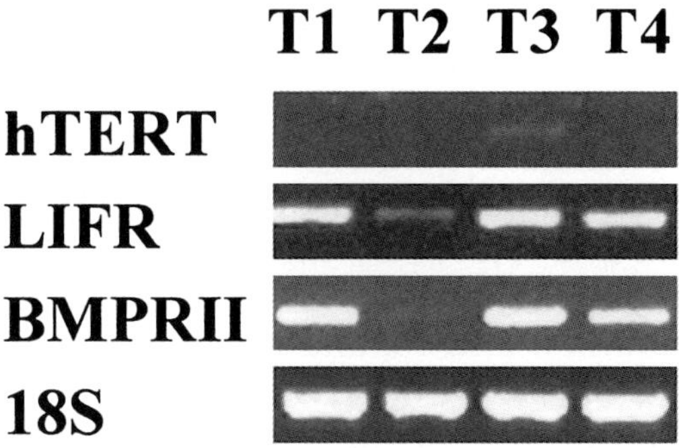

Fig. 2. Expression of stem cell marker genes in cultivated human cytotrophoblastic cell samples. RT-PCR analyses of mRNA expression of human telomerase reverse transcriptase (hTERT), leukemia inhibitory factor receptor (LIFR), and bone morphogenetic protein receptor II (BMPRII). As internal standard the expression of 18S-rRNA (18S) was analysed. T1, T3: trisomy 21; T2, T4: normal karyotype. For details compare Table 2

hTERT was detectable in normal cell samples (F2, F4) and in aneuploid cell samples (F3, T3).

The leukemia inhibitory factor (LIF) is a growth factor necessary for maintaining embryonic stem cells in a proliferative, undifferentiated state (Niwa, 2001; Burdon, 2002; http://www.nih.gov/news/stemcell/scireport.htm). Via RT-PCR we found LIF receptor (LIFR) mRNA to be expressed in all normal and aneuploid human cell samples analysed in this study (Figs. 1 and 2). Whereas in the trisomy 21 samples the expression levels were very similar, independent whether amniotic fluid cell samples or cytotrophoblastic cells were analysed (T1, T3, F1, F3), in three of four analysed cell samples the LIFR expression levels were lower when a normal karyotype was diagnosed (T2, F2, F4) compared to Down Syndrome cell samples.

Bone morphogenetic protein receptors (BMPR) are important for the differentiation of commited mesenchymal cell types from mesenchymal stem and progenitor cells. They are markers for early mesenchymal stem and progenitor cells (Ebendal, 1998; http://www.nih.gov/news/stemcell/scireport.htm). mRNA expression of BMPR II was detectable in all analysed samples (Figs. 1 and 2). Although in one normal cytotrophoblastic cell sample (T2) the expression level was very low, specific differences between normal and trisomy 21 cells could not be detected.

Discussion

Much of the recent excitement surrounding human stem and progenitor cells is connected with the hope of investigators and patients alike that such cells have the potential to treat or cure a myriad of diseases, including Parkinson's, Alzheimer's and heart diseases, diabetes, strokes, spinal cord injuries, burns and genetic diseases with central nervous system damage (Weismann, 2000). The fact that the probably most potent stem cells can be derived from the inner cell mass during embryonic development raises a lot of ethical issues

(Ruiz-Canela, 2002). Accordingly, a large variety of investigations have been initiated to search for different human sources for cells harbouring the potential to differentiate into specific lineages.

In a recent study amniotic fluid cells from pregnant ewes were grown in culture, and a specific subpopulation of morphologically distinct cells was mechanically isolated and selectively expanded under specific culture conditions. Immunocytochemically these cells stained positive for vimentin, smooth muscle actin, cytokeratin 8 and 18, and fibroblast surface protein and negatively for desmin and cluster of differentiation 31. This profile is consistent with a mesenchymal, fibroblast/myofibroblast cell lineage. After expansion, these cells from the amniotic fluid were seeded onto a polyglycolic acid polymer/poly-4-hydroxybutyrate scaffold. Scanning electron microscopy revealed dense, confluent layers of cells surrounding the polymer matrices and firm cell adhesion of both scaffolds (Kaviani, 2001). This is an important demonstration that, in case of prenatal diagnosis of e.g. congenital diaphragmatic hernia or body wall defects, amniotic fluid could serve as a source to create an engineered construct — material containing cells from the fetus, to function as a graft — for implantation immediately after birth. As a consequence of these findings it is interesting and important to further investigate the spectrum of cell types contained in human amniotic fluid.

For example, human amniotic epithelial cells are formed from amnioblasts, separated from the epiblast at about the 8th day after fertilization. They constitute the inner layer of the amnion and have been speculated to form part of the spectrum of cells in human amniotic fluid (Milunsky, 1979; Hoehn, 1982; Gosden, 1983; Sandler, 1995; Prusa, 2002). Amniotic epithelial cells have been demonstrated to express neuronal stem cell markers (Sakuragawa, 1996) and to survive and function in the brain of a rat model of Parkinson's disease (Kakishita, 2000).

Taken together, these data prompted us to investigate the mRNA expression of hTERT, LIFR and BMPR II, genes known to play a role in stem cell proliferation, maintenance and differentiation. Telomerase is an enzyme that helps maintaining telomeres, which protect the ends of chromosomes. Telomerase activity is detectable in human pluripotent stem cells, in germ cells, in most immortalised cell lines, and in tumour samples. Human not-transformed somatic cells, however, do not show telomerase activity and their telomeres are considerably shorter. The leukemia inhibitory factor (LIF) is a growth factor necessary for maintaining embryonic stem cells in a proliferative, undifferentiated state and bone morphogenetic protein receptors (BMPR) are important for the differentiation of commited mesenchymal cell types from mesenchymal stem and progenitor cells (Ebendal, 1998; Niwa, 2001; Burdon, 2002; http://www.nih.gov/news/stemcell/scireport.htm).

In this report we describe for the first time that human amniotic fluid samples and cells from human placental biopsies, cultivated in standard medium, as used for routine prenatal diagnosis, can contain cells expressing hTERT, LIFR and BMPR II. Telomerase activity was already earlier described to be detectable in both uncultured and cultured human amniotic fluid cells from 14 weeks' gestation (Mosquera, 1999). Taken together, these data provide evidence for the presence of specific stem or progenitor cells in such samples. We found stem cell marker gene expression in samples from pregnancies from the 13th up to the 34th week of gestational age. Such cultivated cell samples are available in all prenatal routine diagnosis laboratories. These data provide evidence that human amniotic fluid and cytotrophoblastic cell cultures might provide a new source for research on primary cell systems expressing these stem cell markers. In addition, we found no significant differences in the expression of these stem cell markers between normal and trisomy 21 cell samples, suggesting that early deregulation of the expression of these genes in the here analysed cell sources does not contribute to the molecular development of Down Syndrome.

Acknowledgements

A-R. Prusa is clinically associated with the Children's Hospital of Vienna, Department of Neonatology and Intensive Care, Währinger Gürtel 18–20, A-1090 Vienna. The authors want to thank the Department of Neonatology and Intensive Care for support.

References

Antonarakis SE, Avramopoulos D, Blouin JL, Talbot CC, Schinzel AA (1993) Mitotic errors in somatic cells cause trisomy 21 in about 4.5% of cases and are not associated with advanced maternal age. Nat Genet 3: 146–150

Bahn S, Mimmack M, Ryan M, Caldwell MA, Jauniaux E, Starkey M, Svendsen CN, Emson P (2002) Neuronal target genes of the neuron-restrictive silencer factor in neurospheres derived from fetuses with Down's syndrome: a gene expression study. Lancet 359: 310–315

Benn PA (2002) Advances in prenatal screening for Down syndrome: II first trimester testing, integrated testing, and future directions. Clin Chim Acta 324(1–2): 1–11

Burdon T, Smith A, Savatier P (2002) Signalling, cell cycle and pluripotency in embryonic stem cells. Trends Cell Biol 12(9): 432–438

Cairns NJ (1999) Neuropathology. J Neural Transm [Suppl] 57: 61–74

Ebendal T, Bengtsson H, Soderstrom S (1998) Bone morphogenetic proteins and their receptors: potential functions in the brain. J Neurosci Res 51: 139–146

Engidawork E, Lubec G (2003) Molecular changes in fetal Down syndrome brain. J Neurochem 84: 895–904

Gosden CM (1983) Amniotic fluid cell types and culture. Br Med Bull 39: 348–354

Hernandez D, Fisher EM (1996) Down syndrome genetics: unravelling a multifactorial disorder. Hum Mol Genet 5: 1411–1416

Hoehn H, Salk D (1982) Morphological and biochemical heterogeneity of amniotic fluid cells in culture. Meth Cell Biol 26: 11–34

Kakishita K, Elwan MA, Nakao N, Itakura T, Sakuragawa N (2000) Human amniotic epithelial cells produce dopamine and survive after implantation into the striatum of a rat model of Parkinson's disease: a potential source of donor for transplantation therapy. Exp Neurol 165: 27–34

Kaviani A, Perry TE, Dzakovic A, Jennings RW, Ziegler MM, Fauza DO (2001) The amniotic fluid as a source of cells for fetal tissue engineering. J Pediatr Surg 36: 1662–1665

Lubec G, Engidawork E (2002) The brain in Down syndrome (TRISOMY 21). J Neurol 249: 1347–1356

Miller RJ (2002) The ups and downs of Down's syndrome. Lancet 359: 275–276
Milunsky A (1979) Amniotic fluid cell culture. In: Milunsky A (ed) Genetic disorder and the fetus. Plenum Press, New York
Mosquera A, Fernandez JL, Campos A, Goyanes VJ, Ramiro-Diaz J, Gosalvez J (1999) Simultaneous decrease of telomere length and telomerase activity with ageing of human amniotic fluid cells. J Med Genet 36: 494–496
Niwa H (2001) Molecular mechanism to maintain stem cell renewal of ES cells. Cell Struct Funct 26(3): 137–148
Prusa A-R, Hengstschläger M (2002) Amniotic fluid cells and human stem cell research — a new connection. Med Sci Monit 8: 253–257
Ruiz-Canela M (2002) Embryonic stem cell research: the relevance of ethics in the progress of science. Med Sci Monit 8: SR21–26
Sakuragawa N, Thangavel R, Mizuguchi M, Hirasawa M, Kamo I (1996) Expression of markers for both neuronal and glial cells in human amniotic epithelial cells. Neurosci Lett 209: 9–12
Sandler TW (1995) Langman's medical embryology. Williams and Wilkins, Baltimore, pp 71–80
Wegner R-D (1999) Diagnostic cytogenetics. Springer, Berlin Heidelberg New York Tokyo
Weissman IL (2000) Stem cells: units of development, units of regeneration, and units in evolution. Cell 100: 157–168
Xu C, Inokuma MS, Denham J, Golds K, Kundu P, Gold JD, Carpenter MK (2001) Feeder-free growth of undifferentiated human embryonic stem cells. Nat Biotechnol 19(10): 971–974
Yoon PW, Freeman SB, Sherman SL, Taft LF, Gu Y, Pettay D, Flanders WD, Khoury MJ, Hassold TJ (1996) Advanced maternal age and the risk of Down syndrome characterized by the meiotic stage of the chromosomal error: a population-based study. Am J Hum Genet 58: 628–633

Authors' address: M. Hengstschläger, Ph.D., Prof., Obstetrics and Gynecology, Prenatal Diagnosis and Therapy, University of Vienna, Währinger Gürtel 18-20, A-1090 Vienna, Austria, e-mail: markus.hengstschlaeger@akh-wien.ac.at

BOSTON SCIENTIFIC
HEALTH SCIENCES
LIBRARY
BROOMFIELD

Springer-Verlag and the Environment

WE AT SPRINGER-VERLAG FIRMLY BELIEVE THAT AN international science publisher has a special obligation to the environment, and our corporate policies consistently reflect this conviction.

WE ALSO EXPECT OUR BUSINESS PARTNERS – PRINTERS, paper mills, packaging manufacturers, etc. – to commit themselves to using environmentally friendly materials and production processes.

THE PAPER IN THIS BOOK IS MADE FROM NO-CHLORINE pulp and is acid free, in conformance with international standards for paper permanency.